NUMBER THEORY

Tradition and Modernization

Developments in Mathematics

VOLUME 15

Aims and Scope

Developments in Mathematics is a book series publishing

(i) Proceedings of conferences dealing with the latest research advances,

(ii) Research monographs, and

(iii) Contributed volumes focusing on certain areas of special interest.

Editors of conference proceedings are urged to include a few survey papers for wider appeal. Research monographs, which could be used as texts or references for graduate level courses, would also be suitable for the series. Contributed volumes are those where various authors either write papers or chapters in an organized volume devoted to a topic of special/current interest or importance. A contributed volume could deal with a classical topic that is once again in the limelight owing to new developments.

NUMBER THEORY

Tradition and Modernization

Edited by

WENPENG ZHANG
Northwest University, Xi'an, P.R. China

YOSHIO TANIGAWA
Nagoya University, Nagoya, Japan

Springer

Library of Congress Control Number: 2005937555

ISBN-10: 0-387-30414-2 e-ISBN: 0-387-30829-6

ISBN-13: 978-0387-30414-4

Printed on acid-free paper.

AMS Subject Classifications: 11N99, 11M35, 11F03, 11F85, 11G05

Printed in the United States of America.

9 8 7 6 5 4 3 2 1

springer.com

Contents

Preface vii

About the book and the conference ix

List of participants xi

Positive finiteness of number systems
S. Akiyama 1

On a distribution property of the resudual order of $a \pmod p$ –IV
K. Chinen and L. Murata 11

Diagonalizing "bad" Hecke operators on spaces of cusp forms
Y.-J. Choie and W. Kohnen 23

On the Hilbert-Kamke and the Vinogradov problems in additive number theory
V. N. Chubarikov 27

The Goldbach-Vinogradov theorem in arithmetic progressions
Z. Cui 39

Densities of sets of primes related to decimal expansion of rational numbers
T. Hadano, Y. Kitaoka, T. Kubota and M. Nozaki 67

Spherical functions on p-adic homogeneous spaces
Y. Hironaka 81

On modular forms of weight $(6n+1)/5$ satisfying a certain differential equation
M. Kaneko 97

Some aspects of the modular relation
S. Kanemitsu, Y. Tanigawa, H. Tsukada and M. Yoshimoto 103

Zeros of automorphic L-functions and noncyclic base change
J. Liu and Y. Ye 119

Analytic properties of multiple zeta-functions in several variables
K. Matsumoto 153

Cubic fields and Mordell curves
K. Miyake 175

Towards the reciprocity of quartic theta-Weyl sums, and beyond
Y.-N. Nakai 185

Explicit congruences for Euler polynomials
Z.-W. Sun 205

Square-free integers as sums of two squares
W. Zhai 219

Some applications of L-functions to the mean value of the Dedekind sums and Cochrane sums
W. Zhang 229

Index 239

Preface

This book is a collection of papers contributed by participants of the third China-Japan Seminar on Number Theory held in Xi'an, PRC, during February 12–16, 2004 devoted to "Tradition and Modernization of Number Theory." The volume also assembles those papers which were contributed by invitees who could not attend the seminar. The papers are presented in quite different in depth and cover variety of descriptive details, but the main underlying editorial principle, explained below, enables the reader to have a unified glimpse of the developments of number theory. Thus on the one hand, we cling to the traditional approach presented in greater detail, and on the other, we elucidate its influence on the modernization of the methods in number theory, emphasizing on a few underlying common features such as functional equations for various zeta-functions, modular forms, congruence conditions, exponential sums and algorithmic aspect (see "About the book and the conference," page ix).

It is due to add a few words on our editorial policy. As you may perceive, the general situation surrounding the scientific publishers is now becoming harder, but the new Springer has agreed to publish the present volume. Thanks are due to Professor Dr. Heinze, Ms. Ann Kostant, Messrs. J. Martindale and R. Saley for their generosity and support. In response to their good wishes, we promised to make this book a readable one. To attain this goal, we made a great deal of effort, the pervading principle of our editorial work being to make this not merely an ununified collection of papers presented in mutually disjointed fashion but an organized mathematical volume. All the authors have been so kind as to understand this policy. We are very grateful for their cooperation.

In the course of editing of the book, many people helped us. We wish to express our hearty thanks to them, especially to Drs. Zhefeng Xu, Liping Ding, Jingping Ma, Jie Li, Jing Gao, Huaning Liu and Masami Yoshimoto.

September 22, 2005

The editors
Wenpeng Zhang and Yoshio Tanigawa

About the book and the conference

We can make a brief and rambling review on the papers contained in this volume by grouping them into classes having a few principles in common, where rambling means we do not mind the order of papers.

The first feature is the reduction modulo a prime p or a positive integer. The paper by Cui is a generalization of Vinogradov's three primes theorem with congruence conditions in which the circle method, especially the treatment of exponential integrals on major and minor arcs is exposed in its full details, and even a beginner could grasp the core of the method. Chubarikov's paper on "additive problems" also has a reduction modulo p aspect. The papers by Chinen-Murata and Hadano-Kitaoka-Kubota-Nozaki are concerned with the distribution of quantities connected with primitive roots modulo p. Also, Sun's paper investigating the congruences modulo p between coefficients of the Euler polynomial falls in this category. Kaneko, referred to below, considers reduction mod p of the associated polynomial.

The reduction mod p aspect has been a central theme from the time of Gauss and has created fruitful results in number theory. Thus we may say that considering reduction mod p of various problems leads to a new horizon of research connecting the past to the present.

Secondly, as can be seen from the papers by Akiyama and Hadano-Kitaoka-Kubota-Nozaki, the algorithmic aspect has been an important topic in modern number theory along with the developement of computer science.

Thirdly, the theory of zeta-functions has been a main driving force not only for the developement of number theory and related fields in its applications to varied problems but also in its own right. In particular, the functional equations both local and global have been the main object of research as a manifestation of the modular transformation, to which the paper of Kanemitsu-Tanigawa-Tsukada-Yoshimoto is devoted. Of course, these result from the theory of modular forms, in which field there are two research papers by Choie-Kohnen and Kaneko, which deal with the diagonalization of bad Hecke operators and the modular forms generating differential equations, respecively. The thorough-as-usual survey of Matsumoto of Euler-Zagier sums deals with analytic continuation and may be thought of as the pre-functional equation. Also, in the papers of Hironaka and Liu and Ye an important part is played by the functional equations.

Fourthly, all three principal methods of exponential (trigonometrical) sums due to Weyl, Vinogradov and van der Corput are presented in different fashion and details. In addition to these, there is a paper by Zhang and his school who made full use of the Kloostermann sums. The paper of Nakai is concerned with the structural theory of the theta-Weyl sums—the author's Mittelaltertraum—an interesting point, compared to the treatment of the Kloostermann sum as the functional equation. Zhai's paper deals with the short-interval result on an arithmetic function by the Euler product of its generating zeta function.

Finally, we have been mindful to adopt ideas and methods from other fields from the start. In the volume, we find three modern topics by Hironaka on "Algebraic groups and Prehomogeneous vector spaces," Liu and Ye on "Automorphic L-functions" and by Miyake on "elliptic curves." Hironaka's paper deals with local objects while Liu and Ye's paper, starting from local objects, treats the global objects. These suggest that in the 21st century, analytic number theory is to deal with both aspects.

I would like to state my recollection of the seminar. Looking back, I should say it was a remarkably heartwarming occasion as well as a successful scientific seminar. Indeed, tea service during the session was something that we could not even imagine in Japan. Very genuine and pure-hearted young people, good food, all were great privileges for participants of the seminar. I wholeheartedly thank Zhang Laoshi, the editor of the volume, for conducting this seminar and his enthusiastic students to whom I wish great success in their respective careers; if I would be any help in scientific matter, I would be honored to do my best. Specifically, thanks are due to Huaning Liu (for his efficient support as the official conference correspondent), Jing Gao, Nan Gao, Zhefeng Xu (the leader of the group), Jie Li, Tianping Zhang, Xiaobeng Zhang, Chuan Lv, Liping Ding, Minhui Zhu, Xinwei Lu and Dongmei Ren. Thank you and I wish you a big success.

Now let's meet again in the fourth China-Japan Seminar, "Sailing on the sea of number theory," in which we will continue not only to study the traditional problems in more detail but also try to extend our limit of knowledge on the ground of the hitherto stored rich pile of ideas and principles, challenging new problems in the wider sea of number theory and beyond.

As it is autumn now and one and half year's ago we gathered in Xi'an, nee Chan'an, I am tempted to quote a passage from the most famous poem of Libai (Zi ye wu ge): *Chang an yi pian yue, wan hu dao yi sheng. Qiu feng chui bu jin, zong shi yu guang qing.* The first two lines lead to Erdös' notion of the Book, which is the moon shining in the sky of Xi'an, and on the earth we are struggling through our labor of research.

The series supervisor, Jin Guangzi=S. Kanemitsu

List of participants

Professor Shigeki Akiyama (Niigata University)
Professor Krishnaswami Alladi (University of Florida)
Professor Masaaki Amou (Gunma University)
Dr. Junfeng Chen (Yan'an University)
Professor Yonggao Chen (Nanjing Normal University)
Professor Vladimir N. Chubarikov (Moscow Lomonosov State University)
Dr. Liping Ding (Northwest University)
Professor Shigeki Egami (Toyama University)
Dr. Jing Gao (Xi'an Jiaotong University)
Dr. Nan Gao (Northwest University)
Dr. Dan Ge (Yan'an University)
Professor Jinbao Guo (Yan'an University)
Dr. Yongping Guo (Yan'an University)
Professor Yumiko Hironaka (Waseda University)
Professor Chaohua Jia (Academia Sinica)
Professor Shigeru Kanemitsu (University of Kinki)
Professor Yoshiyuki Kitaoka (Meijo University)
Professor Chao Li (Shangluo Teachers College)
Professor Hailong Li (Weinan Teachers College)
Professor Hongze Li (Shanghai Jiaotong University)
Dr. Jie Li (Northwest University)
Dr. Yansheng Li (Yan'an University)
Professor Guodong Liu (Huizhou University)
Dr. Huaning Liu (Northwest University)
Professor Jianya Liu (Shandong University)
Dr. Chuan Lv (Northwest University)
Professor Kohji Matsumoto (Nagoya University)
Professor Katsuya Miyake (Waseda University)
Professor Kenji Nagasaka (Hosei University)
Professor Yoshinobu Nakai (Yamanashi University)
Dr. Lan Qi (Yan'an University)
Dr. Yan Qu (Shandong University)
Dr. Dongmei Ren (Xi'an Jiaotong University)
Professor Zhiwei Sun (Nanjing University)
Professor Yoshio Tanigawa (Nagoya University)
Professor Xiaoying Wang (Northwest University)
Professor Yang Wang (Nanyang Teachers College)
Professor Yonghui Wang (The Capital Normal University)
Dr. Zhefeng Xu (Northwest University)
Dr. Masami Yoshimoto (University of Kinki)
Dr. Yuan Yi (Xi'an Jiaotong University)
Dr. Weili Yao (Xi'an Jiaotong University)
Dr. Hai Yang (Yan'an University)
Dr. Haiwen Yang (Yan'an University)
Professor Wenguang Zhai (Shandong Normal University)
Dr. Tianping Zhang (Northwest University)
Professor Wenpeng Zhang (Northwest University)
Dr. Xiaobeng Zhang (Northwest University)
Dr. Minhui Zhu (Northwest University)

1 2 3 4 5 6 7 8 9 10 11 12 13 14 15 16 17 18 19 20 21 22 23 24 25 26
27 28 29 30 31 32 33 34 35 36 37 38 39 40 41
42 43 44 45 46 47 48 49 50 51

1. Yonghui Wang, 2. Yongping Guo, 3. Yan Qu, 4. Cuidian Yang, 5. Yansheng Li, 6. Chao Li, 7. Hai Yang, 8. Jinbao Guo, 9. Chuan Lv, 10. Wenguang Zhai, 11. Yonggao Chen, 12. Yoshio Tanigawa, 13. Yumiko Hironaka, 14. Shigeki Egami, 15. Masami Yoshimoto, 16. Masaaki Amou, 17. Shigeki Akiyama, 18. Junfeng Chen, 19. Yoshinobu Nakai, 20. Xiaobeng Zhang, 21. Claus Bauer, 22. Zhengguang Dou, 23. Tianping Zhang, 24. Nan Gao, 25. Zhefeng Xu, 26. Lan Qi, 27. Dan Ge, 28. Yang Wang, 29. Hongze Li, 30. Guodong Liu, 31. Zhiwei Sun, 32. Jianya Liu, 33. Hailong Li, 34. Yoshiyuki Kitaoka, 35. Yuan Yi, 36. Weili Yao, 37. Xiaoying Wang, 38. Liping Ding, 39. Dongmei Ren, 40. Jie Li, 41. Minhui Zhu, 42. Xianzhong Zhao, 43. Chaohua Jia, 44. Jincheng Wang, 45. Kexiao Zhu, 46. Shigeru Kanemitsu, 47. Krishnaswami Alladi, 48. Katsuya Miyake, 49. Kohji Matsumoto, 50. Wenpeng Zhang, 51.Xianlong Xin

POSITIVE FINITENESS OF NUMBER SYSTEMS

Shigeki Akiyama
Department of Mathematics, Faculty of Science, Niigata University
Ikarashi 2-8050, Niigata 950-2181, Japan
akiyama@math.sc.niigata-u.ac.jp

Abstract We characterize the set of β's for which each polynomial in β with non-negative integer coefficients has a finite admissible expression in some number systems.

Keywords: Beta expansion, Canonical number system, Pisot number

2000 Mathematics Subject Classification: 11A63, 37B10

1. Introduction

In this note, we study a certain finiteness property of number systems given as an aggregate of power series in some base β, called *beta-expansion,* the number systems then being called *canonical.*

In relation to symbolic dynamics, an important problem is to determine the set of β's for which each polynomial in base β with non-negative integer coefficients has a finite expression in the corresponding number system. However this problem is rather difficult in general, and we restrict our scope to the set of such β's which does *not* have 'global' finiteness. Let us explain precisely this problem in terms of beta-expansion (cf. [27]).

Let $\beta > 1$ be a real number. Each positive x is uniquely expanded into a *beta-expansion*:

$$x = \sum_{i=M}^{\infty} a_i \beta^{-i} \qquad (M \text{ can be negative})$$

under conditions

$$a_i \in [0, \beta) \cap \mathbb{Z} \quad \text{and} \quad \forall L \geq M \quad 0 \leq x - \sum_{i=M}^{L} a_i \beta^{-i} < \beta^{-L},$$

Number Theory: Tradition and Modernization, pp. 1–10
W. Zhang and Y. Tanigawa, eds.

which is also called a *greedy expansion*. We write this expression as

$$x = x_M x_{M+1} \dots x_0 . x_1 x_2 \dots$$

in an analogy to the usual decimal expansion. If $a_i = 0$ for sufficiently large i, then the expansion is called finite and the tail $00\dots$ is omitted as usual. Let $\mathrm{Fin}(\beta)$ be the set of finite beta expansions. It is obvious that $\mathrm{Fin}(\beta)$ is a subset of $\mathbb{Z}[1/\beta] \cap [0, \infty)$ if β is an algebraic integer[1]. Frougny and Solomyak [14] first studied the property

$$\mathrm{Fin}(\beta) = \mathbb{Z}[1/\beta] \cap [0, \infty)$$

which we call *finiteness* property (F). If β has the property (F), then β is a Pisot number, that is, a real algebraic integer greater than one that all other conjugates of β have modulus less than one.

A polynomial $x^d - a_{d-1}x^{d-1} - \dots - a_0$ with $a_{d-1} \geq a_{d-2} \geq \dots \geq a_0 > 0$ has a Pisot number $\beta > 1$ as a root (cf. [10]). In [14] it is shown that the property (F) holds for this class of β. The complete characterization of β with (F) among algebraic integers (or among Pisot numbers), is a difficult problem when $d \geq 3$ (cf. [2], [8], [4]).

The expansion of 1 is a digit sequence given by an expression $1 = \sum_{i=1}^{\infty} c_i \beta^{-i} = .c_1c_2c_3\dots$ such that $.0c_2c_3\dots$ is the beta expansion of $1 - c_1/\beta$ with $c_1 = \lfloor \beta \rfloor$, with $\lfloor \beta \rfloor$ signifying the integral part of β. This expansion play a crucial role in determining which formal expression can be realized as beta-expansion ([25], [18]). Especially a formal expression

$$1 = \sum_{i=1}^{\infty} d_i \beta^{-i} = .d_1 d_2 \dots$$

coincides with the expansion of 1 if and only if the digit sequence $d_1 d_2 \dots$ is greater than its left shift $d_i d_{i+1} \dots$ for $i > 1$ in natural lexicographical order.

In [14] it is shown that if the expansion of $1 = .c_1c_2\dots$ has infinite decreasing digits (i.e., $c_1 \geq c_2 \geq c_3 \geq \dots$ and $c_i = c_{i+1} > 0$ from some index on), then the set $\mathrm{Fin}(\beta)$ is closed under addition. This is equivalent to the condition:

$$\mathbb{Z}_+[\beta] \subset \mathrm{Fin}(\beta)$$

where $\mathbb{Z}_+ = \mathbb{Z} \cap [0, \infty)$ and $\mathbb{Z}_+[\beta]$ is the set of polynomials in base β as indeterminate with coefficients in $\mathbb{Z}_+$. We call this property *positive*

[1] If β is an algebraic integer, then $\mathbb{Z}[\beta] \subset \mathbb{Z}[1/\beta]$.

finiteness ((PF) for short). The author showed in [3] that (PF) implies *weak finiteness* which has a close connection to Thurston's tiling generated by Pisot unit β (cf. [30], [8]), which fact is one of the motivations to study (PF).

In [9], Ambrož, Frougny, Masáková and Pelantová gave a characterization of (PF) in terms of 'transcription' of minimal forbidden factors. Our problem in this paper is to characterize β with the property (PF) *without* (F). With this restriction of the scope, we can give a complete characterization of such β's:

Theorem 1. *Let $\beta > 1$ be a real number with positive finiteness. Then either β satisfies the finiteness property (F) or β is a Pisot number whose minimal polynomial is of the form:*

$$x^d - (1 + \lfloor \beta \rfloor) x^{d-1} + \sum_{i=2}^{d} a_i x^{d-i}$$

with $a_i \geq 0$ $(i = 2, \dots, d)$, $a_d > 0$ and $\sum_{i=2}^{d} a_i < \lfloor \beta \rfloor$. In the latter case, the expansion of 1 has infinite decreasing digits. Conversely if $\beta > 1$ is a root of the polynomial

$$x^d - B x^{d-1} + \sum_{i=2}^{d} a_i x^{d-i}$$

with $a_i \geq 0$, $a_d > 0$ and $B > 1 + \sum_{i=2}^{d} a_i$, then this polynomial is irreducible and β is a Pisot number with (PF) without (F). We also have $B = 1 + \lfloor \beta \rfloor$.

The study of (PF) is thus reduced to that of (F) by Theorem 1. Unfortunately, we are unable to add any new example of β to those already found in [14].

A parallel problem is solved in another well known number system. Let α be an algebraic integer [2] of degree d having its absolute norm $|N(\alpha)|$. If each element $x \in \mathbb{Z}[\alpha]$ has an expression:

$$x = \sum_{i=0}^{\ell} a_i \alpha^i, \quad a_i \in \mathcal{A} = \{0, 1, \dots, |N(\alpha)| - 1\}$$

then we say that α gives a *canonical number system* (CNS for short). If such an expression exists, then it is unique since $\mathcal{A}$ forms a complete

[2] α is used instead of β to distinguish the difference of number systems.

set of representatives of $\mathbb{Z}[\alpha]/\alpha\mathbb{Z}[\alpha]$ and the digit string is computed from the bottom by successive reduction modulo α. If α gives a CNS, then α must be expanding, that is, all conjugates of α have modulus greater than one ([22]). Assume that α has the minimal polynomial of the form $x^2 + Ax + B$. Then α gives a CNS if and only if $-1 \le A \le B$ and $B \ge 2$ ([19], [20], [15]). When $d \ge 3$, the characterization of α's among expanding algebraic integers is again a difficult question ([6], [28], [7], [11], [12], [5]). It is obvious that CNS is an analogous concept of (F). To pursue this analogy, let us say that α has *positive finiteness* if $\mathbb{Z}_+[\alpha] = \mathcal{A}[\alpha]$, i.e.,

$$\forall 0 \le a_i \in \mathbb{Z} \quad \exists b_j \in \{0, 1, \dots, |N(\alpha)| - 1\} \qquad \sum_{i \ge 0} a_i \alpha^i = \sum_{j \ge 0} b_j \alpha^j.$$

This type of positive finiteness is in fact weaker than CNS and we can show

Theorem 2. *Assume that α has positive finiteness. Then either α gives a CNS or the minimal polynomial of α is given by*

$$\sum_{i=1}^{d} a_i x^i - C \tag{1}$$

with $a_d = 1$, $a_i \ge 0$ and $\sum_{i=1}^{d} a_i < C$. Conversely if α is a root of the irreducible polynomial (1) with the same condition then α has positive finiteness but does not give a CNS.

It is not possible to remove irreducibility in the last statement. For example, $x^2 + x - 12 = (x-3)(x+4)$ but -4 gives a CNS.

In [26], Pethő introduced a more general concept 'CNS polynomial' among expanding polynomials. If the polynomial is irreducible, then the concept coincides with CNS. It is straightforward to generalize above Theorem 2 to this framework. In this wider sense, x^2+x-12 has positive finiteness.

2. Proof of Theorem 1

First we prove the second part of Theorem 1. Assume that $\beta > 1$ is a root of a polynomial:

$$P(x) = x^d - Bx^{d-1} + \sum_{i=2}^{d} a_i x^{d-i} \text{ with } a_i \ge 0,\ a_d > 0 \text{ and } B > 1 + \sum_{i=2}^{d} a_i.$$

By applying Rouché's Theorem, $P(x)$ and $x^d - Bx^{d-1}$ has the same number of roots in the open unit disk. Thus β is a Pisot number and

$P(x)$ is irreducible. In fact, if $P(x)$ is non-trivially decomposed into $P_1(x)P_2(x)$ and $P_1(\beta) = 0$, then the constant term of $P_2(x)$ is less than 1 in modulus, and hence it must vanish. This contradicts $a_d > 0$.

The relation $P(\beta) = 0$ formally gives rise to the expansion

$$1 = .B\ \overline{a_2}\ \overline{a_3}\ \dots\ \overline{a_d}$$

where we put $\overline{x} = -x$ to simplify the notation. Multiplying β^{-j} ($j = 1, 2, \dots$) and summing up we have

$$\begin{aligned} 1 &= .B\ \overline{a_2}\ \overline{a_3}\ \dots\ \overline{a_d} \\ &\quad + .\overline{1}\ B\ \overline{a_2}\ \dots\ \overline{a_{d-1}}\ \overline{a_d} \\ &\quad + .0\ \overline{1}\ B\ \dots\ \overline{a_{d-2}}\ \overline{a_{d-1}}\ \overline{a_d} \\ &\quad + \cdots \\ &= .(B-1)\ (B-1-a_2)\ (B-1-a_2-a_3)\ \dots\ m\ m\ m \dots \end{aligned}$$

with $m = B - 1 - \sum_{i=2}^{d} a_i$. Since the last sequence is lexicographically greater than its left shifts, this gives the expansion of 1 of β with infinite decreasing digits. By the result of [14], this β has the property (PF). Now it is clear that $B = 1 + \lfloor \beta \rfloor$. Since the expansion of 1 is not finite, β does not satisfy (F). This can also be shown in the following way. Since $P(0) < 0$ and $P(1) > 0$, there is a positive conjugate $\beta' \in (0, 1)$. Using Proposition 1 of [1], β does not satisfy the finiteness property (F).

To prove the first part of Theorem 1, we quote two lemmas. First,

Lemma 3 (Theorem 5, Handelman [17]). *Let $\beta > 1$ be an algebraic integer such that other conjugates has modulus less than β and there are no other positive conjugates. Then β is a Perron-Frobenius root of a primitive companion matrix.*

The proof of this lemma relies on the Perron-Frobenius theorem and the fact that for any polynomial $p(x)$ without positive roots, $(1+x)^m p(x)$ have only positive coefficients for sufficiently large m. (A direct proof of this fact will be given in the appendix.) We need another

Lemma 4 (Lemma 2, [14]). *The equality $\mathbb{Z}_+[\beta] = \mathbb{Z}[\beta] \cap [0, \infty)$ holds if and only if β is a Perron-Frobenius root of a primitive companion matrix.*

In the following, we also use the fact that there are only two Pisot numbers less than $\sqrt{2}$. The smallest one, say $\theta \approx 1.32372$, is a positive root of $x^3 - x - 1$ and the second smallest one $\theta_1 \approx 1.38028$ is given by $x^4 - x^3 - 1$ (c.f. [24]). C. L. Siegel [29] was the first to prove that these two are the smallest Pisot numbers. In [1], it is shown that θ has property (F). On the other hand, θ_1 does not satisfy (PF) since $\theta_1 + 1$ has the infinite purely periodic beta expansion $100.0010000100001\ldots$.

Let us assume that $\beta > 1$ has positive finiteness (PF) property but does not have the (F) property. This implies that β is not an integer and greater than $\sqrt{2}$. Since $\mathbb{Z}_+ \subset \mathrm{Fin}(\beta)$, Proposition 1 of [2] implies that β is a Pisot number. We claim that β has a conjugate $\beta' \in (0,1)$. If not, then by Lemma 3, β is a Perron-Frobenius root of a primitive companion matrix. Then by Lemma 4, each element of $\mathbb{Z}[\beta] \cap [0,\infty)$ has a polynomial expression in base β with non-negative integer coefficients. Thus (PF) property implies the (F) property. This is a contradiction, which shows the claim.

By the property (PF), $\kappa = (1 + \lfloor\beta\rfloor)/\beta \in \mathrm{Fin}(\beta)$. Note that $\beta > \sqrt{2}$ implies $\lfloor\beta\rfloor + 1 < \beta^2$ and hence that the beta expansion of κ begins with $a_0 = 1$. Hence, as $\kappa - 1 < \beta^{-1}$, we have a beta expansion:

$$\frac{1 + \lfloor\beta\rfloor}{\beta} = 1.0a_2a_3\ldots a_\ell$$

with $a_\ell \neq 0$. Set $Q(x) = x^\ell - (\lfloor\beta\rfloor + 1)x^{\ell-1} + \sum_{i=2}^{\ell} a_i x^{\ell-i}$. Then $Q(x)$ has two sign changes among its coefficients. By Descartes's law, there exist at most two positive real roots of $Q(x)$, and therefore they must be β and β'. On the other hand, we see $Q(0) = a_\ell > 0$. If $Q(1) > 0$ then there are at least two positive roots of $Q(x)$ in $(0,1)$ which is impossible. Thus we have $Q(1) < 0$ which implies $\sum_{k=2}^{\ell} a_k < \lfloor\beta\rfloor$. But we have already proven under this inequality that $Q(x)$ is irreducible and the expansion of 1 of β has infinite decreasing digits. □

A few lines are due to elucidate the situation. If $\lfloor\beta\rfloor + 1$ has a finite beta expansion in base β, the above procedure yields the same polynomial $Q(x) = x^\ell - (1 + \lfloor\beta\rfloor)x^{\ell-1} + \sum_{k=2}^{\ell} a_k x^{\ell-k}$. Since $\beta > 1$ is a root of $Q(x)$ and $Q(0) > 0$, $Q(x)$ has exactly two positive real roots. If $Q(1) < 0$, then β has (PF) by the same reasoning. If $Q(1) \geq 0$, then there is a root $\eta \geq 1$ other than β. Note that this could happen even if β has property (PF). However in such case, $Q(x)$ must be reducible since β does not have other positive conjugate if it has property (PF). Especially if β satisfies (F), then $Q(x)$ is reducible. For example, $\beta = (1+\sqrt{5})/2$ satisfies (F) and $Q(x) = x^3 - 2x^2 + 1 = (x^2 - x - 1)(x - 1)$. The above proof shows, as a consequence, that $Q(x)$ must be irreducible if β satisfies (PF) without (F).

It is not clear whether the condition $\mathbb{Z}_+ \subset \mathrm{Fin}(\beta)$ implies (PF). We encounter difficulties in proving the existence of a positive conjugate $\beta' \in (0,1)$ under this condition.

3. Proof of Theorem 2

First we recall that if α has positive finiteness, then α is expanding. This was proved in CNS case in [22] and the same proof works in positive finiteness case. (See Lemma 3 and the proof of Theorem 3 in [22].) [3]

Let us assume that α has positive finiteness but does not give a CNS. Let $P(x)$ be the minimal polynomial of α. We claim that there exists a positive conjugate α'. Suppose this is not the case. Then by the remark after Lemma 3, there is a large integer M such that $(1+x)^M P(x)$ has only positive coefficients. This gives a relation of the form $\sum_{i=0}^{\ell} a_i\alpha^i = 0$ with $a_i > 0$. Thus each element of $\mathbb{Z}[\alpha]$ has an equivalent expression in $\mathbb{Z}_+[\alpha]$ which is attained by repeated addition of the above relation. This shows that $\mathbb{Z}_+[\alpha] = \mathbb{Z}[\alpha]$ and positive finiteness of α implies that α gives a CNS. This is a contradiction and the claim is proved. Note that $\alpha' > 1$.

Let $C = |N(\alpha)|$ and let its expression be $C = \sum_{i=0}^{d} a_i\alpha^i$ with $a_i \in \mathcal{A}$. Reducing modulo α, we see that $a_0 = 0$. Set $Q(x) = \sum_{i=1}^{d} a_i x^i - C$. Since $Q(0) < 0$ and there is only one sign change among the coefficients of $Q(x)$, there exists exactly one positive root of $Q(x)$ which is α'. Now $\alpha' > 1$ implies $Q(1) < 0$, i.e., $\sum_{i=1}^{d} a_i < C$. Suppose that $Q(x)$ is not irreducible and $Q(x) = P(x)R(x)$ with $\deg R \geq 1$. From $C = |N(\alpha)|$, we deduce $|R(0)| = 1$ and hence there exists a root η of $Q(x)$ with $|\eta| \leq 1$. Then

$$0 = |Q(\eta)| = \left|\sum_{i=1}^{d} a_i\eta^i - C\right| \geq C - \sum_{i=1}^{d} a_i$$

gives a contradiction. This shows that $Q(x) = P(x)$ and $a_d = 1$.

Finally we prove the converse. Assume that α is a root of the irreducible polynomial $Q(x) = \sum_{i=1}^{d} a_i x^i - C$ with $a_d = 1$, $a_i \geq 0$ and $\sum_{i=1}^{d} a_i < C$. Then $Q(x)$ must be expanding since otherwise there would exist a root η with $|\eta| \leq 1$ of $Q(x)$ and we would have the same contradiction. As $Q(0) < 0$ there exists a positive conjugate α'. Hence α cannot give a CNS, since -1 cannot have a finite expansion (cf. Proposition 6 in [15]). It remains to show that α has positive finiteness. The idea of the present proof can be traced back to [21]. Since α is a root of $Q(x)$, we have an expression

$$a_d a_{d-1} \ \ldots \ a_1 \ \overline{C} = 0. \tag{2}$$

We describe an algorithm to get an equivalent expression in $\mathcal{A}[\beta]$ from each $x = \sum_{i=0}^{\ell} d_i\alpha^i$ with $d_i \in \mathbb{Z}_+$. Adding the relation (2) $\kappa = \lfloor d_0/C \rfloor$

[3]For the later use, it suffices to show an easier fact (Lemma 3 in [22]): 'each conjugate of α has modulus not less than one.'

times, we have an equivalent expression of x in $\mathbb{Z}_+[\alpha]$:

$$d_\ell d_{\ell-1}\ldots d_0 + \kappa \times \left(a_d a_{d-1}\ \ldots\ a_1\ \overline{C}\right) = d'_{\ell'} d'_{\ell'-1}\ldots d'_0$$

whose constant term is $d'_0 = d_0 - \kappa C \in \mathcal{A}$. Repeat the same process on d'_1 to make the coefficients of α^1 fall into $\mathcal{A}$. This process can be continued in a similar manner. In each step, the sum of digits of the expression of x is strictly decreasing. Hence we finally get an expression in $\mathcal{A}[\alpha]$ in finite steps. □

Acknowledgments

The author expresses his most gratitude to Z.Masáková and E.Pelantová. This work was motivated by [9] and the discussion with them during my stay in Prague TU. Thanks are also due to J.Thuswaldner with whom we discussed on positive finiteness of expanding algebraic numbers.

Appendix

Handelman showed in [16], as a special case of his wide theory, that for any polynomial $p(x) \in \mathbb{R}[x]$ having no non-negative roots, there exists a positive integer M such that $(1+x)^M p(x)$ has only positive coefficients (cf. [23] and [13]). This is a crucial fact in proving Lemma 3 and Theorem 2. As the statement itself looks elementary, it may be worthy to note here a direct short proof. To prove this we factorize $p(x)$ into quadratic and linear factors in $\mathbb{R}[x]$. Since a linear factor $(x+a)$ with $a > 0$ does no harm, we prove that for any x^2+bx+c with $b^2 < 4c$ there exists a positive n such that $(1+x)^n(x^2+bx+c)$ has positive coefficients. The k-th coefficient of $(1+x)^n(x^2+bx+c)$ is

$$\binom{n}{k}\left(c\frac{n-k}{k+1} + b + \frac{k-1}{n-k+1}\right).$$

Thus we show that $f(k) = c(n-k)(n-k+1) + b(k+1)(n-k+1) + (k-1)(k+1) > 0$ for $k = 0, 1, \ldots, n$ if n is sufficiently large. From an expression

$$f(k) = -1 + b + bn + cn + cn^2 + (-c + bn - 2cn)k + (1 - b + c)k^2,$$

we see that the minimum is attained when $k = (c - bn + 2cn)/(2 - 2b + 2c)$ as $x^2 + bx + c > 0$ implies $1 - b + c > 0$. Direct computation shows

$$f(k) \geq \frac{-4 + 8b - 4b^2 - 4c + 4bc - c^2 + (4b - 4b^2 + 4c + 2bc)n + (-b^2 + 4c)n^2}{4(1 - b + c)}.$$

As the coefficient of n^2 in the numerator is positive, the assertion is shown. □

References

[1] S. Akiyama, *Pisot numbers and greedy algorithm*, Number Theory (K. Győry, A. Pethő, and V. Sós, eds.), Walter de Gruyter, 1998, pp. 9–21.

[2] S. Akiyama, *Cubic Pisot units with finite beta expansions*, Algebraic Number Theory and Diophantine Analysis (F.Halter-Koch and R.F.Tichy, eds.), de Gruyter, 2000.

[3] S. Akiyama, *On the boundary of self affine tilings generated by Pisot numbers*, J. Math. Soc. Japan **54** (2002), no. 2, 283–308.

[4] S. Akiyama, T. Borbély, H. Brunotte, A. Pethő, and J. M. Thuswaldner, *Generalized radix representations and dynamical systems I*, Acta Math. Hungar (to appear).

[5] S. Akiyama, H. Brunotte, and A. Pethő, *On cubic canonical number systems*, J. Math. Anal.Appl. **281** (2003), 402–415.

[6] S. Akiyama and A. Pethő, *On canonical number systems*, Theor. Comput. Sci. **270** (2002), no. 1-2, 921–933.

[7] S. Akiyama and H. Rao, *New criteria for canonical number systems*, Acta Arith. **111** (2004), no. 1, 5–25.

[8] S. Akiyama, H. Rao, and W. Steiner, *A certain finiteness property of Pisot number systems*, J. Number Theory **107** (2004), no. 1, 135-160.

[9] P. Ambrož, Ch. Frougny, Z. Masáková, and E. Pelantová, *Arithmetics on number systems with irrational bases*, Bull. Belg. Math. Soc. **10** (2003), 1–19.

[10] A. Brauer, *On algebraic equations with all but one root in the interior of the unit circle*, Math. Nachr. **4** (1951), 250–257.

[11] H. Brunotte, *On trinomial bases of radix representations of algebraic integers*, Acta Sci. Math. (Szeged) **67** (2001), no. 3-4, 521–527.

[12] H. Brunotte, *Characterization of CNS trinomials*, Acta Sci. Math. (Szeged) **68** (2002), 673–679.

[13] V. de Angelis and S. Tuncel, *Handelman's theorem on polynomials with positive multiples*, Codes, systems, and graphical models, IMA Vol. Math. Appl. vol. 123, Springer, 2001, pp. 439–445.

[14] Ch. Frougny and B. Solomyak, *Finite beta-expansions*, Ergod. Th. and Dynam. Sys. **12** (1992), 713–723.

[15] W. J. Gilbert, *Radix representations of quadratic fields*, J. Math. Anal. Appl. **83** (1981), 264–274.

[16] D. Handelman, *Positive polynomials and product type actions of compact groups*, Mem. Amer. Math. Soc. **54** (1985), no. 320, xi+79 pp.

[17] D. Handelman, *Spectral radii of primitive integral companion matrices and log concave polynomials*, Contemp. Math. (P.Walters, ed.), vol. 135, 1992, Symbolic Dynamics and its Applications, pp. 231–237.

[18] Sh. Ito and Y. Takahashi, *Markov subshifts and realization of β-expansions*, J. Math. Soc. Japan **26** (1974), 33–55.

[19] I. Kátai and B. Kovács, *Kanonische Zahlensysteme in der Theorie der quadratischen Zahlen*, Acta Sci. Math. (Szeged) **42** (1980), 99–107.

[20] I. Kátai and B. Kovács, *Canonical number systems in imaginary quadratic fields*, Acta Math. Acad. Sci. Hungar. **37** (1981), 159–164.

[21] I. Kátai and J. Szabó, *Canonical number systems for complex integers*, Acta Sci. Math. (Szeged) **37** (1975), 255–260.

[22] B. Kovács and A. Pethő, *Number systems in integral domains, especially in orders of algebraic number fields*, Acta Sci. Math. (Szeged) **55** (1991), 286–299.

[23] E. Meissner, *Über positive Darstellung von Polynomen*, Math.Ann. **70** (1911), 223–235.

[24] M. Grandet-Hugot M. Pathiaux-Delefosse M.J. Bertin, A. Decomps-Guilloux and J.P. Schreiber, *Pisot and Salem numbers*, Birkhäuser, 1992.

[25] W. Parry, *On the β-expansions of real numbers*, Acta Math. Acad. Sci. Hungar. **11** (1960), 401–416.

[26] A. Pethő, *On a polynomial transformation and its application to the construction of a public key cryptosystem*, Computational number theory (Debrecen, 1989) (Berlin), de Gruyter, Berlin, 1991, pp. 31–43.

[27] A. Rényi, *Representations for real numbers and their ergodic properties*, Acta Math. Acad. Sci. Hungar. **8** (1957), 477–493.

[28] K. Scheicher and J. M. Thuswaldner, *On the characterization of canonical number systems*, Osaka Math. J. **41** (2004), 327 - 351

[29] C.L. Siegel, *Algebraic numbers whose conjugates lie in the unit circle*, Duke Math. J. **11** (1944), 597-602.

[30] W. Thurston, *Groups, tilings and finite state automata*, AMS Colloquium Lecture Notes, 1989.

ON A DISTRIBUTION PROPERTY OF THE RESIDUAL ORDER OF $a \pmod{p}$– IV

Koji Chinen[1] and Leo Murata[2]

[1] *Department of Mathematics, Faculty of Engineering, Osaka Institute of Technology, Omiya, Asahi-ku, Osaka 535-8585, Japan*
YHK03302@nifty.ne.jp

[2] *Department of Mathematics, Faculty of Economics, Meiji Gakuin University, 1-2-37 Shirokanedai, Minato-ku, Tokyo 108-8636, Japan*
leo@eco.meijigakuin.ac.jp

Abstract Let a be a positive integer and $Q_a(x;k,l)$ be the set of primes $p \leq x$ such that the residual order of $a \pmod{p}$ in $\mathbf{Z}/p\mathbf{Z}$ is congruent to l modulo k. In this paper, under the assumption of the Generalized Riemann Hypothesis, we prove that for any residue class $l \pmod{k}$ the set $Q_a(x;k,l)$ has the natural density $\Delta_a(k,l)$ and the values of $\Delta_a(k,l)$ are effectively computable. We also consider some number theoretical properties of $\Delta_a(k,l)$ as a number theoretical function of k and l.

Keywords: Residual order, Artin's conjecture for primitive roots

2000 Mathematics Subject Classification: 11N05, 11N25, 11R18

1. Introduction

Let a be a positive integer which we assume is not a perfect b-th power with $b \geq 2$ and p a prime number not dividing a. We define $D_a(p) = \sharp\langle a \pmod{p}\rangle$ — the multiplicative order of $a \pmod{p}$ in $(\mathbf{Z}/p\mathbf{Z})^\times$, and for an arbitrary residue class $l \pmod{k}$ with $k \geq 2$, we consider the set

$$Q_a(x;k,l) = \{p \leq x \ : D_a(p) \equiv l \pmod{k}\}$$

and denote its natural density by $\Delta_a(k,l)$, to be precise,

$$\Delta_a(k,l) = \lim_{x\to\infty} \frac{\sharp Q_a(x;k,l)}{\pi(x)},$$

where $\pi(x) = \sum_{p\leq x} 1$.

Number Theory: Tradition and Modernization, pp. 11–22
W. Zhang and Y. Tanigawa, eds.

In [1] and [7], we studied the case $k = 4$. There assuming the Generalized Riemann Hypothesis (GRH), we proved that any $Q_a(x; 4, l)$ has the natural density $\Delta_a(4, l)$, and determined its explicit value. In [2], we extended our previous result to the case $k = q^r$, a prime power. On the basis of these results, we succeeded in revealing the relation between the natural density of $Q_a(x; q^{r-1}, l)$ and that of $Q_a(x; q^r, l)$. It is clear that, for any $r \geq 1$,

$$\Delta_a(q^{r-1}, l) = \sum_{t=0}^{q-1} \Delta_a(q^r, l + tq^{r-1})$$

and we were able to verify that, when r is not "very small", we have

$$\Delta_a(q^r, j + tq^{r-1}) = \frac{1}{q}\Delta_a(q^{r-1}, j),$$

for any t, — "equi-distribution property" — for details, see [2].

In this paper we study the most general case — k being composite.

Our main result is :

Theorem 1.1. *We assume GRH, and assume a is not a perfect b-th power with $b \geq 2$. Then, for any residue class $l \pmod{k}$, the set $Q_a(x; k, l)$ has the natural density $\Delta_a(k, l)$, and the values of $\Delta_a(k, l)$ are effectively computable.*

From this result, we find some interesting relationships between $\Delta_a(k, l)$ and $\Delta_a(k', l')$ with $k'|k$ and $l' \equiv l \pmod{k'}$.

In order to prove Theorem 1.1, we make use of two combined methods.

Let $I_a(p)$ be the **residual index** of $a \pmod{p}$, i.e. $I_a(p) = |(\mathbf{Z}/p\mathbf{Z})^\times : \langle a \pmod{p}\rangle|$. The first method is the one we already used in [1] and [7], and consists of the following: in order to calculate the density $\Delta_a(4, 1)$, first we decompose the set $Q_a(x; 4, 1)$, which reads in terms of cardinarity:

$$\begin{aligned}
&\sharp Q_a(x; 4, 1) \\
&= \sum_{f\geq 1}\sum_{l\geq 0} \sharp\{p \leq x \,:\, I_a(p) = 2^f + l\cdot 2^{f+2}, p \equiv 1 + 2^f \pmod{2^{f+2}}\} \\
&+ \sum_{f\geq 1}\sum_{l\geq 0} \sharp\{p \leq x \,:\, I_a(p) = 3\cdot 2^f + l\cdot 2^{f+2}, p \equiv 1 + 3\cdot 2^f \pmod{2^{f+2}}\}
\end{aligned} \tag{1.1}$$

(cf. [1] formula (3.4)). We calculate all cardinal numbers on the right hand side. In the process the calculations of the extension degree $[\tilde{G}_{k,n,d} : \mathbf{Q}]$ and the coefficients $c_r(k, n, d)$ $(r = 1, 3)$ play crucial roles (for details, see [1]). The technique used here is a generalization of that of Hooley [5], in which under GRH he obtained a quantitative result on Artin's conjecture for primitive roots. This method is feasible again in this paper (Section 2).

Let

$$k = p_1^{e_1} \dots p_r^{e_r}$$

be the prime power decomposition of k, where p_i's are distinct primes and $e_i \geq 1$. If l satisfies the condition $p_i^{e_i} \nmid l$ for any i, $1 \leq i \leq r$, we can apply the above method to such $Q_a(x; k, l)$'s. Then we can prove the existence of its natural density and can calculate it directly (Theorem 2.2).

Our second method is more elementary. For $Q_a(x; k, l)$ such that $p_i^{e_i} | l$ for some i, we can prove in Theorem 3.1 that the natural density of such $Q_a(x; k, l)$ is written as a linear combination of the densities of

$$Q_a(x; k, l') \quad \text{with } p_i^{e_i} \nmid l' \quad \text{for any } i, 1 \leq i \leq r,$$

and those of

$$Q_a(x; k', l'') \quad \text{with } k'|k.$$

Then, by Theorem 2.2, we can prove inductively the existence of the natural density of $Q_a(x; k, l)$ and determine simultaneously its explicit value.

Here we remark that, if $p_i^{e_i} | l$ for all i, then $l = 0$ and we already have a similar result in Hasse [3], [4] and Odoni [8].

2. Existence of the Density — l not divisible by any $p_i^{e_i}$

Let $k = \prod_{i=1}^r p_i^{e_i}$ as above and put $l = h \prod_{i=1}^r p_i^{f_i}$ ($(h, k) = 1$). In this section, we assume $0 \leq f_i \leq e_i - 1$ for all i. For $g_i \geq f_i$, let

$$k' = k'(g_1, \dots, g_r) = \prod_{i=1}^r p_i^{g_i} \quad \text{and} \quad k'' = k''(g_1, \dots, g_r) = \prod_{i=1}^r p_i^{e_i + g_i}.$$

Then under GRH, we can prove the existence of the density $\Delta_a(k, l)$ in a similar manner to that of [2, Section 2]. In fact, we can decompose the set $Q_a(x; k, l)$ which reads, in terms of cardinality,

Lemma 2.1. *Under the above notations, we have*

$$\sharp Q_a(x; k, l) = \sum_{g_1 \geq f_1} \cdots \sum_{g_r \geq f_r} \sum_{\substack{0 < v < k \\ (v,k)=1}} \sum_{t \geq 0} \sharp N_a(x; m; 1 + vk' \pmod{k''}), \tag{2.1}$$

where

$$N_a(x;m;1+vk' \pmod{k''}) = \left\{p \le x \,:\, I_a(p) = m,\ p \equiv 1+vk' \pmod{k''}\right\},$$

$$m = \left\{\overline{h}v \pmod{\prod_{i=1}^{r} p_i^{e_i-f_i}} + t\prod_{i=1}^{r} p_i^{e_i-f_i}\right\} \cdot \prod_{i=1}^{r} p_i^{g_i-f_i} \tag{2.2}$$

and where $\overline{h}h \equiv 1 \pmod{\prod_{i=1}^{r} p_i^{e_i-f_i}}$.

Proof. The proof goes on the same lines as in [2, Lemma 2.2] and is omitted. □

This decomposition turns out to yield the existence of the density $\Delta_a(k,l)$. Before stating the main theorem of this section, we introduce some notations. For $k \in \mathbf{N}$, let $\zeta_k = \exp(2\pi i/k)$. We denote Euler's totient by $\varphi(k)$. We define the following two types of number fields:

$$\begin{aligned} G_{m,n,d} &= \mathbf{Q}(a^{1/mn}, \zeta_{md}, \zeta_n), \\ \tilde{G}_{m,n,d} &= G_{m,n,d}(\zeta_{kk'}). \end{aligned}$$

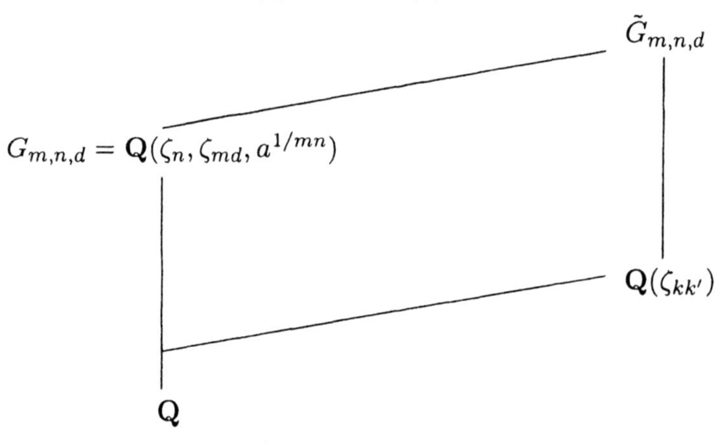

We take an automorphism $\sigma_v \in \mathrm{Gal}(\mathbf{Q}(\zeta_{kk'})/\mathbf{Q})$ determined uniquely by the condition $\sigma_v : \zeta_{kk'} \mapsto \zeta_{kk'}^{1+vk'}$ $(0 < v < k,\ (v,k) = 1)$, and we consider the automorphism $\sigma_v^* \in \mathrm{Gal}(\tilde{G}_{m,n,d}/G_{m,n,d})$ which satisfies $\sigma_v^*|_{\mathbf{Q}(\zeta_{kk'})} = \sigma_v$. We can verify that such a σ_v^* is unique if it exists (see [1, Lemma 4.3]).

Theorem 2.2. *Let k and l be as above. Then under GRH, we have*

$$\sharp Q_a(x;k,l) = \Delta_a(k,l)\,\mathrm{li}\ x + O\left(\frac{x}{\log x \log\log x}\right)$$

as $x \to \infty$, where

$$\Delta_a(k,l) = \sum_{g_1 \geq f_1} \cdots \sum_{g_r \geq f_r} \sum_{\substack{0<v<k \\ (v,k)=1}} \sum_{t \geq 0} \frac{m}{\varphi(m)} \sum_{d|m} \frac{\mu(d)}{d} \sum_{n=1}^{\infty} \frac{\mu(n) c_v(m,n,d)}{[\tilde{G}_{m,n,d} : \mathbf{Q}]}. \tag{2.3}$$

The series on the right hand side always converges, the number m is defined by (2.2) and

$$c_v(m,n,d) = \begin{cases} 1, & \textit{if } \sigma_v^* \textit{ exists}, \\ 0, & \textit{otherwise}. \end{cases}$$

Remark 1. When $(1+vk',k) > 1$, we define $c_v(m,n,d) = 0$ (in this case σ_v does not exist).

Proof. We can prove this theorem similarly to [1, Section 4], and so we state the outline only (see also [2, Section 2]). From (2.1) we have

$$\begin{aligned} &\sharp N_a\Big(x; m; 1+vk' \ (\mathrm{mod}\ k'')\Big) \\ &= \frac{1}{[K_m : \mathbf{Q}]} \frac{m_0}{\varphi(m_0)} \sum_{d|m_0} \frac{\mu(d)}{d} \sharp B\Big(x; K_m; a^{1/m}; md; 1+vk' \ (\mathrm{mod}\ k'')\Big), \end{aligned}$$

where $K_m = \mathbf{Q}(\zeta_{m_0}, a^{1/m})$, $m_0 = \prod_{p|m,\ p:\text{prime}} p$,

$$\begin{aligned} &B(x; K_m; a^{1/m}; N; s \ (\mathrm{mod}\ t)) \\ &= \left\{ \mathfrak{p} : \begin{array}{l} \text{a prime ideal in } K_m,\ N\mathfrak{p} = p^1 \leq x, p \equiv 1 \ (\mathrm{mod}\ N), \\ p \equiv s \ (\mathrm{mod}\ t),\ a^{1/m} \text{ is a primitive root mod}\,\mathfrak{p} \end{array} \right\} \end{aligned}$$

and $N\mathfrak{p}$ is the (absolute) norm of $\mathfrak{p}$. Next we define

$$\begin{aligned} &P(x; K_m; a^{1/m}; md; s \ (\mathrm{mod}\ t); n) \\ &= \left\{ \mathfrak{p} : \begin{array}{l} \text{a prime ideal in } K_m \text{ s.t. } N\mathfrak{p} = p^1 \leq x, p \equiv 1 \ (\mathrm{mod}\ md), \\ p \equiv s \ (\mathrm{mod}\ t), \text{ and the equation } X^q \equiv a^{1/m} \ (\mathrm{mod}\ \mathfrak{p}) \\ \text{is solvable in } O_{K_m} \text{ for any } q|n. \end{array} \right\}. \end{aligned}$$

Then we have

$$\begin{aligned} &\sharp B\Big(x; K_m; a^{1/m}; md; 1+vk' \ (\mathrm{mod}\ k'')\Big) \\ &= \sum_n{}' \mu(n) \sharp P\Big(x; K_m; a^{1/m}; md; 1+vk' \ (\mathrm{mod}\ k''); n\Big) \\ &\quad + O\left(\frac{x(\log\log x)^3}{\log^2 x}\right), \end{aligned}$$

where $\sum'_n$ means the sum over such $n \le x$ which are either 1 or a square free positive integer composed entirely of prime factors not exceeding $(1/8)\log x$, and the constant implied by the O-symbol depends only on a, k and l (see Propositions 4.1 and 4.2 of [1]).

By the uniqueness of σ^*, we can prove similarly as in [1, Proposition 4.4],

$$\begin{aligned}&\sharp P\Big(x; K_m; a^{1/m}; md; 1+vk' \pmod{k''}; n\Big)\\ &= \pi(x; \tilde{G}_{m,n,d}/K_m, \{\sigma^*\}) + O(m^2\sqrt{x}),\end{aligned}$$

where

$$\pi(x; L/K, C) = \sharp\left\{\mathfrak{p} : \begin{array}{l}\text{a prime ideal in } K\text{, unramified in } L,\\ (\mathfrak{p}, L/K) = C,\ N\mathfrak{p} \le x\end{array}\right\}$$

for a finite Galois extension L/K and a conjugacy class C in $\mathrm{Gal}(L/K)$, $(\mathfrak{p}, L/K)$ being the Frobenius symbol. The constant implied by the O-symbol depends only on a, k and l.

We can estimate $[\tilde{G}_{m,n,d} : K_m]$ and the discriminant $d_{\tilde{G}_{m,n,d}}$ of $\tilde{G}_{m,n,d}$ as follows:

$$[\tilde{G}_{m,n,d} : K_m] = \delta \frac{d}{m_0\varphi((n, m_0))} \cdot mn\varphi(n)$$

and

$$\log |d_{\tilde{G}_{m,n,d}}| \ll (mnd)^3 \log(mnd),$$

where δ and the constant implied by $\ll$ depends only on a, k and l (the proof is similar to [1, Lemma 4.6]). This estimate is based on Lagarias-Odlyzko [6].

By Lemma 2.1, $\sharp Q_a(x; k, l)$ is the infinite sum of $\sharp N_a\big(x; m; 1 + vk' \pmod{k''}\big)$'s, and the above results show that each $\sharp N_a\big(x; m; 1 + vk' \pmod{k''}\big)$ is the sum of $\pi(x; \tilde{G}_{m,n,d}/K_m, \{\sigma^*\})$ plus error terms. The sum of these main terms gives rise to the main term $\Delta_a(k, l) \times \mathrm{li}(x)$. And, in a similar way as in [1], we can estimate the sum of the error terms by $O(x \log^{-1} x \log\log^{-1} x)$, completing the proof. □

3. Existence of the Density — l being divisible by some $p_i^{e_i}$

In this section, we shall prove the following result:

Theorem 3.1. *If l is divisible by some $p_i^{e_i}$, then $Q_a(x; k, l)$ has the natural density $\Delta_a(k, l)$ and we can calculate it effectively.*

We prove this theorem by induction on r — the number of distinct prime factors of k.

For $k = p_1^{e_1}$, our assertion is true by [2].

For $k = \prod_{i=1}^{r} p_i^{e_i}$ — the general case — we assume, without loss of generality,

$$\prod_{p_i^{e_i} \| l} p_i^{e_i} = p_1^{e_1} \dots p_s^{e_s}.$$

If $s = 0$ (i.e. l is not divisible by any $p_i^{e_i}$), our assertion is true by Theorem 2.2, and so we assume $s \geq 1$ and put $l_0 = p_1^{e_1} \dots p_s^{e_s}$.

Write

$$l = m_0 + n_0 \frac{k}{l_0}, \; 0 \leq m_0 < \frac{k}{l_0}$$

and consider the decomposition:

$$Q_a\Big(x; \frac{k}{l_0}, m_0\Big) = \bigcup_{n=0}^{l_0-1} Q_a\Big(x; k, m_0 + n\frac{k}{l_0}\Big). \tag{3.1}$$

For $j \geq s+1$, since $p_j^{e_j} | \frac{k}{l_0}$ and $p_j^{e_j} \nmid l$, then for any n, $p_j^{e_j}$ does not divide $m_0 + n\frac{k}{l_0}$. So we have, for any n,

$$\sharp\Big\{j \,:\, p_j^{e_j} \,\Big|\, m_0 + n\frac{k}{l_0}\Big\} \leq s.$$

Moreover the condition

$$\sharp\Big\{j \,:\, p_j^{e_j} \,\Big|\, m_0 + n\frac{k}{l_0}\Big\} = s$$

is satisfied, if and only if,

$$\Big\{j \,:\, p_j^{e_j} \,\Big|\, m_0 + n\frac{k}{l_0}\Big\} = \{1, 2, \dots, s\} \quad \text{and } n = n_0.$$

In fact, $\{j \,:\, p_j^{e_j} | m_0 + n_0\frac{k}{l_0}\} = \{1, 2, \dots, s\}$ is clear, and if $\sharp\{j \,:\, p_j^{e_j} | m_0 + n'\frac{k}{l_0}\} = s$ for some n', then l_0 divides both $m_0 + n'\frac{k}{l_0}$ and $m_0 + n_0\frac{k}{l_0}$, thus $n' = n_0$. Therefore, except for $Q_a(x; k, l)$, all other $Q_a(x; k, m_0 + n\frac{k}{l_0})$ appearing in (3.1) satisfy

$$\sharp\Big\{j \,:\, p_j^{e_j} \,\Big|\, m_0 + n\frac{k}{l_0}\Big\} < s,$$

and for those $Q_a(x; k, m_0 + n\frac{k}{l_0})$ we know the existence of its natural density, from the induction hypothesis. And, also from the induction hypothesis, the set on the left hand side of (3.1) has its density. Then we can conclude that $Q_a(x; k, l)$ has its natural density.

This proof provides an *algorithm* to determine the density $\Delta_a(x;k,l)$, but it is difficult to write down the pervading formula in general. In the next section, we will present a numerical example and clarify the contents of Theorem 3.1.

4. Some numerical examples

We take $a = 5$ and $k = 12 = 2^2 \cdot 3^1$. Unconditionally, we have $\Delta_5(12,0) = 1/4$.

For such an l with $2^2 \nmid l$ and $3 \nmid l$, we can apply Theorem 2.2, and get the densities :

$$\Delta_5(12,1), \Delta_5(12,2), \Delta_5(12,5), \Delta_5(12,7), \Delta_5(12,10), \Delta_5(12,11).$$

First we state the results for these densities. We can determine the value $c_v(m,n,d)$ in Theorem 2.2 similarly to [2, Section 3]:

Proposition 4.1. *We assume GRH and let $a = 5$, $k = 12$. Then we have the following:*
(I) *When $l = 1, 5, 7, 11$, the value $c_v(m,n,d)$ in (2.3) is given as follows:*
(i) *If $g_1 \geq 1$ and $g_2 \geq 1$,*

$$c_v(m,n,d) = \begin{cases} 1, & \text{if } 2 \nmid d \text{ and } 3 \nmid d, \\ 0, & \text{otherwise.} \end{cases}$$

(ii) *If $g_1 \geq 1$ and $g_2 = 0$, then $c_v(m,n,d) = 1$ if and only if*
(a) *$2 \nmid d$, $3 \nmid n$, g_1: odd, $v \equiv 5 \pmod 6$ or*
(b) *$2 \nmid d$, $3 \nmid n$, g_1: even, $v \equiv 1 \pmod 6$,*
and $c_v(m,n,d) = 0$ in all other cases.

(II) *When $l = 2, 10$, the value $c_v(m,n,d)$ in (2.3) is given as follows:*
(i) *If $g_1 \geq 1$ and $g_2 \geq 1$,*

$$c_v(m,n,d) = \begin{cases} 1, & \text{if } 3 \nmid d, \\ 0, & \text{otherwise.} \end{cases}$$

(ii) *If $g_1 \geq 1$ and $g_2 = 0$, then $c_v(m,n,d) = 1$ if and only if*
(a) *$3 \nmid n$, g_1: odd, $v \equiv 5 \pmod 6$ or*
(b) *$3 \nmid n$, g_1: even, $v \equiv 1 \pmod 6$,*
and $c_v(m,n,d) = 0$ in all other cases.

We can also calculate the extension degree $[\tilde{G}_{m,n,d} : \mathbf{Q}]$ (see [2, Lemma 3.3]). In the following lemma, $\langle m_1, \cdots, m_n \rangle$ means the least common multiple of $m_1, \cdots, m_n$.

Lemma 4.2.

$$[\tilde{G}_{m,n,d} : \mathbf{Q}] = \begin{cases} mn\varphi(\langle md, n, 2^{g_1+2}3^{g_2+1}\rangle), \\ \frac{1}{2}mn\varphi(\langle md, n, 2^{g_1+2}3^{g_2+1}\rangle), \end{cases}$$

where the latter case happens if and only if mn *is even and* $5|\langle md, n\rangle$.

Now we can transform the series (2.3) for $a = 5$, $k = 12$ and $l = 1, 2, 5, 7, 10, 11$ into an expression involving some Euler products. The proof is similar to [7, Section 5] (see also [2, Section 4]):

Theorem 4.3. *Let* χ *be a nontrivial character of* $(\mathbf{Z}/6\mathbf{Z})^\times$. *We define the constant* C_χ *by*

$$\begin{aligned} C_\chi &= \prod_{\substack{p:\text{prime}\\ p\neq 2,3}} \left(1 - \frac{p(1-\chi(p))}{(p-1)(p^2-\chi(p))}\right) \\ &= \prod_{\substack{p:\text{prime}\\ p\equiv 5 \pmod 6}} \left(1 - \frac{-2p}{(p-1)(p^2+1)}\right) \approx 0.86989. \end{aligned}$$

Then under GRH, we have the following:
(I) *For* $l = 1,\ 5,\ 7,\ 11$,

$$\Delta_5(12, l) = \frac{1}{96} + \frac{1}{120}\left(5 - \chi(l)\frac{126}{47}C_\chi\right).$$

(II) *For* $l = 2,\ 10$,

$$\Delta_5(12, 6 \pm 4) = \frac{5}{48} \pm \frac{109}{1880}C_\chi.$$

Theoretical approximate values are

$$\begin{aligned} \Delta_5(12,1) = \Delta_5(12,7) &= \frac{5}{96} - \frac{21}{940}C \approx 0.03265, \\ \Delta_5(12,5) = \Delta_5(12,11) &= \frac{5}{96} + \frac{21}{940}C \approx 0.07151, \\ \Delta_5(12,2) &= \frac{5}{48} - \frac{109}{1880}C \approx 0.053732, \\ \Delta_5(12,10) &= \frac{5}{48} + \frac{109}{1880}C \approx 0.154602. \end{aligned}$$

For the remaining values of l, i.e. for $l = 3$, 4, 6, 8 and 9, we have by Theorem 3.1,

$$\Delta_5(12,3) = \Delta_5(4,3) - \Delta_5(12,7) - \Delta_5(12,11) = \frac{1}{16},$$
$$\Delta_5(12,6) = \Delta_5(4,2) - \Delta_5(12,2) - \Delta_5(12,10) = \frac{1}{8},$$
$$\Delta_5(12,9) = \Delta_5(4,1) - \Delta_5(12,1) - \Delta_5(12,5) = \frac{1}{16},$$

and

$$\Delta_5(12,4) = \Delta_5(3,1) - \Delta_5(12,1) - \Delta_5(12,7) - \Delta_5(12,10) = \frac{1}{6} - \frac{5}{376}C,$$
$$\Delta_5(12,8) = \Delta_5(3,2) - \Delta_5(12,2) - \Delta_5(12,5) - \Delta_5(12,11) = \frac{1}{24} + \frac{5}{376}C.$$

Consequently, we can determine all densities.

Numerical data seem to be well-matched with these theoretical densities. In the table below, $\Delta_5(x;12,l) = \sharp Q_5(x;12,l)/\pi(x)$ at $x = 179424673$ (10^7 th prime).

Table 1. Experimental densities $\Delta_5(x;12,l)$.

j	0	1	2	3	4	5
theoretical	0.125000	0.032650	0.053732	0.062500	0.155099	0.071517
experimental	0.124955	0.032617	0.053689	0.062416	0.154655	0.071531

j	6	7	8	9	10	11
theoretical	0.125000	0.032650	0.053234	0.062500	0.154601	0.071517
experimental	0.125067	0.032665	0.053736	0.062595	0.154542	0.071532

Remark 2. When considering $\Delta_a(12,l)$, one may expect that one would encounter the multiplicative characters mod 12, but in the above example, only the character mod 6 appeared. This is caused by the fact that $c_v(m,n,d)$ is determined by the condition of v (mod 6). We have already come across similar phenomena in our previous papers. For example, in [7], we needed the nontrivial character mod 4 in general, which give rise to the absolute constant C (see [7, Theorem 1.2]), but in some cases, we obtained the densities $\Delta_a(4,1) = \Delta_a(4,3) = 1/6$ (under GRH) and C did not appear. We can explain this "vanishing" of the absolute constant from the same viewpoint. Thus, if we take $a = 10$ for example, then $c_v(m,n,d)$ is *not* determined by the condition of v (mod 6). Indeed, when $l = 1, 5, 7, 11$, $c_v(m,n,d) = 1$ happens in the following cases:

(I) If $g_2 \geq 1$,

(i) $g_1 \geq 3$; $2, 3 \nmid d$,

(ii) $g_1 = 2;\ 2, 3 \nmid d;\ 5 \nmid \langle md, n\rangle$,

(iii-a) $g_1 = 1;\ 2, 3 \nmid d;\ 5 \nmid \langle md, n\rangle$,

(iii-b) $g_1 = 1;\ 2, 3 \nmid d;\ 5|\langle md, n\rangle$, $\begin{cases} g_2 : \text{odd}, & r = 1, 5, \\ g_2 : \text{even}, & r = 7, 11. \end{cases}$

(II) If $g_2 = 0$,

(i) $g_1 \geq 3,\ 2 \nmid d,\ 3 \nmid n$, $\begin{cases} g_1 : \text{odd}, & r \equiv 5 \pmod 6, \\ g_1 : \text{even}, & r \equiv 1 \pmod 6, \end{cases}$

(ii) $g_1 = 2,\ 2 \nmid d,\ 3 \nmid n,\ 5 \nmid \langle md, n\rangle,\ r \equiv 1 \pmod 6$,

(iii-a) $g_1 = 1,\ 2 \nmid d,\ 3 \nmid n,\ 5 \nmid \langle md, n\rangle,\ r \equiv 5 \pmod 6$,

(iii-b) $g_1 = 1,\ 2 \nmid d,\ 3 \nmid n,\ 5|\langle md, n\rangle,\ r = 11$.

In such cases, it happens that $\Delta_a(12, l)$'s are indeed determined mod 12. We can observe it from the following experimental results:

Table 2. Experimental densities $\Delta_{10}(x; 12, l)$.

l	0	1	2	3	4	5
$\Delta_{10}(x; 12, l)$	0.124991	0.028384	0.062410	0.062506	0.148836	0.075796

l	6	7	8	9	10	11
$\Delta_{10}(x; 12, l)$	0.125061	0.034013	0.059552	0.062442	0.145885	0.070125

Remark 3. We notice that the distribution property of $\Delta_5(12, j)$ are complicated.

When

$$j \pmod{12} = j_1 \pmod 4 \times j_2 \pmod 3$$

in $\mathbf{Z}/12\mathbf{Z} \cong \mathbf{Z}/4\mathbf{Z} \times \mathbf{Z}/3\mathbf{Z}$, we naïvely expect

$$\Delta_5(12, j) = \Delta_5(4, j_1)\Delta_5(3, j_2)$$

— local multiplicity —, but the following examples show that the distribution is not so simple.

1

$$\Delta_5(3, 0) = \frac{3}{8}$$

$$\Delta_5(3, 1) = \frac{3}{8}$$

$$\Delta_5(3, 2) = \frac{1}{4}$$

$$\Delta_5(4,1) = \frac{1}{6} \quad \begin{cases} \Delta_5(12,9) = \dfrac{1}{16} \\[2mm] \Delta_5(12,1) = \dfrac{5}{96} - \dfrac{21}{940}C \\[2mm] \Delta_5(12,5) = \dfrac{5}{96} + \dfrac{21}{940}C \end{cases}$$

References

[1] K. Chinen and L. Murata, *On a distribution property of the residual order of* a (mod$p$), J. Number Theory **105** (2004), 60–81.

[2] K. Chinen and L. Murata, *On a distribution property of the residual order of* a (mod$p$) — *III*, preprint.

[3] H. Hasse, *Über die Dichte der Primzahlen* p*, für die eine vorgegebene ganzrationale Zahl* $a \neq 0$ *von durch eine vorgegebene Primzahl* $l \neq 2$ *teilbarer bzw. unteilbarer Ordnung mod* p *ist*, Math. Ann. **162** (1965), 74–76.

[4] H. Hasse, *Über die Dichte der Primzahlen* p*, für die eine vorgegebene ganzrationale Zahl* $a \neq 0$ *von gerader bzw. ungerader Ordnung mod* p *ist*, Math. Ann. **166** (1966), 19–23.

[5] C. Hooley, *On Artin's conjecture*, J. Reine Angew. Math. **225** (1967), 209–220.

[6] J. C. Lagarias and A. M. Odlyzko, *Effective versions of the Chebotarev density theorem*, in : Algebraic Number Fields (Durham, 1975), Acadeic Press, London, 1977, 409-464.

[7] L. Murata and K. Chinen, *On a distribution property of the residual order of* a (mod$p$) — *II*, J. Number Theory **105** (2004), 82–100.

[8] R. W. Odoni, *A conjecture of Krishnamurthy on decimal periods and some allied problems*, J. Number Theory **13** (1981), 303–319.

DIAGONALIZING "BAD" HECKE OPERATORS ON SPACES OF CUSP FORMS

YoungJu Choie [1] and Winfried Kohnen [2]

[1] *Department of Mathematics, Pohang Institute of Science and Technology, Pohang 790-784, Korea*
yjc@postech.ac.kr

[2] *Universität Heidelberg, Mathematisches Institut, INF 288, D-69120 Heidelberg, Germany*
winfried@mathi.uni-heidelberg.de

Abstract We show that "bad" Hecke operators on space of newforms "often" can be diagonalized.

Keywords: newforms, Hecke operators

2000 Mathematics Subject Classification: 11F33

1. Introduction

For an even integer $k \geq 2$ and $M \in \mathbf{N}$ let $S_k(M)$ be the space of cusp forms of weight k with respect to the usual Hecke congruence subgroup $\Gamma_0(M) = \{\begin{pmatrix} a & b \\ c & d \end{pmatrix} \in SL_2(\mathbf{Z}) \,|\, c \equiv 0 \pmod{M}\}$. As is well-known, there is a splitting $S_k(M) = S_k^{new}(M) \oplus S_k^{old}(M)$ where $S_k^{old}(M)$ is the space of old forms "coming from lower levels" and the space of newforms $S_k^{new}(M)$ is the orthogonal complement of $S_k^{old}(M)$ in $S_k(M)$ with respect to the Petersson scalar product [1].

Recall that one can write

$$S_k^{old}(M) = \bigoplus_{dt|M, t \neq M} S_k^{new}(t)|V_d \tag{1}$$

Number Theory: Tradition and Modernization, pp. 23–26
W. Zhang and Y. Tanigawa, eds.

where for $f = \sum_{n\geq 1} a(n)q^n \in S_k(M)$ we have put

$$(f|V_d)(z) := \sum_{n\geq 1} a(n)q^{nd} \qquad (z \in \mathcal{H}).$$

Here as usual $\mathcal{H}$ denotes the complex upper half-plane and $q = e^{2\pi iz}$ for $z \in \mathcal{H}$.

If p is a prime, then there is a Hecke operator T_p ($M \not\equiv 0 \pmod p$) resp. U_p ($M \equiv 0 \pmod p$) on $S_k(M)$. Recall that for $f(z) = \sum_{n\geq 1} a(n)q^n \in S_k(M)$ one has

$$(f|T_p)(z) = \sum_{n\geq 1} (a(pn) + p^{k-1}a(\frac{n}{p}))q^n$$

(with the convention that $a(\frac{n}{p}) = 0$ if $n \not\equiv 0 \pmod p$) and

$$(f|U_p)(z) = \sum_{n\geq 1} a(pn)q^n.$$

The T_p generate a commutative $\mathbf{C}$-algebra of hermitian operators on $S_k(M)$ and hence can be (simultaneously) diagonalized. The "bad" Hecke operators U_p are in general hermitian only on $S_k^{new}(M)$, not on $S_k(M)$.

Now fix $N \in \mathbf{N}$ and suppose that N is squarefree. The purpose of this paper is to show that U_p can be diagonalized on $S_k(pN)$ for all prime numbers p with $p \nmid N$ up to a finite number r of exceptions. The number r can be bounded by an explicit constant depending only on k and N.

Note that in [2] certain "bad" Hecke operators (with index a prime dividing the level M) are constructed and it is shown that $S_k(M)$ has a basis consisting of eigenfunctions of *all* Hecke operators (including the "bad" ones). However, those "bad" Hecke operators are different from the U_p-operators.

2. Statement of result and proof

Theorem. *Fix a squarefree $N \in \mathbf{N}$. Then U_p is diagonalizable on $S_k(pN)$ for all primes $p \nmid N$ up to a finite number r of exceptions. One has $r \leq C_{k,N}$ where*

$$C_{k,N} := \sum_{t|N} [\frac{k}{12}\sigma_1(t)] \cdot 2^{\sum_{\mu\geq 1}\left[\frac{[\frac{k}{12}\sigma_1(t)]}{2^\mu}\right]}$$

and $\sigma_1(t)$ is the sum of the positive divisors of t.

Remark. The bound $C_{k,N}$ can be slightly improved, see the arguments below.

Proof. In (1) put $M = pN$ where p is a prime with $N \not\equiv 0 \pmod p$.

On the right-hand side of (1), suppose that $p|t$. Then since N is squarefree and $N \not\equiv 0 \pmod p$, it follows that p does not divide d. Hence since U_p and V_d $(d \not\equiv 0 \pmod p)$ commute and U_p is hermitian on $S_k^{new}(t)$, we infer that U_p can be diagonalized on $S_k^{new}(t)|V_d$.

Let us now consider the subspace

$$\begin{aligned} S_k^{old,p}(pN) :&= \bigoplus_{dt|pN,\, t\neq pN,\, t\not\equiv 0 \pmod p} S_k^{new}(t)|V_d \\ &= \bigoplus_{t|N,\, d|p\frac{N}{t}} S_k^{new}(t)|V_d. \end{aligned}$$

Since p does not divide N, we have clearly

$$S_k^{old,p}(pN) = \bigoplus_{t|N,\, d|\frac{N}{t}} \left(S_k^{new}(t)|V_d \oplus S_k^{new}(t)|V_{pd}\right). \tag{2}$$

Let f be a normalized Hecke eigenform in $S_k^{new}(t)$ where $t|N$, with eigenvalue λ_p under the Hecke operator T_p acting on $S_k(t)$. Then

$$f|U_p = \lambda_p f - p^{k-1} f|V_p.$$

Since on the right-hand side of (2), p does not divide d, we conclude that U_p transforms $S_k^{new}(t)|V_d$ to $S_k^{new}(t)|V_{pd}$. Also V_pU_p is the identity operator and hence U_p maps $S_k^{new}(t)|V_{pd}$ to $S_k^{new}(t)|V_d$.

In fact, we see that U_p acts on the two-dimensional subspace $W_f := \mathbf{C}f|V_d \oplus \mathbf{C}f|V_{pd}$ through the matrix

$$\begin{pmatrix} \lambda_p & -p^{k-1} \\ 1 & 0 \end{pmatrix}$$

which has eigenvalues

$$\mu_p^{\pm} = \frac{\lambda_p \pm \sqrt{\lambda_p^2 - 4p^{k-1}}}{2}.$$

By Deligne's theorem, we have $\lambda_p^2 \leq 4p^{k-1}$, hence $\mu_p^+ \neq \mu_p^-$ and so U_p is diagonalizable on W_f unless $\lambda_p^2 = 4p^{k-1}$. In the latter case, however, since k is even it follows that $\sqrt{p}$ is contained in the number field K_f obtained from $\mathbf{Q}$ by adjoining all the Hecke eigenvalues λ_n $(n \geq 1)$ of f.

Let $p_1, \ldots, p_s$ be different primes. Then the degree of the extension

$$\mathbf{Q}(\sqrt{p_1}, \ldots, \sqrt{p_s})/\mathbf{Q}$$

is 2^s and we see that the number r of exceptional primes is bounded by a constant depending only on k and N.

To get an explicit bound, recall that $Gal(\mathbf{C}/\mathbf{Q})$ acts on the set of normalized Hecke eigenforms f in $S_k^{new}(t)$ by sending $f = \sum_{n\geq 1} a(n)q^n$ to $f^\sigma := \sum_{n\geq 1} a(n)^\sigma q^n$ $(\sigma \in Gal(\mathbf{C}/\mathbf{Q}))$. If

$$K_{t,k} := \prod_f K_f$$

(with f running over the normalized Hecke eigenforms in $S_k^{new}(t)$) is the composite field, it therefore follows that we can view $Gal(K_{t,k}/\mathbf{Q})$ as a subgroup of the symmetric group $\mathcal{S}_{g_{t,k}}$ where $g_{t,k}$ is the number of normalized Hecke eigenforms in $S_k^{new}(t)$.

Hence we conclude that $[K_{t,k} : \mathbf{Q}] \mid g_{t,k}!$ and hence also $[K_f : \mathbf{Q}] \mid g_{t,k}!$.

We clearly have $g_{t,k} \leq \dim S_k(t)$ and as is well-known, since t is square-free

$$\dim S_k(t) \leq \frac{k}{12}[SL_2(\mathbf{Z}) : \Gamma_0(t)] = \frac{k}{12}\sigma_1(t).$$

Therefore $[K_f : \mathbf{Q}] \mid [\frac{k}{12}\sigma_1(t)]!$ and we infer that the contribution to the number r coming from each $t|N$ is bounded by

$$g_{t,k} \cdot 2^{s_{t,k}} \leq [\frac{k}{12}\sigma_1(t)]2^{s_{t,k}}$$

where $2^{s_{t,k}}$ is the highest power of 2 that divides $[\frac{k}{12}\sigma_1(t)]!$. By an elementary argument, if $m \in \mathbf{N}$ then the highest 2-power that divides $m!$ is given by 2^ν with $\nu = \sum_{\mu\geq 1}[\frac{m}{2^\mu}]$. Hence we see that the total number r of exceptions is bounded by $C_{k,N}$.

Acknowledgments

This research was partially supported by the grant KOSEF R01-2003-00011596-0.

References

[1] S. Lang, *Introduction to modular forms*, Grundl. d. Math. Wiss. 222, Springer, Berlin Heidelberg New York, 1976.

[2] A. Pizer, *Hecke operators for* $\Gamma_0(N)$, J. Algebra **83** (1983), 39–64.

ON THE HILBERT-KAMKE AND THE VINOGRADOV PROBLEMS IN ADDITIVE NUMBER THEORY

Vladimir. N. Chubarikov
Faculty of Mechanics and Mathematics, M. V. Lomonosov Moscow State University, Vorobjovy Gory, 119992, Moscow, Russia
chubarik@mech.math.msu.su

Abstract This paper is survey on recent results on the system of equations $x_1^s + \cdots + x_k^s = \beta_s$, $1 \le s \le n$, in natural number unknowns — the Hilbert-Kamke system, and one on prime unknowns — the Vinogradov system of equations. The main problem is to determine or estimate the Hardy-Littlewood function $G(m)$ through (mean values of) trigonometrical sums and find the exponent of convergence of the associated singular integrals. We shall also state the corresponding results on the multivariate version of these problems.

Keywords: Basis, Hilbert-Kamke problem, Hilbert-Kamke problem in prime unknowns, Singular integrals, Mean value of trigonometrical sums

2000 Mathematics Subject Classification: Primary 11P05, 11D72

In order to treat the simultaneous systems of equations, in addition to a single one, we extend the notion of the set addition (cf. H. Halberstam-K. F. Roth [10]) to the case of n-dimensional vectors.

Let A consist of an infinite set of n-dimensional vectors $\bar{a} = (a_1, \ldots, a_n)$ with natural number components, and of the zero vector $(0, \ldots, 0)$, where the dimension n is fixed.

Then as in [10], A is said to be a basis of $(\mathbb{N} \cup \{0\})^n$ if at a finite step k we have

$$W(k) = \underbrace{A + \cdots + A}_{k} = (\mathbb{N} \cup \{0\})^n.$$

It is quite natural to consider the following two types of problems.

(A) To find the minimal number k_1 for which $W(k_1) = (\mathbb{N} \cup \{0\})^n$.

Number Theory: Tradition and Modernization, pp. 27–37
W. Zhang and Y. Tanigawa, eds.

(B) To find the minimal number k_2 such that the compliment of $W(k_2)$ in $W(k_1)$, $W(k_1)\backslash W(k_2)$ is finite.

We refer to these as the basis problems. We illustrate the above notions by the Waring problem with exponent m (the dimension being 1), in which $k_1 = g(m)$ signifies the least number k for which all natural numbers are expressed as the sum of k, m-th powers of natural numbers and $k_2 = G(m)$ is the least number for which sufficiently large natural numbers are expressed in the same way; $g(m)$ and $G(m)$ being known as the Hardy-Littlewood functions.

The existence of $G(m)$ was proved by D. Hilbert [11], who posed a further problem of simultaneous representation of several numbers by finite sums of powers of natural numbers with exponents $1, 2, \ldots, n$, or in other words, the solvability of the system of diophantine equations

$$\begin{cases} x_1 + \cdots + x_k = N_1, \\ \qquad \cdots \\ x_1^n + \cdots + x_k^n = N_n \end{cases}$$

in natural number unknowns $x_1, \ldots, x_k$. This problem subsequently became known as the "Hilbert-Kamke problem", with the corresponding function $G_1(n)$.

If we replace the natural numbers by primes, then we have the Goldbach problem and the Hilbert-Kamke problem in primes, respectively (to the latter of which we refer later). It should be noted that, the initial sequence A being formed by the multiplicative principle, the basis problems ask for finding connections between additive and multiplicative structure of the natural number sequences.

We note that the simultaneous basis problems have special features which distinguish them from the one-dimensional basis problem, i.e. we have to impose additional conditions on $N_1, \ldots, N_n$.

The first type of conditions, called *order conditions*, arise quite naturally from the growth conditions

$$(x_1^m + \cdots + x_k^m)^{r/m} \geq x_1^r + \cdots + x_k^r \geq k^{1-r/m}(x_1^m + \cdots + x_k^m)^{r/m},$$

whence for $m \leq r$

$$N_m^{r/m} \geq N_r \geq k^{1-r/m} N_m^{r/m}, \quad 1 \leq m, r \leq n \tag{1}$$

The order conditions in (1) are not sufficient as shown by G. I. Arkhipov [2], who found the necessary and sufficient order conditions.

The second type of necessary conditions is of *arithmetical nature* found by K. K. Mardzhanischvili [13], who, using Vinogradov's method [15]–[22], showed that for an appropriate $k = k(n)$, these conditions are also sufficient. G. I. Arkhipov [2] showed that these arithmetical conditions can be expressed as the solvability of the system of linear equations

$$t_1 \cdot 1^s + t_2 \cdot 2^s + \cdots + t_k \cdot k^s = N_s, \quad 1 \le s \le n$$

in integers $t_1, \ldots, t_k$.

The Hilbert-Kamke system of equations may be reduced in the form

$$x_1^s + \cdots + x_k^s = \beta_s, \quad 1 \le s \le n, \tag{2}$$

in unknowns x_m in the range $0 \le x_m \le 1$, $1 \le m \le k$, where $\beta_s = N_s P^{-s}$, $P > 1$.

Let γ be the singular integral of the system (2)

$$\gamma = \gamma_m(\beta_1, \cdots, \beta_n) = \int_{-\infty}^{\infty} \cdots \int_{-\infty}^{\infty} I^k e^{-2\pi i(\alpha_1\beta_1 + \cdots + \alpha_n\beta_n)} d\alpha_1 \cdots \alpha_n, \tag{3}$$

where

$$I = I(\alpha_1, \ldots, \alpha_n) = \int_0^1 e^{2\pi i f(x)} dx, \tag{4}$$

with $f(x) = \alpha_1 x + \cdots + \alpha_n x^n$. We note that γ admits another interpretation as follows. We denote by $\Omega = \Omega(h)$, the subset of the k-dimensional unit interval such that

1) $0 \le x_m \le 1, \quad m = 1, \ldots, k,$

2) $\left| \sum_{m=1}^{k} x_m^s - \beta_s \right| \le h, \quad h > 0, s = 1, \ldots, n,$

and set

$$\mu(h) = \int \cdots \int_{\Omega(h)} dx_1 \cdots dx_k.$$

Then for $k > 0.5n(n+1) + 1$ we have

$$\gamma = \lim_{h \to 0+} 2^{-n} h^{-n} \mu(h).$$

Now we shall state a result of Arkhipov on the estimate of γ. To this end we need the notion of the characteristic $\Delta = \Delta(y_1, \cdots, y_l)$ of vectors $(y_1, \ldots, y_l)$, $l \ge n$, which is defined as follows. By some method, say κ, we choose n entries with all indices distinct and label them by $z_1, \ldots, z_n$. We augment this by adding two more numbers $z_0 = 0, z_{n+1} = 1$. Then we set

$$\Delta(y_1, \ldots, y_l) = \max_{\kappa} \min_{\substack{0 \le i,j \le n+1 \\ i \ne j}} |z_i - z_j|.$$

Theorem 1 (G. I. Arkhipov [6]). *Let ε be the largest value of the characteristic of solution $(x_1, \dots, x_k)$ of the Hilbert-Kamke system (2) of equations. Then we have*

$$2^{2n(n-k)}k^{n-k}n^{-k-n}\varepsilon^{n(k-n)} \le \gamma \le 2^{2n^2}k^{2n}n^{k-2n}\varepsilon^{k-3n}.$$

It follows that $\varepsilon = 0$ if and only if $\gamma = 0$.

The following list gives main estimates for the function $G_1(n)$.

1. $n^2 \le G_1(n) \le 2^{n^2-n-2}n^4$ (K. K. Mardzhanischvili [13]),
2. $2^n - 1 < G_1(n) \le 3n^3 2^n - n$ (G. I. Arkhipov [2]),
3. $G(n) \sim 2^n$ (D. A. Mit'kin [14]).

We note, in passing, that in his investigation on the Hilbert-Kamke problem, G. I. Arkhipov gave a negative answer to a version of the Artin hypotheses on the system of forms. Subsequently he and A. A. Karatsuba obtained a lower bound of exponential type for the number of variables in the Artin hypotheses on the representation of the zero by forms of even degree.

We now turn to the Hilbert-Kamke problems in primes mentioned above, concerning the Hilbert-Kamke type system of equations

$$\begin{cases} p_1 + \cdots + p_k = N_1, \\ \quad \cdots \\ p_1^n + \cdots + p_k^n = N_n \end{cases} \tag{5}$$

in prime unknowns $p_1, \dots, p_k$. We call (5) the Vinogradov system of equations and the solvability of the system or the existence of the Hardy-Littlewood function $H_1(n)$ the Vinogradov problem.

Relying essentially on Vinogradov's estimates on trigonometric sums in primes, K. K. Mardzhanischvili and L.-K. Hua obtained the asymptotic formulae for the number of solutions of the Vinogradov system of equations with the number of summands $k \ll n^2 \log$. In 1985 the author [8] established the existence of, and obtained estimates on, $H_1(n)$ for the first time.

We say that the n-tuples $(N_1, \dots, N_n)$, $N_1 \to \infty$, belongs to the $(\bar{\gamma}, \varepsilon)$-cone if its entries satisfy the conditions

$$N_m = N_1^m(\gamma_m + O(N_1^{-\varepsilon}), \quad 1 \le m \le n,$$

where $\bar{\gamma} = (\gamma_1, \dots, \gamma_n)$, $\gamma_1, \dots, \gamma_n$ are some positive constants and ε is a small positive constant. An n-tuple belonging to the $(\bar{\gamma}, \varepsilon)$-cone is said to satisfy *the real solvability condition* if there exists a number $P = P(\bar{\gamma})$ such that for each $N_1 \geq P$, the system of equations

$$\begin{cases} x_1 + \cdots + x_k = \gamma_1, \\ \quad \cdots \\ x_1^n + \cdots + x_k^n = \gamma_n \end{cases} \tag{6}$$

is solvable in real numbers x_i, $0 \leq x_i \leq 1$, $1 \leq i \leq k$, and the Jacobi matrix

$$\|mx_s^{m-1}\|, \quad (1 \leq m \leq n, 1 \leq s \leq k)$$

of the solution $x_1, \ldots, x_k$ of (6) has a maximal rank.

We may now state a theorem on the value of the singular integral.

Theorem 2. *The value of the singular integral in the asymptotic formula for the Vinogradov system* (5) *of equations is positive or* 0 *according as the n-tuple $(N_1, \ldots, N_n)$ satisfies the real solvability condition or not. Furthermore, in the former case, if $N_m = N_1^m \gamma_m$, $1 \leq m \leq n$, then the system* (5) *has only finitely many solutions.*

Now we introduce another solvability condition for the Vinogradov system (4) of equations, of arithmetical nature, which, properly formulated, allows one to establish the existence of $H_1(n)$, in conjunction with the order condition.

We say that an n-tuple $(N_1, \ldots, N_n)$ satisfies the p-solvability condition if for any fixed prime number p, the system of congruences

$$\sum_{r=1}^{pl_p+r} t_{\nu,p}\nu^m \equiv N_m \pmod{p^{\delta_p}}, \quad 0 \leq m \leq n$$

is solvable in residue class unknowns $t_{\nu,p}$ to the modulus p^{δ_p}, where $n+1 = (p-1)l_p + r$, $0 \leq r \leq p-1$, i.e. $l_p = [\frac{n+1}{p-1}]$ and $p^{\delta_p} \| (n+l_p)\,!$, the highest power of p which divides $(n+l_p)\,!$. We say $(N_1, \ldots, N_n)$ satisfies the arithmetical solvability condition if the above system of congruences is solvable for each prime $p \leq n+1$.

The p-solvability condition may be stated in an equivalent form:

Theorem 3. *The n-tuple $(N_1, \ldots, N_n)$ satisfies the p-solvability condition if and only if the system of congruences*

$$\begin{cases} x_1 + \cdots + x_k \equiv N_1 \pmod{p^\alpha} \\ \quad \cdots \\ x_1^n + \cdots + x_k^n \equiv N_n \pmod{p^\alpha} \end{cases}$$

is solvable in x_s, for each natural number α, where $1 \leq x_s \leq p^\alpha$, $(x_s, p) = 1$, $1 \leq s \leq k$.

An n-tuple $(N_1, \dots, N_n)$ satisfying both real and arithmetical conditions is called *admissible*. For admissible n-tuples we have a complete description of the number k of summands given in the following

Theorem 4. *For an admissible n-tuple $(N_1, \dots, N_n)$, the number k of summands in the representatives*

$$\begin{cases} N_1 = p_1 + \cdots + p_k, \\ \qquad \cdots \\ N_n = p_1^n + \cdots + p_k^n, \end{cases} \tag{7}$$

has the form

$$k = k_0 + b(n)s, \tag{8}$$

where s is 0 or 1 to be determined by the arithmetical conditions (see below),

$$b(n) = \prod_{p \le n+1} p^{\delta_p}, \; p^{\delta_p} \| \left(n + \left[\frac{n+1}{p-1} \right] \right)!,$$

and k_0 is the least positive integer that satisfies the congruences

$$k_0 \equiv k_p \pmod{p^{\delta_p}}, \quad k_p = \sum_{\nu=1}^{pl_p + r} t_{\nu,p},$$

for all $p \le n+1$.

Let a denote the number of summands in the representation (7) of the admissible n-tuple $(N_1, \dots, N_n)$ for which the singular integral has a positive value. If $k_0 \ge a$, then we may take $s = 0$ in (8), but if $k_0 < a-1$, then for each such k_0, there exists an admissible n-tuple $(N_1, \dots, N_n)$ which are not represented as the sum of k_0 terms in the prescribed form, but represented as the sum of $k_0 + b(n)$ terms.

We turn to the multi-dimensional case. As the one-dimensional Hilbert-Kamke and Vinogradov problems are studied through single trigonometric sums, so are the multivariate Hilbert-Kamke and Vinogradov problems through multiple trigonometric sums; especially we need mean value theorems for them which we developed during 1974-1987 [8, 9, 4, 5] and state as Theorem 5 below.

The multiple (r-ple) trigonometric sum $S = S(\bar{\alpha})$ is defined by

$$S = S(\bar{\alpha}) = \sum_{x_1=1}^{P_1} \cdots \sum_{x_r=1}^{P_r} e^{2\pi i F(x_1, \dots, x_r)},$$

where $P_1, \ldots, P_r$ are natural numbers with $P_1 = \min(P_1, \ldots, P_r)$, $F(x_1, \ldots, x_r)$ is a polynomial in $x_1, \ldots, x_r$ with real coefficients,

$$F(x_1, \ldots, x_r) = \sum_{t_1=0}^{n_1} \cdots \sum_{t_r=0}^{n_r} \alpha(t_1, \ldots, t_r) x_1^{t_1} \cdots x_r^{t_r}$$

with $\alpha(0, \ldots, 0) = 0$ and $0 \leq \alpha(t_1, \ldots, t_r) \leq 1$.

Theorem 5. *Let*

$$J = J(P_1, \ldots, P_r; n_1, \ldots, n_r; k, r) = \int \underset{E}{\cdots} \int |S(\bar{\alpha})|^{2k} d\bar{\alpha},$$

where the integral is taken over the m*-dimensional unit cube* $E \subset \mathbb{R}^m$, $m = (n_1 + 1) \cdots (n_r + 1)$*:*

$$\bar{\alpha} = (0, \ldots, \alpha(t_1, \ldots, t_r), \ldots, \alpha(n_1, \ldots, n_r)) \in E,$$

i.e. J *equals to the number of solutions of the system of equations*

$$\sum_{j=1}^{2k} (-1)^{\nu} x_{1,j}^{t_1} \cdots x_{r,j}^{t_r} = 0 \quad (0 \leq t_1 \leq n_1, \ldots, 0 \leq t_r \leq n_r, t_1 + \cdots + t_r \geq 1)$$

in unknowns $x_{i,j}$ *in the range*

$$1 \leq x_{1,j} \leq P_1, \ldots, 1 \leq x_{r,j} \leq P_r, \quad j = 1, \ldots, 2k.$$

Then for any integer $\tau \geq 0$*, we have the estimate*

$$J \leq k^{2m\tau} \kappa^{4\kappa^2 \Delta(\tau)} 2^{8m\kappa\tau} (P_1 \cdots P_r)^{2k} P^{-\kappa \Delta(\tau)},$$

where $\kappa = n_1 \nu_1 + \cdots + n_r \nu_r$ *with natural numbers* $\nu_1, \ldots, \nu_r$ *such that*

$$-1 < \frac{\ln P_s}{\ln P_1} \leq 0, \quad 1 \leq s \leq r,$$

and other quantities are defined by

$$\kappa\gamma = 1, \Delta(\tau) = 0.5m(1 - (1 - \gamma)^{\tau}), P = (P_1^{n_1} \cdots P_r^{n_r})^{\gamma}.$$

From Theorem 5 we may deduce a sharp estimate for the multiple trigonometric sum $S = S(\bar{\alpha})$ for $\bar{\alpha}$ belonging to a principal part E_2 of E, which we now define. Let $q = q(t_1, \ldots, t_r)$ denote the denominator in Dirichlet's approximation to $\alpha = \alpha(t_1, \ldots, t_r) = \frac{a}{q} + \frac{\theta}{q^{\tau}}$ with $|\theta| \leq 1$ and $\tau = \tau(t_1, \ldots, t_r) = P_1^{t_1} \cdots P_r^{t_r} P^{-1/3}$. Let Q_0 be the least common multiple of all $q = q(t_1, \ldots, t_r)$, $t_1 + \ldots + t_r \geq 2$, and let E_2 be the set of all vectors $\bar{\alpha}$ whose components $\alpha(t_1, \ldots, t_r)$ satisfy the inequality $Q_0 \geq P^{1/6}$.

We may now state

Theorem 6. *For each $\bar{\alpha} \in E_2$ we have*

$$|S(\bar{\alpha})| \leq 2^{32\kappa} P_1 \cdots P_r P^{-\rho},$$

where $\rho = (32m\kappa \ln(8m\kappa))^{-1}$ and other quantities have the same meaning as in Theorem 5.

Since the study on the asymptotic formula for the members of solutions of the Hilbert-Kamke problems is closely related to Hua's problem on the exponent of convergence of singular integrals in Tarry's problem ([7]), we state at this point some results for the singular integral of the form (cf. the definition of $\gamma(\beta_1, \ldots, \beta_n)$)

$$\theta_0 = \int_{-\infty}^{\infty} \cdots \int_{-\infty}^{\infty} |I|^{2k} d\alpha_n \cdots d\alpha_1,$$

where I means the integral in (4).

Theorem 7. *The singular integral θ_0 converges for $2k > \frac{n^2+n}{2} + 1$ and diverges for $2k \leq \frac{n^2+n}{2} + 1$. More generally, let $r, \ldots, m$ and n be integers such that $1 \leq r < \cdots < m < n$, $r + \cdots + m + n < \frac{n^2+n}{2}$ and*

$$\theta_0' = \int_{-\infty}^{\infty} \cdots \int_{-\infty}^{\infty} \left| \int_0^1 \exp\left(2\pi i(\alpha_n x^n + \alpha_m x^m + \cdots + \alpha_r x^r)\right) dx \right|^{2k} d\alpha_n d\alpha_m \cdots d\alpha_r.$$

Then the integral θ_0' converges for $2k > n + m + \cdots + r$ and diverges for $2k \leq n + m + \cdots + r$.

The convergence criteria in Theorem 7 are based on the estimate of the trigonometric integral I_1, similar to I in (3).

Theorem 8. *Let I_1 denote the integral*

$$I_1 = \int_a^b e^{2\pi i f(x)} dx,$$

where $f(x) = \alpha_n x^n + \cdots + \alpha_1 x$ with real coefficients $\alpha_1, \ldots, \alpha_n$, $n \geq 1$. Let

$$\beta_r(x) = \frac{1}{r!} f^{(r)}(x), \quad r = 1, \ldots, n$$

and let

$$H = H(\alpha_n, \dots, n_1) = \min_{a \le x \le b} \sum_{r=1}^{n} |\beta_r(x)|^{1/r}.$$

Then the estimate

$$|I_1| \le \min(b-a, 6en^3 H^{-1})$$

holds.

In the multivariate Tarry problem, the singular integral $\theta = \theta_r(k)$ corresponding to θ_0 is

$$\theta = \int_{-\infty}^{\infty} \cdots \int_{-\infty}^{\infty} \left| \int_0^1 \cdots \int_0^1 \exp(2\pi i F_1(x_1, \dots, x_r)) dx_1 \cdots dx_r \right|^{2k} d\alpha(0, \dots, 1) \cdots d\alpha(n_1, \dots, n_r),$$

where F_1 corresponds to F in Theorem 5, with an additional condition $t_1 + \cdots + t_r \ge 1$:

$$F_1(x_1, \dots, x_r) = \sum_{\substack{t_1=0 \\ t_1+\cdots+t_r \ge 1}}^{n_1} \cdots \sum_{t_r=0}^{n_r} \alpha(t_1, \dots, t_r) x_1^{t_1} \cdots x_r^{t_r}.$$

The counterpart of Theorem 7 for the multivariate Tarry problem is that the singular integral θ_r converges for $2k > rC_{n+r}^{r+1} + r$, where r is the number of variables, and n does not exceed any of n_i $1 \le i \le r$.

The multivariate Hilbert-Kamke problem refers to the solvability and the estimate of the number $G_1(n;r)$ of summands of the system of equations

$$x_{1,1}^{t_1} \cdots x_{r,1}^{t_r} + \cdots + x_{1,k}^{t_1} \cdots x_{r,k}^{t_r} = N(t_1, \dots, t_r)$$
$$(0 \le t_i \le n_i,\ 1 \le i \le r,\ t_1 + \cdots + t_r \ge 1)$$

in natural number unknowns.

We obtained the estimate [5]

$$G_1(n;r) \ll (rn)^{r+1} 2^n.$$

Similarly, the multivariate Vinogradov problem refers to the solvability, and the estimate of the number $H_1(n;r)$ of summands, of the system of equations

$$p_{1,1}^{t_1} \cdots p_{r,1}^{t_r} + \cdots + p_{1,k}^{t_1} \cdots p_{r,k}^{t_r} = N(t_1, \dots, t_r)$$
$$(0 \le t_i \le n_i,\ 1 \le i \le r,\ t_1 + \cdots + t_r \ge 1)$$

in prime number unknowns.

For the study of this problem we developed the theory of multiple trigonometric sum $\tilde{S}$ with primes [8, 9], where $\tilde{S}$ is similar to the ordinary multiple trigonometric sum , with summation being restricted to primes:

$$\tilde{S} = \sum_{p_1 \le P_1} \cdots \sum_{p_r \le P_r} e^{2\pi i F(p_1,\dots,p_r)}.$$

We obtained the estimate whose precision corresponds to that of S in Theorem 5, and as one of its two applications, we gave the asymptotic for the number k of summands to the expected accuracy:

$$H_1(n;r) \sim (b(n))^r.$$

Finally, for comparison's sake, we restate the estimates [4, 5, 14]:

$$G_1(n) \sim 2^n(1+O(1)), \qquad G_1(n;r) \ll (rn)^{r+1}2^n,$$
$$H_1(n) \sim b(n), \quad b(n) > n!, \quad H_1(n;r) \sim (b(n))^r.$$

References

[1] G. I. Arkhipov, *Multiple trigonometric sums*, Dokl. Akad. Nauk SSSR **219** (1974), 1036–1037; English transl. in Soviet Math. Dokl. **15** (1974), 1702–1704.

[2] G. I. Arkhipov, *The values of a singular series in a Hilbert – Kamke problem*, ibid. **259** (1981), no.2, 265–267; English transl. in Soviet Math. Dokl. **24** (1981), no.1, 49–51.

[3] G. I. Arkhipov and V. N. Chubarikov, *Multiple trigonometric sums*, Izv. Akad. Nauk SSSR Ser. Mat. **40** (1976), 209–220; English transl. in Math. USSR Izv. **10** (1976), 200–210.

[4] G. I. Arkhipov and V. N. Chubarikov, *On the number of summands in the Hilbert – Kamke problem in prime numbers*, Dokl. Akad. Nauk SSSR **330** (1993), no.4, 407–408; English transl. in Soviet Math. Dokl. **47** (1993), no. 3, 485–488.

[5] G. I. Arkhipov and V. N. Chubarikov, *On the asymptotics of the number of summands in a multidimensional additive problem with prime numbers*, Dokl. Akad. Nauk SSSR **331** (1993), no.1, 5–6; English transl. in Soviet Math. Dokl. **48** (1994), no.1, 1–3.

[6] 14. G. I. Arkhipov, A. A. Karatsuba and V. N. Chubarikov, *The theory of multiple trigonometric sums*, Nauka, 1987, pp. 368.

[7] G. I. Arkhipov, A. A. Karatsuba and V. N. Chubarikov, *Exponent of convergence of singular integral in the Tarry problem*, Dokl. Akad. Nauk SSSR **248** (1979), no.2, 268–272; English transl. in Soviet Math. Dokl. **20** (1979), no.5, 978–981.

[8] V. N. Chubarikov, *Multiple trigonometric sums over primes*, Dokl. Akad. Nauk SSSR **278** (1984), no.2, 302–304; English transl. in Soviet Math. Dokl. **30** (1984), 391–393.

[9] V. N. Chubarikov, *Estimates of multiple trigonometric sums with primes*, Izv. Akad. Nauk SSSR Ser. Mat. **49** (1985) no.5, 1031–1067; English transl. in Math USSR Izv. **27** (1986), 323–357.

[10] H. Halberstam and K. F. Roth, *Sequences*, Oxford Univ. Press, Oxford, 1966.

[11] D. Hilbert, *Beweis für die Darstellbarkeit der ganzen Zahlen durch eine feste Anzahl n-ter Potenzen*, Math. Ann. **67** (1909), 281–300.

[12] Loo-Keng Hua, *Additive theory of prime numbers*, Proc. of Steklov Math. Institute of Acad. of Sciences of USSR, 22 1947 (In Russian); English transl. of Chinese rev. ed. Amer. Math. Soc. Province, RI 1965.

[13] K. K. Mardzhanishvili, *On simultaneous representation of numbers by sums of complete first, second, ..., nth powers*, Izv. Akad. Nauk SSSR Ser. Mat. **1** (1937), 609-631. (In Russian).

[14] D. A. Mit'kin, *Estimate for the number of summands in the Hilbert - Kamke problem*, Mat. Sb. (N. S.) **129** (171) (1986), no.4, 549–577; English transl. in Math. USSR Sb. **57** (1987) 561-590.

[15] I. M. Vinogradov, *Selected works*, Springer-Verlag, Berlin, 1985.

[16] I. M. Vinogradov, *An introduction to the Theory of Numbers*, Pergamon Press. London & New York, 1955, 155pp.

[17] I. M. Vinogradov, *A general Waring theorem*, Mat. Sb. **31** (1924), 490-507.

[18] I. M. Vinogradov, *On Waring theorem*, Izv. Akad. Nauk SSSR, ser. phiz.-matem. no.4-5 (1928), 393-400; English transl. in [15], 101–106.

[19] I. M. Vinogradov, *A new estimate for $G(n)$ in Waring's problem*, Dokl. Akad. Nauk SSSR **5** (1934), 249–253; English transl. in [15], 107–109.

[20] I. M. Vinogradov, *On an upper bound for $G(n)$ in Waring's problem*, Izv. Akad. Nauk SSSR, ser. phiz.-matem. no.10 (1934), 1455-1469; English transl. in [15], 110–123.

[21] I. M. Vinogradov, *A new variant of the proof of Waring's theorem*, Trudy Matem. Instituta im. V. A. Steklova **9** (1935), 5–15.

[22] I. M. Vinogradov, *The representation of an odd number as a sum of three prime numbers*, Dokl. Akad. Nauk SSSR **15** (1937), 291–294; English transl. in [15], 129–132.

THE GOLDBACH-VINOGRADOV THEOREM IN ARITHMETIC PROGRESSIONS

Zhen Cui
Department of Mathematics, Shanghai Jiao Tong University, Shanghai, 200240, P. R. China
cuizhen@sjtu.edu.cn

Abstract We prove that the ternary Goldbach problem can be solved with two of the prime variables in different arithmetic progressions.

Keywords: The Goldbach-Vinogradov Theorem (the Three Primes Theorem), the circle method, the singular series, the mean value theorem, the explicit formula

2000 Mathematics Subject Classification: 11P32

1. Introduction

In 1937, I. M. Vinogradov [16, 17] proved that there exists an absolute constant $N_0 > 0$ such that every odd integer N larger than N_0 is a sum of three primes. This result gave a decisive answer to the famous ternary Goldbach problem and is nowadays called the Goldbach-Vinogradov Theorem (Theorem G-V) or Three Primes Theorem. More precisely, the theorem reads

Theorem G-V. *For any large odd integer N, let $R(N)$ denote the number of solutions of the Diophantine equation*

$$N = p_1 + p_2 + p_3$$

in prime variables p_j, $j = 1, 2, 3$. Then we have

$$R(N) = \sigma(N)\frac{N^2}{2\log^3 N}\left(1 + O(\frac{1}{\log N})\right), \qquad as \quad N \to \infty,$$

Project supported partially by the National Natural Science Foundation of China(Grant No. 10471090).

Number Theory: Tradition and Modernization, pp. 39–65
W. Zhang and Y. Tanigawa, eds.

where $\sigma(N)$ is the so-called singular series defined by

$$\sigma(N)=\prod_{p|N}\left(1-\frac{1}{(p-1)^2}\right)\prod_{p\nmid N}\left(1+\frac{1}{(p-1)^3}\right)>\frac{1}{2}$$

and the constant implied by the O-symbol is absolute. Here, and in what follows, p runs through odd primes.

After Vinogradov's work, a large number of generalizations and refinements of Theorem G-V have been obtained by many authors. Most of them imposed some restrictions on the prime variables p_j $(j=1,2,3)$ to get refinements while some of them have considered the generalization with all or some of the three primes lying in arithmetic progressions. In fact, such problems were investigated even before Vinogradov's work. For instance, Rademacher [13] established a theorem in this direction in 1926. To state his theorem we first introduce some notation.

For positive integers m and N, we define

$$\mathcal{B}(N,m)=\left\{\boldsymbol{b}=(b_1,b_2,b_3)\in\mathbb{N}^3\ \middle|\ \begin{array}{c}1\le b_j\le m,\ (b_j,m)=1\ \text{and}\\ b_1+b_2+b_3\equiv N(\bmod m)\end{array}\right\}.$$

Then one has (see [9] for example)

$$\sharp\mathcal{B}(N,m)=m^2\prod_{\substack{p|m\\ p|N}}\frac{(p-1)(p-2)}{p^2}\prod_{\substack{p|m\\ p\nmid N}}\frac{p^2-3p+3}{p^2}.$$

Let $J(N,m,\boldsymbol{b})$ be the number of solutions of the Diophantine equation in prime variables p_j $(1\le j\le 3)$

$$\begin{cases} N &= p_1+p_2+p_3,\\ p_j &\equiv b_j \pmod m, \quad j=1,2,3.\end{cases} \tag{1.1}$$

Rademacher [13] proved

Theorem R. *Let m be a given positive integer. Then under the assumption of GRH, we have*

$$J(N,m,\boldsymbol{b})=\sigma(N,m)\frac{N^2}{2\log^3 N}(1+o(1)) \tag{1.2}$$

for odd N and all $\boldsymbol{b}\in\mathcal{B}(N,m)$. Here $\sigma(N,m)$ is the singular series *associated with this problem, satisfying*

$$\sigma(N,m) = \frac{C(m)}{m^2}\prod_{p|m}\frac{p^3}{(p-1)^3+1}\prod_{\substack{p|N\\ p\nmid m}}\frac{(p-1)((p-1)^2-1)}{(p-1)^3+1}$$
$$\times\prod_{p>2}\left(1+\frac{1}{(p-1)^3}\right) \gg 1,$$

where $C(m) = 2$ *for odd* m, *and* $C(m) = 8$ *for even* m.

Since this result was proved before Vinogradov's pioneering works [16, 17], the appearance of GRH was somehow inevitable. The GRH was removed by Zulauf [20] in 1952, and independently by Ayoub [1] in 1953 by different methods. Their arguments with some minor modifications can actually give (1.2) for every $r \le \log^A N$ for arbitrary $A > 0$. Here a natural question arises: Is (1.1) still solvable for large r (for example, for r up to a certain power of N)? Recently, M. C. Liu and T. Zhan [10] showed

Theorem L-Z. *Let* N *denote a large odd integer. There exists an absolute and computable constant* $\delta > 0$ *such that (1.1) is solvable for every* $m \le N^\delta$ *and every* $\boldsymbol{b} \in \mathcal{B}(N,m)$.

This partly answered the above question. On average in m, in 1993, Wolke [18] first broke the bound $\log^A N$ for m. In fact, he proved that Rademacher's formula (1.2) is true for almost all prime moduli $m = p \ll N^{1/11}$ by establishing the following Bombieri-Vinogradov type mean-value theorem concerning exponential sums over primes:

$$\sum_{q\le Q}\max_{y\le x}\max_{(a,q)=1}\max_{|\lambda|\le\theta}\left|\sum_{n\le y}\Lambda(n)e\left(n(\frac{a}{q}+\lambda)\right) - \frac{\mu(q)}{\varphi(q)}\sum_{n\le y}e(n\lambda)\right|$$
$$\ll x\log^{-A} x, \tag{1.3}$$

provided that Q and θ satisfy the conditions

$$Q = x^{1/4}, \qquad \theta = \min(Q^{-4}, (\log x)^{-8(A+21)}). \tag{1.4}$$

Recently, T. Zhan and J. Y. Liu [19, 7] relaxed the ranges of Q and θ in (1.4), so that the mean-value estimate (1.3) holds for

$$Q = x^{1/3}\log^{-B} x, \qquad \theta = Q^{-3}\log^{-B} x, \tag{1.5}$$

where $B > 0$ is a constant depending only on A. These ranges of Q and θ are wide enough to get the results which were obtained under GRH. On the basis of this improvement, it was shown that Rademacher's formula (1.2) holds true for almost all prime moduli $m = p \leq N^{3/20} \log^{-B} N$ by J. Y. Liu [6], and for almost all positive moduli $m \leq N^{1/8-\varepsilon}$ by J. Y. Liu and T. Zhan [7]. Very recently, Z. Cui [2] showed that Rademacher's formula (1.2) is true for almost all prime moduli $m = p \leq N^{1/6-\varepsilon}$ by another approach.

All the above-mentioned results are concerned with the case of all the prime variables lying in arithmetic progressions to the same modulus. In the present paper, we are concerned with the prime variables in arithmetic progressions to different moduli. Let $R/2 < r_1 \neq r_2 \leq R$ be primes and b_1, b_2 be two integers such that $(r_1, b_1) = (r_2, b_2) = 1$. Let also N be a large odd integer. Denote by $J(N, r_1, b_1, r_2, b_2)$ the number of solutions of the Diophantine equation in prime variables p_j

$$\begin{cases} N = p_1 + p_2 + p_3, \\ p_j \equiv b_j \pmod{r_j}, \quad j = 1, 2. \end{cases} \tag{1.6}$$

Then we have the following

Theorem 1. *Let N be a fixed large odd integer, $\varepsilon > 0$ arbitrarily small and*

$$N^{\varepsilon} \leq R \leq N^{1/8-\varepsilon}.$$

Let $A > 0$ be arbitrary. Then for all pairs of primes (r_1, r_2) such that $R/2 < r_1 \neq r_2 \leq R$, the Diophantine equation (1.6) in prime variables is solvable for all integers $(r_1, b_1) = (r_2, b_2) = 1$ with at most $O(R^2 \log^{-A} N)$ exceptions. As for the number of its solutions, we have

$$J(N, r_1, b_1, r_2, b_2) = \sigma(N, r_1, r_2, b_1, b_2) \frac{N^2}{2 \log^3 N} \left(1 + O\left(\frac{1}{\log N}\right)\right), \tag{1.7}$$

where the singular series $\sigma(N, r_1, r_2, b_1, b_2)$ is defined in (6.4).

The above result is a consequence of the following

Theorem 2. *Let notations be as in Theorem 1. Then*

$$\sum_{R/2 < r_j \leq R} \max_{(b_j, r_j) = 1} \left| \sum_{\substack{N = p_1 + p_2 + p_3 \\ p_j \equiv b_j (\bmod r_j)}} (\log p_1)(\log p_2)(\log p_3) - \frac{\sigma(N, r_1, r_2, b_1, b_2) N^2}{2} \right| \ll N^2 \log^{-A} N. \tag{1.8}$$

Deduction of Theorem 1 from Theorem 2. Let $E(R)$ be the set of pairs of primes (r_1, r_2) such that $R/2 < r_j \le R$ $(j = 1, 2)$ for which

$$\max_{b_1,b_2}\left|\sum_{\substack{N=p_1+p_2+p_3\\ p_j\equiv b_j(\bmod r_j)}}(\log p_1)(\log p_2)(\log p_3)-\frac{\sigma(N,r_1,r_2,b_1,b_2)N^2}{2}\right| > \frac{N^2}{r_1r_2\log N}.$$

Then one deduces from Theorem 2 that

$$\sum_{(r_1,r_2)\in E(R)}\frac{1}{r_1r_2}\le \log^{-A}N$$

for arbitrary $A > 0$, and consequently

$$\sharp E(R)=\sum_{(r_1,r_2)\in E(R)}1\le R^2\sum_{(r_1,r_2)}\frac{1}{r_1r_2}\le R^2\log^{-A}N. \tag{1.9}$$

Since

$$\frac{1}{r_1r_2}\ll \sigma(N,r_1,r_2,b_1,b_2)\ll\frac{1}{r_1r_2}, \tag{1.10}$$

one sees that (1.7) is true for all $(r_1, r_2) \notin E(R)$ and all $(b_1, r_1) = (b_2, r_2) = 1$. This completes the proof of Theorem 1.

Remark 1. Our result is not a trivial generalization of the former ones in that new difficulty arises and the former methods do not work well in dealing with it. We will also explain the difference between ours and other cases.

Remark 2. If we do not concern ourselves with the numerical values of δ, Theorem 1 for $R \le N^\delta$ can be derived trivially from the result in the same moduli case.

In the remaining part of the paper we shall prove Theorem 2.

2. Notations and outline of the method.

We will use standard notation in number theory such as the von Mangoldt function $\Lambda(n)$, the Euler function $\varphi(n)$, etc. In addition, the letter r with or without subscripts always stands for a prime number, so that $r \ll \varphi(r)$. Let $L = \log N$. The letters δ and ε denote sufficiently small positive numbers; whose values may vary at each occurrence. For example, we can write $N^\delta L^5 \ll N^\delta$, $N^\delta N^\varepsilon \ll N^\varepsilon$. The letter c is reserved for absolute constants. The expression $r \sim R$ means $R/2 < r \le R$, and

$\sum_{r\sim R}$ means the sum over r in this range, where $N^\varepsilon \le R \le N^{1/8-\varepsilon}$. Let $B = A + 100$ and let

$$M = NL^{-12}, \quad P = RL^{2B}, \quad Q = NR^{-2}L^{-4B}. \tag{2.1}$$

It follows from Dirichlet's Theorem on rational approximations that each $\alpha \in [-1/Q, 1-1/Q]$ can be written in the form

$$\alpha = a/q + \lambda, \quad q \le Q, \quad 1 \le a \le q, \quad (a,q) = 1, \quad |\lambda| \le 1/qQ, \tag{2.2}$$

where α and λ are used as variables of integration in this context throughout. Denote by $\mathfrak{M}(a,q)$ the set of α in (2.2). We define the major arcs $\mathfrak{M}_1$, $\mathfrak{M}_2$, $\mathfrak{M}_2'$ with their union $\mathfrak{M}$ and the minor arcs $\mathfrak{m}_1$, $\mathfrak{m}_2$, $\mathfrak{m}_3$, $\mathfrak{m}_4$ with their union $\mathfrak{m}$ as follows;

$$\mathfrak{M}_1 = \bigcup_{\substack{q\le P\\ r_1\nmid q,\, r_2\nmid q}} \bigcup_{(a,q)=1} \mathfrak{M}(a,q), \qquad \mathfrak{M}_2 = \bigcup_{\substack{q\le P\\ r_1\nmid q,\, r_2\mid q}} \bigcup_{(a,q)=1} \mathfrak{M}(a,q),$$

$$\mathfrak{M}_2' = \bigcup_{\substack{q\le P\\ r_1\mid q,\, r_2\nmid q}} \bigcup_{(a,q)=1} \mathfrak{M}(a,q), \qquad \mathfrak{m}_1 = \bigcup_{\substack{P<q\le R^2L^{2B}\\ r_1\nmid q}} \bigcup_{(a,q)=1} \mathfrak{M}(a,q),$$

$$\mathfrak{m}_2 = \bigcup_{\substack{P<q\le R^2L^{2B}\\ r_1\mid q,\, r_2\nmid q}} \bigcup_{(a,q)=1} \mathfrak{M}(a,q), \quad \mathfrak{m}_3 = \bigcup_{\substack{k\le L^{2B}\\ q=r_1r_2k}} \bigcup_{(a,q)=1} \mathfrak{M}(a,q)$$

and

$$\mathfrak{m}_4 = [-1/Q, 1-1/Q] \backslash (\mathfrak{M} \cup \mathfrak{m}_1 \cup \mathfrak{m}_2 \cup \mathfrak{m}_3).$$

We have clearly $\mathfrak{m} = [-1/Q, 1-1/Q]\backslash\mathfrak{M}$.

Let $e(\alpha) = e^{2\pi i\alpha}$ as usual, and define the exponential sum and the exponential sum in arithmetic progression by

$$S(\alpha) = \sum_{M<p\le N} (\log p) e(p\alpha), \tag{2.3}$$

$$S(\alpha, r, b) = \sum_{\substack{M<p\le N\\ p\equiv b \bmod r}} (\log p) e(p\alpha), \tag{2.4}$$

respectively, where p runs through primes.

The assertion of Theorem 2 is equivalent to

$$\sum_{r_1\sim R}\sum_{r_2\sim R} \max_{(b_j,r_j)=1} \left| \int_{-1/Q}^{1-1/Q} S(\alpha,r_1,b_1)S(\alpha,r_2,b_2)S(\alpha)e(-N\alpha)d\alpha \right.$$
$$\left. - \sigma(N, r_1, b_1, r_2, b_2)\frac{N^2}{2} \right| \ll N^2L^{-A}.$$

To prove Theorem 2 it thus suffices to show

$$\sum_{r_1\sim R}\sum_{r_2\sim R}\max_{(b_j,r_j)=1}\left|\int_{\mathfrak{M}} S(\alpha,r_1,b_1)S(\alpha,r_2,b_2)S(\alpha)e(-N\alpha)d\alpha \right. \\ \left. - \sigma(N,r_1,b_1,r_2,b_2)\frac{N^2}{2}\right| \ll N^2L^{-A}, \quad (2.5)$$

and

$$\sum_{r_1\sim R}\sum_{r_2\sim R}\max_{(b_j,r_j)=1}\left|\int_{\mathfrak{m}} S(\alpha,r_1,b_1)S(\alpha,r_2,b_2)S(\alpha)e(-N\alpha)d\alpha\right| \ll N^2L^{-A}. \quad (2.6)$$

To estimate the latter integral over the minor arc $\mathfrak{m}$, we need the following two lemmas.

Lemma 2.1 ([15]). *For any α in the form of (2.2), we have*

$$S(\alpha) \ll NL^4(q^{-1/2} + q^{1/2}N^{-1/2} + N^{-1/5}). \quad (2.7)$$

Lemma 2.2 ([2]). *For any α satisfying*

$$r\alpha = a_1/q_1 + \lambda_1,\ |\lambda_1| \le \frac{1}{q_1^2}, \quad and \quad r^2\alpha = a_2/q_2 + \lambda_2,\ |\lambda_2| \le \frac{1}{q_2^2},$$

we have

$$S(\alpha,r,b) \ll \left(q_1 + \frac{N}{rq_1} + \frac{N}{rq_2^{1/2}} + \frac{N^{5/6}}{r^{1/2}} + N^{1/2}q_2^{1/2}\right)L^3. \quad (2.8)$$

Proof of (2.6). It follows from (2.7) that

$$S(\alpha) \ll \frac{N}{R}L^{-A-4} \quad (2.9)$$

for $\alpha \in \mathfrak{m}_4$, since $q \ge R^2L^B$ in this case. From (2.8), we also have

$$S(\alpha,r_j,b_j) \ll \frac{N}{R^{3/2}}L^{-A-4} \quad (2.10)$$

for $\alpha \in \mathfrak{m}_j,\ \ j = 1,2$. By the Schwarz inequality, we have

$$\left|\int_{\mathfrak{m}\setminus\mathfrak{m}_3}\right| = \left|\int_{\mathfrak{m}\setminus\mathfrak{m}_3} S(\alpha,r_1,b_1)S(\alpha,r_2,b_2)S(\alpha)e(-N\alpha)d\alpha\right|$$

$$\ll \max_{\alpha\in\mathfrak{m}_1} |S(\alpha,r_1,b_1)| \left(\int_0^1 |S(\alpha,r_2,b_2)|^2 d\alpha\right)^{1/2} \left(\int_0^1 |S(\alpha)|^2 d\alpha\right)^{1/2}$$

$$+ \max_{\alpha\in\mathfrak{m}_2} |S(\alpha,r_2,b_2)| \left(\int_0^1 |S(\alpha,r_1,b_1)|^2 d\alpha\right)^{1/2} \left(\int_0^1 |S(\alpha)|^2 d\alpha\right)^{1/2}$$

$$+ \max_{\alpha\in\mathfrak{m}_4} |S(\alpha)| \left(\int_0^1 |S(\alpha,r_1,b_1)|^2 d\alpha\right)^{1/2} \left(\int_0^1 |S(\alpha,r_2,b_2)|^2 d\alpha\right)^{1/2}$$

Since we have

$$\int_0^1 |S(\alpha,r_j,b_j)|^2 d\alpha = \sum_{\substack{M<p\le N\\ p\equiv b_j \bmod r_j}} \log^2 p < \sum_{\substack{M<n\le N\\ n\equiv b_j \bmod r_j}} \log^2 n \ll \frac{NL^2}{r_j}$$

and

$$\int_0^1 |S(\alpha)|^2 d\alpha \sim NL,$$

we infer from (2.9) and (2.10) that

$$\begin{aligned}\left|\int_{\mathfrak{m}\setminus\mathfrak{m}_3}\right| &\ll \frac{N}{R^{3/2}} L^{-A-2} \left(\left(\frac{N}{r_2}\right)^{1/2} + \left(\frac{N}{r_1}\right)^{1/2}\right) N^{1/2} \\ &\quad + \frac{N}{R} L^{-A-2} \left(\frac{N}{r_1}\right)^{1/2} \left(\frac{N}{r_2}\right)^{1/2} \\ &\ll \frac{N^2}{r_1 r_2} L^{-A-2}. \end{aligned} \tag{2.11}$$

uniformly for $r_1 \sim R$ and $r_2 \sim R$.

The above upper bound over $\mathfrak{m}\setminus\mathfrak{m}_3$ is admissible for every individual pair (r_1, r_2). But over $\mathfrak{m}_3$, the individual bound is not under control, and we need another treatment. First note that in the form of (2.2),

$$R^2/4 \le q \le R^2 L^{2B},$$

holds for any $\alpha \in \mathfrak{m}_3$, and so we extend the domain of the integral and take the mean-value as follows

$$\sum_{r_1\sim R} \sum_{r_2\sim R} \max_{(b_j,r_j)=1} \left| \int_{\mathfrak{m}_3} S(\alpha,r_1,b_1) S(\alpha,r_2,b_2) S(\alpha) e(-N\alpha) d\alpha \right|$$

$$\ll \sum_{r_1\sim R} \sum_{r_2\sim R} \max_{(b_j,r_j)=1} \sum_{R^2/4\le q\le R^2 L^{2B}}$$

$$\times \sum_{\substack{a=1\\(a,q)=1}}^{q} \int_{\mathfrak{M}(a,q)} |S(\alpha,r_1,b_1)S(\alpha,r_2,b_2)S(\alpha)|d\alpha$$

$$\ll \sum_{\frac{R^2}{4}\le q\le R^2L^{2B}} \sum_{\substack{a=1\\(a,q)=1}}^{q} \sum_{r_2\sim R} \max_{(b_2,r_2)=1}$$

$$\int_{\mathfrak{M}(a,q)} \left(\sum_{r_1\sim R} \max_{(b_1,r_1)=1} |S(\alpha,r_1,b_1)|\right) |S(\alpha,r_2,b_2)S(\alpha)|d\alpha$$

$$\ll \max_{q,a} \left(\sum_{r_1\sim R} \max_{(b_1,r_1)=1} |S(a/q+\lambda,r_1,b_1)|\right)$$

$$\times \sum_{r_2\sim R} \max_{(b_2,r_2)=1} \int_0^1 |S(\alpha,r_2,b_2)S(\alpha)|d\alpha.$$

To estimate the sum in the first bracket above, we distinguish two cases. In the case $r_1|q$, which can happen at most two times, we use the trivial estimate; and in the case $(r_1,q)=1$ (recall that r_1 is a prime), we have

$$r^2\alpha = ar^2/q + r^2\lambda = m + a_2/q + \lambda_2$$

by (2.2). Here m and a_2 are integers not contributing the exponential sum $S(\alpha,r,b)$, and $\lambda_2 = r^2\lambda$. The benefit is that $q=q_2$ remain unchanged, and

$$|\lambda_2| = |r^2\lambda| \le r^2/qQ = 1/q(Q/r^2).$$

Here

$$Q/r^2 = N/(R^2r^2L^{4B}) > R^{4+\varepsilon} > q,$$

where we have used (2.1) and $R \ll N^{1/8-\varepsilon}$; by the similar arguments we also have $q_1 = q$ in the notation of Lemma 2.2. Hence we have

$$\sum_{r_1\sim R} |S(a/q+\lambda,r_1,b_1)|$$

$$\ll \sum_{\substack{r_1\sim R\\ r_1\nmid q}} |S(a/q+\lambda,r_1,b_1)| + \max_{r_1\sim R} |S(a/q+\lambda,r_1,b_1)|$$

$$\ll \sum_{r_1\sim R} \frac{NL^3}{r_1q^{1/2}} + \max_{r_1\sim R} \frac{NL}{r_1(q/r_1)^{1/2}}$$

$$\ll \frac{NL}{R}, \tag{2.12}$$

by (2.8). Applying (2.12) and estimating as above to obtain

$$\sum_{r_1\sim R}\sum_{r_2\sim R}\max_{(b_j,r_j)=1}\left|\int_{\mathfrak{m}_3}\right| \ll \frac{N}{R}\sum_{r_2\sim R}\int_0^1 |S(\alpha,r_2,b_2)S(\alpha)|d\alpha$$
$$\ll \frac{N}{R}\sum_{r_2\sim R}\left(\frac{N}{r_2}\right)^{1/2} N^{1/2}L^2 \ll N^2L^{-A}.$$

Now, using (2.11) for the first sum, we conclude that

$$\sum_{r_1\sim R}\sum_{r_2\sim R}\max\left|\int_{\mathfrak{m}}\right| \ll \sum_{r_1\sim R}\sum_{r_2\sim R}\frac{N^2}{r_1r_2}L^{-A-2} + N^2L^{-A} \ll N^2L^{-A},$$

which proves (2.6).

3. Gaussian Sums and preliminaries for the major arcs.

Let us begin with the definitions of the general Gaussian sums. For a character χ mod q, we define the classical Gaussian sum $G(\chi,a)$ by

$$G(\chi,a) = \sum_{\substack{c=1\\(c,q)=1}}^{q} \bar{\chi}(c)e\left(\frac{ac}{q}\right) \tag{3.1}$$

and its generalization by

$$C(\chi,q,b,m,a) = \sum_{\substack{c=1\\(c,q)=1\\c\equiv b \bmod m}}^{q} \bar{\chi}(c)e\left(\frac{ac}{q}\right), \tag{3.2}$$

where we further write

$$C(q,b,m,a) = C(\chi^0,q,b,m,a),$$

χ^0 signifying the principal character. Suppose χ mod q be induced by the primitive character χ^* mod q^*. Then for $(a,q)=1$, the classical estimates give

$$|G(\chi^0,a)| \le 1, \quad |G(\chi,a)| \le (q^*)^{1/2}, \tag{3.3}$$

while for $m|q$ and $(a,q)=(b,q)=1$, Liu and Zhan [7] showed that

$$C(\chi,q,b,m,a) \le (q^*)^{1/2}. \tag{3.4}$$

Define

$$V(\lambda) = \sum_{M<n\le N} e(n\lambda), \quad W(\chi,\lambda) = \sum_{M<p\le N} (\log p)e(\lambda p) - \delta_\chi V(\lambda), \tag{3.5}$$

where $\delta_\chi = 1$ or 0 according as χ is principal or not, respectively. From the definition, we have $W(\chi,\lambda) = W(\chi^*,\lambda)$, where χ is induced by the primitive character χ^*. Note that for $(m,q)=1$, any character χ mod mq can be written uniquely as $\chi = \xi\eta$ with ξ mod m and η mod q. Then, following the method of Liu and Zhan [7] and using the orthogonality of characters, we have for $m \nmid q$,

$$\begin{aligned} S(\alpha,m,b) &= \frac{G(\chi^0,a)}{\varphi(mq)}V(\lambda) + \frac{1}{\varphi(mq)}\sum_{\xi \bmod m}\bar{\xi}(b)\sum_{\eta \bmod q} G(\eta,a)W(\xi\eta,\lambda) \\ &=: S_1(m,a,q,\lambda) + S_2(m,a,b,q,\lambda), \end{aligned} \tag{3.6}$$

say, while for $m|q$, we have

$$\begin{aligned} S(\alpha,m,b) &= \frac{C(q,b,m,a)}{\varphi(q)}V(\lambda) + \frac{1}{\varphi(q)}\sum_{\eta \bmod q} C(\eta,q,b,m,a)W(\eta,\lambda) \\ &=: U_1(m,a,b,q,\lambda) + U_2(m,a,b,q,\lambda), \end{aligned} \tag{3.7}$$

say. We also have

$$\begin{aligned} S(\alpha) &= \frac{G(\chi^0,a)}{\varphi(q)}V(\lambda) + \frac{1}{\varphi(q)}\sum_{\chi \bmod q} G(\chi,a)W(\chi,\lambda) \\ &=: S_1(a,q,\lambda) + S_2(a,q,\lambda), \end{aligned} \tag{3.8}$$

say.

Finally we introduce quantities which will be used in the estimation of integrals on major arcs and whose estimates will be given in §5.

$$J_0 = \sum_{q\le P}\sideset{}{^*}\sum_{\chi \bmod q} \frac{1}{q^{1/6}} \max_{|\lambda|\le 1/qQ} |W(\chi,\lambda)|, \tag{3.9}$$

$$J_1 = \sum_{r\sim R}\sum_{\substack{q\le P \\ r\nmid q}}\sideset{}{^*}\sum_{\xi \bmod r}\sideset{}{^*}\sum_{\eta \bmod q} \frac{1}{rq^{1/6}} \max_{|\lambda|\le 1/qQ} |W(\xi\eta,\lambda)|, \tag{3.10}$$

$$K_0 = \sum_{q\le P}\sideset{}{^*}\sum_{\chi \bmod q} \frac{1}{q^{1/6}} \left\{\int_{-1/qQ}^{1/qQ} |W(\chi,\lambda)|d\lambda\right\}^{1/2}, \tag{3.11}$$

$$K_1 = \sum_{r\sim R}\sum_{\substack{q\le P \\ r\nmid q}}\sideset{}{^*}\sum_{\xi \bmod r}\sideset{}{^*}\sum_{\eta \bmod q} \frac{1}{rq^{1/6}} \left\{\int_{-1/qQ}^{1/qQ} |W(\xi\eta,\lambda)|d\lambda\right\}^{1/2}, \tag{3.12}$$

$$K_2 = \sum_{k\le L^{2B}} \sideset{}{^*}\sum_{\eta \bmod k} \left\{ \int_{-2/RQ}^{2/RQ} |W(\eta,\lambda)|d\lambda \right\}^{1/2}, \tag{3.13}$$

$$K_3 = \sum_{r\sim R} \sum_{k\le L^{2B}} \sideset{}{^*}\sum_{\eta_1 \bmod r} \sideset{}{^*}\sum_{\eta_2 \bmod k} \frac{1}{R} \left\{ \int_{-2/RQ}^{2/RQ} |W(\eta_1\eta_2,\lambda)|d\lambda \right\}^{1/2}, \tag{3.14}$$

$$K_4 = \sum_{r\sim R} \max_{\substack{r'\sim R\\ r'\ne r}} \sum_{k\le L^{2B}} \sideset{}{^*}\sum_{\xi \bmod r} \sideset{}{^*}\sum_{\eta \bmod k} \sideset{}{^*}\sum_{\eta' \bmod r'} \frac{1}{R} \left\{ \int_{-2/RQ}^{2/RQ} |W(\xi\eta\eta',\lambda)|d\lambda \right\}^{1/2}, \tag{3.15}$$

$$K_5 = \max_{r'\sim R} \sum_{k\le L^{2B}} \sideset{}{^*}\sum_{\eta \bmod k} \sideset{}{^*}\sum_{\eta' \bmod r'} \left\{ \int_{-2/RQ}^{2/RQ} |W(\eta\eta',\lambda)|d\lambda \right\}^{1/2}. \tag{3.16}$$

Here and in the sequel $\sum^*_{\chi \bmod q}$ means the sum taken over all the primitive characters $\chi \bmod q$.

4. On the major arcs.

In this section, we will calculate the integrals on the major arcs $\mathfrak{M}_1$, $\mathfrak{M}_2$ and $\mathfrak{M}_2'$. Let us begin with $\mathfrak{M}_1$. We have

$$\begin{aligned} &\int_{\mathfrak{M}_1} S(\alpha,r_1,b_1)S(\alpha,r_2,b_2)S(\alpha)e(-N\alpha)d\alpha \\ &= \sum_{\substack{q\le P\\ r_1\nmid q, r_2\nmid q}} \sum_{\substack{a=1\\ (a,q)=1}}^{q} \int_{-1/qQ}^{1/qQ} (S_1(r_1,a,q,\lambda) + S_2(r_1,a,b_1,q,\lambda)) \\ &\quad \times \big(S_1(r_2,a,q,\lambda) + S_2(r_2,a,b_2,q,\lambda)\big)\big(S_1(a,q,\lambda) + S_2(a,q,\lambda)\big) \\ &\quad \times e\left(-N(a/q+\lambda)\right) d\lambda =: I_{10} + I_{11} + I_{12} + I_{13}, \end{aligned} \tag{4.1}$$

where

$$I_{10} = \sum_{\substack{q\le P\\ r_1\nmid q, r_2\nmid q}} \sum_{\substack{a=1\\ (a,q)=1}}^{q} \int_{-1/qQ}^{1/qQ} S_1(r_1,a,q,\lambda)S_1(r_2,a,q,\lambda)S_1(a,q,\lambda) \times e\big(-N(a/q+\lambda)\big)d\lambda,$$

$$I_{11} = \sum_{\substack{q\le P\\ r_1\nmid q, r_2\nmid q}} \sum_{\substack{a=1\\ (a,q)=1}}^{q} \int_{-1/qQ}^{1/qQ} (S_1(r_1,a,q,\lambda)S_2(r_2,a,b_2,q,\lambda)S(a/q+\lambda) + S_2(r_1,a,b_1,q,\lambda)S_1(r_2,a,q,\lambda)S(a/q+\lambda))e\big(-N(a/q+\lambda)\big)d\lambda,$$

$$I_{12} = \sum_{\substack{q\le P\\ r_1\nmid q, r_2\nmid q}} \sum_{\substack{a=1\\(a,q)=1}}^{q} \int_{-1/qQ}^{1/qQ} (S_2(r_1,a,b_1,q,\lambda)S_2(r_2,a,b_2,q,\lambda)S_1(a,q,\lambda)$$
$$+ S_2(r_1,a,b_1,q,\lambda)S_2(r_2,a,b_2,q,\lambda)S_2(a,q,\lambda))e(-N(a/q+\lambda))d\lambda$$
$$=: I_{121} + I_{122},$$

say, and

$$I_{13} = \sum_{\substack{q\le P\\ r_1\nmid q, r_2\nmid q}} \sum_{\substack{a=1\\(a,q)=1}}^{q} \int_{-1/qQ}^{1/qQ} S_1(r_1,a,q,\lambda)S_1(r_2,a,q,\lambda)S_2(a,q,\lambda)$$
$$\times e(-N(a/q+\lambda))d\lambda.$$

Substituting (3.6) and (3.8) into I_{13}, we have

$$|I_{13}| \ll \sum_{\substack{q\le P\\ r_1\nmid q, r_2\nmid q}} \frac{1}{r_1 r_2 \varphi^3(q)} \sum_{\substack{a=1\\(a,q)=1}}^{q} \max_{|\lambda|\le 1/qQ} \left| \sum_{\chi \bmod q} G(\chi,a)G^2(\chi^0,a)W(\chi,\lambda)\right|$$
$$\times \int_{-1/qQ}^{1/qQ} \left|V^2(\lambda)\right| d\lambda$$
$$\ll \frac{N}{r_1 r_2} \sum_{q\le P} \frac{1}{\varphi(q)^3} \sum_{\substack{a=1\\(a,q)=1}}^{q} \max_{|\lambda|\le 1/qQ} \left| \sum_{\chi \bmod q} G(\chi,a)W(\chi,\lambda)\right|$$

on using $|G^2(\chi^0,a)|\le 1$, $\int_{-1/qQ}^{1/qQ} |V^2(\lambda)|d\lambda \ll \int_0^1 |V^2(\lambda)|d\lambda \ll N.$

Now applying (3.3) and the estimate $\varphi(q) \gg qL^{-c}$ and reducing to primitive characters, we obtain

$$|I_{13}| \ll \frac{N}{r_1 r_2} L^c \sum_{q\le P} \sideset{}{^*}\sum_{\chi \bmod q} \frac{1}{q^{3/2}} \max_{|\lambda|\le 1/qQ} |W(\chi,\lambda)| \ll J_0 \frac{N}{r_1 r_2} L^c.$$

Hence, summing over r_1 and r_2, we have

$$\sum_{r_1\sim R} \sum_{r_2\sim R} |I_{13}| \ll L^c N J_0. \tag{4.2}$$

Similarly as above, we have

$$\sum_{r_1\sim R} \sum_{r_2\sim R} |I_{11}| \ll L^c N (J_0 + J_1). \tag{4.3}$$

We also have by (3.6) and (3.8)

$$|I_{122}| \ll \sum_{q\le P}\sum_{\xi_1 \bmod r_1}\sum_{\xi_2 \bmod r_2}\sum_{\eta_1 \bmod q}\sum_{\eta_2 \bmod q}\sum_{\chi_3 \bmod q}\frac{1}{\varphi^3(q)r_1r_2}$$
$$\times \sum_{\substack{a=1\\(a,q)=1}}^{q} e(-aN/q)G(\eta_1,a)G(\eta_2,a)G(\chi_1,a)\bar{\xi}_1(b_1)\bar{\xi}_2(b_2)$$
$$\times \int_{-1/qQ}^{1/qQ} W(\xi_1\eta_1,\lambda)W(\xi_2\eta_2,\lambda)W(\chi_3,\lambda)e(-N\lambda)d\lambda.$$

Reducing characters $\xi_j \bmod r_j$, $\eta_j \bmod q$ and $\chi \bmod q$ to primitive characters $\xi_j \bmod r_j^*$, $\eta_j \bmod q_j$ $(j=1,2)$ and $\chi_3 \bmod q_3$, we have divisibility conditions $r_j^*|r_j$ and $q_j|q$, where the former reduces to $r_j^*=1$ or $r_j^*=r_j$ and the latter, to $q_0=[q_1,q_2,q_3]|q$, $[a,b,c]$ denoting the least common multiple of a,b and c. Hence the right-side of the above inequality is majorized by the sums over $r_j^*\le R$, $\xi_j \bmod r_j$, $q_j\le P, \eta_j \bmod q_j (j=1,2), q_3\le P, \chi_3 \bmod q_3$ and $q_0|q$ with summands

$$\frac{(q_0)^{3/2}L^3}{q^2r_1r_2}\int_{-1/qQ}^{1/qQ}|W(\xi_1\eta_1,\lambda)W(\xi_2\eta_2,\lambda)W(\chi_3,\lambda)|d\lambda.$$

Distributing the factor suitably (collecting those with the same suffices into one), we have

$$\sum_{r_1\sim R}\sum_{r_2\sim R}|I_{122}| \ll L^cJ_0(K_0+K_1)^2. \tag{4.4}$$

Similarly as above, we have, on noting $|V(\lambda)|\le N$,

$$\sum_{r_1\sim R}\sum_{r_2\sim R}|I_{121}| \ll L^cN(K_0^2+K_0K_1+K_1^2). \tag{4.5}$$

Now let us go on to the estimation on $\mathfrak{M}_2$, the estimation on $\mathfrak{M}_2'$ being similar.

$$\int_{\mathfrak{M}_2} S(\alpha,r_1,b_1)S(\alpha,r_2,b_2)S(\alpha)e(-N\alpha)d\alpha$$
$$= \sum_{\substack{q\le P\\ r_1\nmid q, r_2|q}}\sum_{\substack{a=1\\(a,q)=1}}^{q}\int_{-1/qQ}^{1/qQ}(S_1(r_1,a,q,\lambda)+S_2(r_1,a,b_1,q,\lambda))$$
$$\times \big(U_1(r_2,a,q,\lambda)+U_2(r_2,a,b_2,q,\lambda)\big)$$
$$\times \big(S_1(a,q)+S_2(a,q)\big)e\left(-N(a/q+\lambda)\right)d\lambda$$

$$=: I_{20} + I_{21} + I_{22} + I_{23}, \tag{4.6}$$

where

$$I_{20} = \sum_{\substack{q \le P \\ r_1 \nmid q, r_2 | q}} \sum_{\substack{a=1 \\ (a,q)=1}}^{q} \int_{-1/qQ}^{1/qQ} S_1(r_1, a, q, \lambda) U_1(r_2, a, q, \lambda) S_1(a, q) \times e(-N(a/q + \lambda)) d\lambda,$$

$$I_{21} = \sum_{\substack{q \le P \\ r_1 \nmid q, r_2 | q}} \sum_{\substack{a=1 \\ (a,q)=1}}^{q} \int_{-1/qQ}^{1/qQ} \Big(S_1(r_1, a, q, \lambda) U_2(r_2, a, b_2, q, \lambda) S(a/q + \lambda)$$

$$+ S_2(r_1, a, b_1, q, \lambda) U_1(r_2, a, b_2 q, \lambda) S(a/q + \lambda) \Big) e(-N(a/q + \lambda)) d\lambda$$

$$=: I_{211} + I_{212}, \quad \text{say},$$

$$I_{22} = \sum_{\substack{q \le P \\ r_1 \nmid q, r_2 | q}} \sum_{\substack{a=1 \\ (a,q)=1}}^{q} \int_{-1/qQ}^{1/qQ} S_2(r_1, a, b_1, q, \lambda) U_2(r_2, a, b_2, q, \lambda) S(a/q + \lambda) \times e(-N(a/q + \lambda))\, d\lambda$$

and

$$I_{23} = \sum_{\substack{q \le P \\ r_1 \nmid q, r_2 | q}} \sum_{\substack{a=1 \\ (a,q)=1}}^{q} \int_{-1/qQ}^{1/qQ} S_1(r_1, a, q, \lambda) U_1(r_2, a, b_2, q, \lambda) S_2(a, q, \lambda) \times e(-N(a/q + \lambda))\, d\lambda.$$

By (3.3) and (3.4), arguing as in the proof of (4.2), we have

$$|I_{23}| \ll \sum_{\substack{q \le P \\ r_1 \nmid q, r_2 | q}} \frac{1}{r_1 \varphi^3(q)} \sum_{\substack{a=1 \\ (a,q)=1}}^{q} \max_{|\lambda| \le 1/qQ} \left| \sum_{\chi_3 \bmod q} G(\chi^0, a) C(q, b, r_2, a) G(\chi_3, a) W(\chi_3, \lambda) \right| \int_{-1/qQ}^{1/qQ} |V(\lambda)|^2 d\lambda$$

$$\ll N L^c \sum_{\substack{q \le P \\ r_1 \nmid q, r_2 | q}} \frac{1}{r_1 \varphi^2(q)} \sideset{}{^*}\sum_{\chi_3 \bmod q} \max_{|\lambda| \le 1/qQ} |W(\chi_3, \lambda)|$$

$$\ll L^c \frac{N}{r_1 r_2} \sum_{\substack{q \le P \\ r_1 \nmid q, r_2 | q}} \sideset{}{^*}\sum_{\chi_3 \bmod q} \frac{1}{q} \max_{|\lambda| \le 1/qQ} |W(\chi_3, \lambda)|.$$

on replacing q^{-1} by r_2^{-1}. Hence

$$\sum_{r_1 \sim R} \sum_{r_2 \sim R} |I_{23}| \ll L^c N J_0. \tag{4.7}$$

Substituting (3.6) and (3.7), we deduce that

$$|I_{22}| = \sum_{\substack{q\le P \\ r_1 \nmid q, r_2 | q}} \frac{1}{r_1\varphi^2(q)} \sum_{\substack{a=1 \\ (a,q)=1}}^{q} \max_{|\lambda|\le 1/qQ} |S(a/q+\lambda)| \int_{-1/qQ}^{1/qQ} \Bigg| \sum_{\xi_1 \bmod r_1} \bar{\xi}(b)$$

$$\sum_{\eta_1 \bmod q} G(\eta_1, a) \sum_{\eta_2 \bmod q} C(\eta_2, q, b_2, r_2, a) W(\xi_1\eta_1, \lambda) W(\eta_2, \lambda) \Bigg| d\lambda.$$

Instead of (2.9) and (3.4) we use the fact that for $r_2|q$, $S(a/q+\lambda) \ll L^3N/q^{1/2}$, and $C(\eta, q, b, r_2, a) \ll q/r_2$, respectively as well as (3.3) to obtain

$$\begin{aligned}
|I_{22}| \ll & \frac{L^cN}{r_1r_2} \sum_{\substack{q\le P \\ r_1 \nmid q, r_2|q}} \sum_{\xi_1 \bmod r_1} \sum_{\eta_1 \bmod q} \sum_{\eta_2 \bmod q} \int_{-1/qQ}^{1/qQ} |W(\xi_1\eta_1, \lambda) W(\eta_2, \lambda)|\, d\lambda \\
\ll & \frac{L^cN}{r_1r_2} \sum_{k\le L^{2B}} \sum_{\xi_1 \bmod r_1} \sum_{\eta_{11} \bmod r_2} \sum_{\eta_{12} \bmod k} \sum_{\eta_{21} \bmod r_2} \sum_{\eta_{22} \bmod k} \\
& \times \int_{-1/r_2Q}^{1/r_2Q} |W(\xi_1\eta_{11}\eta_{12}, \lambda) W(\eta_{21}\eta_{22}, \lambda)|\, d\lambda
\end{aligned}$$

on decomposing $q = r_2k, k \le L^{2B}, \eta_j = \eta_{j1}\eta_{j2}, (j = 1, 2)$. Reducing to primitive characters mod k_j, mod r_j^*, $j = 1, 2$, we have

$$\begin{aligned}
& \sum_{r_1\sim R} \sum_{r_2\sim R} |I_{22}| \\
& \ll NL^{c+8B} \sum_{\substack{r_1\sim R \\ r_1^*|r_1}} \sum_{\substack{r_2\sim R \\ r_2^*|r_2}} \frac{1}{r_1r_2} \sum_{k_1\le L^{2B}} \sum_{k_2\le L^{2B}} \sideset{}{^*}\sum_{\xi_1 \bmod r_1^*} \sideset{}{^*}\sum_{\eta_{11} \bmod r_2^*} \sideset{}{^*}\sum_{\eta_{12} \bmod k_1} \\
& \quad \times \sideset{}{^*}\sum_{\eta_{21} \bmod r_2^*} \sideset{}{^*}\sum_{\eta_{22} \bmod k_2} \int_{-2/RQ}^{2/RQ} |W(\xi_1\eta_{11}\eta_{12}, \lambda) W(\eta_{21}\eta_{22}, \lambda)| d\lambda.
\end{aligned}$$

As in the proof of (4.4), we note that $r_j^* = 1$ or $r_j^* = r_j$ for $j = 1, 2$ and that if $r_1 \neq r_2$, then $(r_1^*, r_2^*) = 1$. Hence for primitive characters $\xi_1 \bmod r_1^*$ and $\eta_{11} \bmod r_2^*$, $\xi_1\eta_{11}$ is also a primitive character mod$r_1^*r_2^*$. Hence, by the Schwarz inequality,

$$\begin{aligned}
\sum_{r_1\sim R} \sum_{r_2\sim R} |I_{22}| \ll NL^{c+8B} & \sum_{r_2^*\le R} \frac{1}{r_2^*} \sum_{k_2\le L^{2B}} \sum_{\eta_{21} \bmod r_2^*} \\
& \times \sideset{}{^*}\sum_{\eta_{22} \bmod k_2} \left\{ \int_{-2/RQ}^{2/RQ} |W(\eta_{21}\eta_{22}, \lambda)|^2 d\lambda \right\}^{1/2}
\end{aligned}$$

$$\times \sum_{r_1^* \le R} \frac{1}{r_1^*} \sum_{k_1 \le L^{2B}} \sideset{}{^*}\sum_{\xi_1 \bmod r_1^*}$$

$$\times \sideset{}{^*}\sum_{\eta_{11} \bmod r_2^*} \sideset{}{^*}\sum_{\eta_{12} \bmod k_1} \left\{ \int_{-2/RQ}^{2/RQ} |W(\xi_1 \eta_{11} \eta_{12}, \lambda)|^2 d\lambda \right\}^{1/2}$$

$$\ll NL^{c+8B}(K_2 + K_3)(K_2 + K_3 + K_4 + K_5). \tag{4.8}$$

Similarly, we have

$$\sum_{r_1 \sim R} \sum_{r_2 \sim R} |I_{211}| \ll NL^c J_0, \tag{4.9}$$

and

$$\sum_{r_1 \sim R} \sum_{r_2 \sim R} |I_{212}| \ll NL^c J_1. \tag{4.10}$$

Now, collecting the formulas (4.1)–(4.10), we have

$$\sum_{r_1 \sim R} \sum_{r_2 \sim R} \left| \int_{\mathfrak{M}} S(\alpha, r_1, b_1) S(\alpha, r_2, b_2) S(\alpha) d\alpha - (I_{10} + I_{20} + I'_{20}) \right|$$

$$\ll L^{c+8B} \Big\{ N(J_0 + J_1) + N(K_0 + K_1 + K_2 + K_3)$$

$$\times (K_0 + K_1 + K_2 + K_3 + K_4 + K_5) \Big\}, \tag{4.11}$$

where I'_{20} is defined in a similar way to I_{20} from the integral over $\mathfrak{M}'_2$. We will show that the estimates of J and K are admissible in the next section.

5. The estimates of J and K.

We will need some lemmas in the following discussion. Let u satisfy $M < u \le N$, and let $M_1, \cdots, M_{10}$ be positive integers such that

$$2^{-10} M \le M_1 \cdots M_{10} < u, \quad \text{and} \quad 2M_6, \ldots, 2M_{10} \le u^{1/5}. \tag{5.1}$$

We write $\boldsymbol{M} = (M_1, M_2, \cdots, M_{10})$. For $j = 1, \ldots, 10$, let

$$a_j(m) = \begin{cases} \log m, & \text{if} \quad j = 1, \\ 1, & \text{if} \quad j = 2, 3, 4, 5, \\ \mu(m), & \text{if} \quad j = 6, 7, 8, 9, 10. \end{cases}$$

We define the following functions of a complex variable s:

$$f_j(s) = f_j(s,\chi) = \sum_{m\sim M_j} \frac{a_j(m)\chi(m)}{m^s}, \quad F(s) = F(s,\chi) = f_1(s)\cdots f_{10}(s).$$

Lemma 5.1 ([8], Lemma 5.2). *Let $F(s,\chi)$ be defined as above. Then for any $S \ge 1$ and $0 < T_3 \le N$,*

$$\sum_{k\sim S}\sum_{\chi \bmod k}{}^{*} \int_{T_3}^{2T_3} \left|F\left(\frac{1}{2}+it,\chi\right)\right| dt \ll (S^2T_3 + ST_3^{1/2}N^{3/10} + N^{1/4})L^c, \tag{5.2}$$

where $k \sim S$ means $S/2 < k \le S$.

Lemma 5.2. *For $T \ge 2$, let $N^*(\alpha,q,T)$ denote the number of zeros of all the L-functions $L(s,\chi)$ with primitive characters $\chi \bmod q$ in the region $\mathrm{Re}\, s \ge \alpha, |\mathrm{Im}\, s| \le T$. Then*

$$N^*(\alpha,q,T) \ll (qT)^{12(1-\alpha)/5}\log^c(qT).$$

Lemma 5.3. *Let $T \ge 2$. There is an absolute constant $c_1 > 0$ such that $\prod_{\chi \bmod q} L(s,\chi)$ has no zeros in the region*

$$\mathrm{Re} s \ge 1 - c_1/\max\{\log q, \log^{4/5} T\}, \quad |\mathrm{Im}\, s| \le T,$$

except a possible Siegel zero.

Lemmas 5.2 and 5.3 are well-known results in number theory. For the proof of Lemma 5.2, see for example pp. 640 and 642, 634, and 669 in Pan and Pan [11]. For the proof of Lemma 5.3, see Satz VIII.6.2 in Prachar [12].

In order to approximate $W(\chi,\lambda)$ by a sum over integers, we introduce

$$\hat{W}(\chi,\lambda) = \sum_{M<m\le N} (\Lambda(m)\chi(m) - \delta_\chi)e(m\lambda). \tag{5.3}$$

Since $W(\chi,\lambda) - \hat{W}(\chi,\lambda)$ is a sum of $\log p$ over p^j, $M < p^j \le N, j \ge 2$, we have

$$W(\chi,\lambda) = \hat{W}(\chi,\lambda) + O(N^{1/2}).$$

Hence in what follows we will use $\hat{W}(\chi,\lambda)$ in place of $W(\chi,\lambda)$. Our task is to prove Lemma 5.4 and Lemma 5.5 which give the estimates for J_j, K_j; the estimates for J_0, K_0 over $S \le L^{2B}$ are given in the former, and those over $L^{2B} < S \le P$ are given in the latter. The proofs depend on Lemmas 5.1–5.3, Gallagher's lemma plus the explicit formula in the former, and Heath-Brown's identity in the latter.

It is customary and convenient to work with the truncated Chebyshev function with χ

$$\psi(\chi, X, Y) = \sum_{X<n\leq Y} \Lambda(n)\chi(n) = \psi(\chi, Y) - \psi(\chi, X) \qquad (5.4),$$

where

$$\psi(\chi, X) = \psi(\chi, 0, X) \qquad (5.5).$$

Then we have

$$\hat{W}(\chi, \lambda) = \int_M^N e(u\lambda)d\left\{\psi(\chi, u) - \delta_\chi u\right\}. \qquad (5.6)$$

We now prove

Lemma 5.4. *Let $A_1 > 0$ be arbitrary. Then for any $B_1 > 0$, we have*

$$K_2 \ll N^{1/2}L^{-A_1}, \qquad (5.7)$$

$$\max_{S\leq L^{B_1}} \sum_{k\sim S} \frac{1}{k^{1/6}} \max_{|\lambda|\leq 1/kQ} \sum_{\chi \bmod k}^{*} |\hat{W}(\chi, \lambda)| \ll NL^{-A_1}, \qquad (5.8)$$

and

$$\max_{S\leq L^{B_1}} \sum_{k\sim S} \sum_{\chi \bmod k}^{*} \frac{1}{k^{1/6}} \left\{\int_{-1/kQ}^{1/kQ} |\hat{W}(\chi, \lambda)|^2 d\lambda\right\}^{1/2} \ll N^{1/2}L^{-A_1}. \quad (5.9)$$

Proof. We use the explicit formula (see [3], p. 109 and 120, or [11], p. 313).

$$\psi(\chi, u) = \delta_\chi u - \sum_{|\gamma|\leq T} \frac{u^\rho}{\rho} + O\left\{\left(\frac{u}{T} + 1\right)\log^2(quT)\right\}, \qquad (5.10)$$

where the sum is over non-trivial zeros $\rho = \beta + i\gamma$ of the L-function $L(s, \chi)$ such that $|\gamma| \leq T$, and T is a parameter, $2 \leq T \leq u$.

Substituting (5.10) with $T = N^{1/3}$ in (5.6), we have

$$\hat{W}(\chi, \lambda) \ll NL^3 \sum_{|\gamma|\leq N^{1/3}} N^{\beta-1} + N^{2/3}R^2L^c.$$

Now, by Lemma 5.3, $\prod_{\chi \bmod k} L(s, \chi)$ has no zero in the region $\sigma \geq 1 - \eta(T), |t| \leq T$ except for a possible Siegel zero, where we write $\eta(T) = c_1 \log^{-4/5} T$. But in our present situation, moduli k do not exceed L^{2B+B_1}, whence Siegel's theorem (see e.g. [3], §21) asserts that there is no such zero, and so $\beta \le 1-\eta(N^{1/3})$, and a fortiori $\beta - 1 \le 0$. Hence

$$\sum_{|\gamma|\le N^{1/3}} N^{\beta-1} \ll \int_0^{1-\eta(N^{1/3})} N^{\frac{\alpha-1}{2}} dN^*(\alpha, q, N^{1/3})$$

$$\ll L^c \int_0^{1-\eta(N^{1/3})} N^{-\frac{1}{10}(1-\alpha)} d\alpha \ll L^c N^{-\eta(N^{1/3})/10}$$

$$\ll \exp(-c_2 L^{1/5}). \tag{5.11}$$

by Lemma 5.2. Summing over $k \le L^{2B+B_1}$ and the primitive characters χ mod k, leads to (5.8).

On the other hand, by Gallagher's lemma [4] it follows that

$$\int_{-1/Q}^{1/Q} |\hat{W}(\chi,\lambda)|^2 d\lambda \ll Q^{-2} \int_{-\infty}^{\infty} \left| \sum_{\substack{t<n\le t+Q \\ M<n\le N}} \Lambda(n)\chi(n) - \delta_\chi \sum_{\substack{t<n\le t+Q \\ M<n\le N}} 1 \right|^2 dt$$

$$\ll Q^{-2} \int_M^N \left| \psi(\chi, X, Y) - \delta_\chi \sum_{X<n\le Y} 1 \right|^2 dt \tag{5.12}$$

on writing $X = \max(t, M)$ and $Y = \min(t+Q, N)$. By (5.10), the integrand on the right of (5.12) is

$$\ll \left| \sum_{|\gamma|\le N^{1/3}} \left(\frac{Y^\rho}{\rho} - \frac{X^\rho}{\rho} \right) + N^{2/3} L^c \right|^2 \ll \left(\sum_{|\gamma|\le N^{1/3}} \int_X^Y u^{\beta-1} du + N^{2/3} L^c \right)^2$$

$$\ll \left(L^c Q \sum_{|\gamma|\le N^{1/3}} N^{\beta-1} + N^{2/3} L^c \right)^2.$$

Hence, applying (5.11), we obtain

$$\left(\int_{-1/Q}^{1/Q} |\hat{W}(\chi,\lambda)|^2 d\lambda \right)^{1/2} \ll Q^{-1} \int_M^N \left(\left(Q \exp(-c_2 L^{1/5}) + N^{2/3} \right) L^c \right)^2 dt$$

$$\ll Q^{-1} N^{1/2} \left(Q \exp(-c_2 L^{1/5}) + N^{2/3} \right) L^c$$

$$\ll N^{1/2} \exp(-c_2 L^{1/5}).$$

Summing over primitive characters χ mod k and $k \le L^{2B+B_1}$, we deduce (5.8) and (5.10). □

Lemma 5.5. *For an arbitrary constant $A_1 > 0$ and for a constant $B_1 = B_1(A) > 0$ large enough, we have*

$$J_1 \ll N^{1/2}L^{-A_1}, \tag{5.13}$$

$$\max_{L^{B_1}\le S\le P}\sum_{k\sim S}\frac{1}{k^{1/6}}\max_{|\lambda|\le 1/kQ}\sum_{\chi \bmod k}{}^{*}|\hat{W}(\chi,\lambda)| \ll NL^{-A_1}, \tag{5.14}$$

$$K_1, K_3, K_4 \ll N^{1/2}L^{-A_1}, \tag{5.15}$$

$$\max_{L^{B_1}\le S\le P}\sum_{k\sim S}\frac{1}{k^{1/6}}\sum_{\chi \bmod k}{}^{*}\left\{\int_{-1/kQ}^{1/kQ}|\hat{W}(\chi,\lambda)|^2 d\lambda\right\}^{1/2} \ll N^{1/2}L^{-A_1}, \tag{5.16}$$

and

$$K_5 \ll N^{1/2}L^{c}. \tag{5.17}$$

Proof. We only give proofs of (5.14) and (5.17), other cases being similar.

We apply Heath-Brown's identity (Lemma 1 [5] with $k = 5$)

$$\Lambda(m) = \sum_{j=1}^{5}\binom{5}{j}(-1)^{j-1}\sum_{\substack{m_1\cdots m_{2j}=m\\ m_{j+1}\cdots m_{2j}\le u}}(\log m_1)\mu(m_{j+1})\cdots\mu(m_{2j})$$

to the sum $\psi(\chi, M, u) = \sum_{M<n\le u}\Lambda(m)\chi(m)$ to find that it is a linear combination of $O(L^{10})$ terms, each of which is of the form

$$\sigma(u;\boldsymbol{M}) = \sum_{\substack{m_1\sim M_1\\ M<m_1\cdots m_{10}\le u}}\cdots\sum_{m_{10}\sim M_{10}} a_1(m_1)\chi(m_1)\cdots a_{10}(m_{10})\chi(m_{10}).$$

By Perron's formula (see e.g. p. 60, Lemma 3.12 in [14])

$$\sigma(u;\boldsymbol{M}) = \frac{1}{2\pi i}\int_{1+1/L-iN}^{1+1/L+iN} F(s,\chi)\frac{u^s - M^s}{s}ds + O(L^2).$$

Shifting the line of integration to $s = 1/2 + it$, we obtain

$$\sigma(u;\boldsymbol{M}) = \frac{1}{2\pi}\int_{-N}^{N} F(\frac{1}{2}+it,\chi)\frac{u^{1/2+it} - M^{1/2+it}}{1/2+it}dt + O(L^2). \tag{5.18}$$

For $S > R \ge N^{\varepsilon}$ we have $\chi \ne \chi^0$ and (5.3) becomes

$$\hat{W}(\chi,\lambda) = \int_M^N e(u\lambda)d\psi(\chi, M, u),$$

and so $\hat{W}(\chi, \lambda)$ is a linear combination of $O(L^{10})$ terms, each of which is of the form

$$\int_M^N e(u\lambda)d\sigma(u; M) = \frac{1}{2\pi}\int_{-N}^N F\left(\frac{1}{2}+it,\chi\right)\int_M^N u^{-1/2+it}e(u\lambda)dudt$$
$$+O\left(L^2(1+|\lambda|N)\right),$$

whence

$$\begin{aligned}\hat{W}(\chi, \lambda)& \\ \ll L^{10}\max_M &\left|\int_{-N}^N F\left(\frac{1}{2}+it,\chi\right)\int_M^N u^{-1/2}e\left(\frac{t}{2\pi}\log u+\lambda u\right)dudt\right| \\ &+R^2P^2L^{12}. \end{aligned} \tag{5.19}$$

In view of

$$\frac{d}{du}\left(\frac{t}{2\pi}\log u+\lambda u\right)=\frac{t}{2\pi u}+\lambda, \quad \frac{d^2}{du^2}\left(\frac{t}{2\pi}\log u+\lambda u\right)=-\frac{t}{2\pi u^2},$$

Lemmas 4.4 and 4.3 in [14, p.71] show that the inner integral in (5.19) can be estimated as

$$\begin{aligned}&\ll M^{-1/2}\min\left\{\frac{N}{(|t|+1)^{1/2}}, \frac{N}{\min_{M<u\le N}|t+2\pi\lambda u|}\right\} \\ &\ll \begin{cases} N^{1/2}L^6/(|t|+1)^{1/2}, & \text{if } \ |t|\le T_0, \\ N^{1/2}L^6/|t|, & \text{if } \ T_0<|t|\le T, \end{cases}\end{aligned} \tag{5.20}$$

where T_0 is a parameter such that $|t+2\pi\lambda u|>|t|/2$ whenever $|t|>T_0$.

It is clear that (5.14) is a consequence of

$$\sum_{k\sim S}\sum_{\chi \bmod k}\max_{|\lambda|\le 1/kQ}|\hat{W}(\chi,\lambda)|\ll NS^{1/6}L^{-A_1}. \tag{5.21}$$

Substituting (5.20) in (5.19) with $T_0=8\pi N/SQ$, we see that (5.21) is a consequence of the following two estimates: For $0<T_1\le T_0$, we have

$$\sum_{k\sim S}\sideset{}{^*}\sum_{\chi \bmod k}\int_{T_1}^{2T_1}\left|F\left(\frac{1}{2}+it,\chi\right)\right|dt\ll S^{1/6}N^{1/2}(T_1+1)^{1/2}L^{-A_1}, \tag{5.22}$$

while for $T_0 < T_2 \le N$, we have

$$\sum_{k\sim S}\sum_{\chi \bmod k}{}^{*}\int_{T_2}^{2T_2}\left|F\left(\frac{1}{2}+it,\chi\right)\right| dt \ll R^{1/6}N^{1/2}T_2L^{-A_1}. \tag{5.23}$$

Both (5.22) and (5.23) follow from lemma 5.1 provided $L^{B_1} \le S \le P \le N^{3/22-\varepsilon}$. This completes the proof of (5.14).

Now we turn to the proof of (5.17), which is rather parallel to that of Lemma 5.4 and (5.14). We therefore suppress the details and indicate necessary modifications of the proof of (5.14).

Correspondingly to (5.12), we have

$$\int_{-2/RQ}^{2/RQ}|\hat{W}(\chi,\lambda)|^2 d\lambda \ll (RQ)^{-2}\int_M^N |\psi(\chi,X,Y)|^2\, dt, \tag{5.24}$$

where $X = \max(t, M)$ and $Y = \min(t + RQ/2, N)$. The counterpart of (5.18) then reads

$$\sigma(Y;M) = \sigma(Y,X;M) = \frac{1}{2\pi}\int_{-N}^{N} F\left(\frac{1}{2}+it,\chi\right)\tau(t)dt + O(L^2),$$

where

$$\tau(t) = \tau(t;Y,X) = \frac{Y^{\frac{1}{2}+it} - X^{\frac{1}{2}+it}}{\frac{1}{2}+it}.$$

Similarly as in the proof of (5.18), we have

$$\begin{aligned}\tau(t) &= \frac{1}{2}\int_X^Y u^{-1/2}e\left(\frac{t}{2\pi}\log u\right)du\\ &\ll Y^{1/2} - X^{1/2}\\ &\ll (t+RQ)^{1/2} - t^{1/2}\\ &\ll t^{1/2}\{(1+RQ/t)^{1/2} - 1\},\end{aligned}$$

the last term being $\ll t^{-1/2}RQ$, since $M - RQ/2 \le t \le N$. Using the data $RQ/2 \le 2N/RL^{-2B} \le M/R$ in the above and the trivial estimate

$$\tau(t) \ll \frac{Y^{1/2}}{|t|} \ll \frac{N^{1/2}}{|t|},$$

we deduce that

$$\tau(t) \ll \min\left(M^{-1/2}RQ, \frac{N^{1/2}}{|t|}\right) \ll L^6 \min\left(\frac{RQ}{N^{1/2}}, \frac{N^{1/2}}{|t|}\right).$$

Equating $RQ/N^{1/2} = N^{1/2}/|t|$, we have $|t| = T_0 = N/(RQ)$, so that

$$\sigma(Y;M) \ll \frac{RQL^6}{N^{1/2}} \int_{|t|\le T_0} \left| F\left(\frac{1}{2}+it,\chi\right)\right| dt + N^{1/2}L^6 \int_{T_0<|t|\le N} \left| F\left(\frac{1}{2}+it,\chi\right)\right| \frac{dt}{|t|} + O(L^2).$$

Substituting these $O(L^{10})$ terms in (5.24), we obtain

$$\int_{-2/(RQ)}^{2/(RQ)} |\hat{W}(\chi,\lambda)|^2 d\lambda \ll L^{32} \max_M \left(\int_{|t|\le T_0} \left| F\left(\frac{1}{2}+it,\chi\right)\right| dt\right)^2 + \frac{N^2L^{32}}{(RQ)^2} \max_M \left(\int_{T_0<|t|\le N} \left| F\left(\frac{1}{2}+it,\chi\right)\right| \frac{dt}{|t|}\right)^2 + \frac{NL^{30}}{(RQ)^2}.$$

Thus, as in the proof of (5.14), it suffices for the proof of (5.17) to have

$$\sum_{k\sim S} \sum_{\chi \bmod k}^{*} \int_{T_1}^{2T_1} \left| F\left(\frac{1}{2}+it\right)\right| dt \ll N^{1/2}L^c \tag{5.25}$$

holds for $R/2 \le S \le RL^{2B}$ and $0 < T_1 \le T_0$, and

$$\sum_{k\sim S} \sum_{\chi \bmod k}^{*} \int_{T_2}^{2T_2} \left| F\left(\frac{1}{2}+it\right)\right| dt \ll RQN^{-1/2}T_2L^c \tag{5.26}$$

holds for $R/2 \le S \le RL^{2B}$ and $T_0 < T_2 \le N$. (5.25) and (5.26) again follows from Lemma 5.1 provided $R \le N^{2/15-\varepsilon}$. This completes the proof of (5.17). □

Collecting the estimates (5.7)–(5.9) and (5.13)–(5.17), it follow from (4.11) that

$$\sum_{r_1\sim R} \sum_{r_2\sim R} \left| \int_{\mathfrak{M}} S(\alpha,r_1,b_1)S(\alpha,r_2,b_2)S(\alpha)e(-N\alpha)d\alpha - (I_{10}+I_{20}+I'_{20})\right| \ll N^2L^{-A}. \tag{5.27}$$

6. The main terms and the singular series

Now, it remains to compute I_{10} and I_{20}, I'_{20}. Recalling the definition of I_{10} (given right after (4.1)) and using (3.6) and (3.8), we have

$$I_{10} = \sum_{\substack{q\le P \\ r_1\nmid q,\, r_2\nmid q}} \sum_{\substack{a=1 \\ (a,q)=1}}^{q} \frac{G^3(\chi_0,a)}{\varphi(r_1r_2)\varphi^3(q)} e\left(-\frac{aN}{q}\right) \int_{-1/qQ}^{1/qQ} V^3(\lambda)e(-N\lambda)d\lambda$$

$$= \sum_{\substack{q \le P \\ r_1 \nmid q, r_2 \nmid q}} \sum_{\substack{a=1 \\ (a,q)=1}}^{q} \frac{G^3(\chi_0, a)}{\varphi(r_1 r_2)\varphi^3(q)} e\left(-\frac{aN}{q}\right) \left(\frac{1}{2}N^2 + O(N^2 R^{-2} L^{-4B})\right). \tag{6.1}$$

Similarly, (recalling the definition right after (4.6)), we have

$$I_{20} = \sum_{\substack{q \le P \\ r_1 \nmid q, r_2 | q}} \sum_{\substack{a=1 \\ (a,q)=1}}^{q} \frac{G^2(\chi_0, a) C(q, b_2, r_2, a)}{\varphi(r_1)\varphi^3(q)} e\left(-\frac{aN}{q}\right) \times \left(\frac{1}{2}N^2 + O(N^2 R^{-2} L^{-4B})\right) \tag{6.2}$$

and

$$\begin{aligned} I'_{20} &= \sum_{\substack{q \le P \\ r_1 | q, r_2 \nmid q}} \sum_{\substack{a=1 \\ (a,q)=1}}^{q} \int_{-1/qQ}^{1/qQ} S_1(r_2, a, q, \lambda) U_1(r_1, a, b_1, q, \lambda) S_1(a, q) \\ &\qquad \times e\,(-N(a/q + \lambda))\, d\lambda \\ &= \sum_{\substack{q \le P \\ r_1 | q, r_2 \nmid q}} \sum_{\substack{a=1 \\ (a,q)=1}}^{q} \frac{G^2(\chi_0, a) C(q, b_1, r_1, a)}{\varphi(r_2)\varphi^3(q)} e\left(-\frac{aN}{q}\right) \\ &\qquad \times \left(\frac{1}{2}N^2 + O(N^2 R^{-2} L^{-4B})\right). \end{aligned} \tag{6.3}$$

Now we may give the postponed definition of the singular series

$$\begin{aligned} \sigma(N, r_1, r_2, b_1, b_2) = &\sum_{\substack{q \le P \\ r_1 \nmid q, r_2 \nmid q}} \sum_{\substack{a=1 \\ (a,q)=1}}^{q} \frac{G^3(\chi_0, a)}{\varphi(r_1 r_2)\varphi^3(q)} e\left(-\frac{aN}{q}\right) \\ &+ \sum_{\substack{q \le P \\ r_1 \nmid q, r_2 | q}} \sum_{\substack{a=1 \\ (a,q)=1}}^{q} \frac{G^2(\chi_0, a) C(q, b_2, r_2, a)}{\varphi(r_1)\varphi^3(q)} e\left(-\frac{aN}{q}\right) \\ &+ \sum_{\substack{q \le P \\ r_1 | q, r_2 \nmid q}} \sum_{\substack{a=1 \\ (a,q)=1}}^{q} \frac{G^2(\chi_0, a) C(q, b_1, r_1, a)}{\varphi(r_2)\varphi^3(q)} e\left(-\frac{aN}{q}\right). \end{aligned} \tag{6.4}$$

From the well-known result $|G(\chi^0)| \le 1$ and $|C(q, b, r, a)| \le 1$, it is easily checked that

$$\frac{1}{\varphi(r_1 r_2)} \ll \sigma(N, r_1, r_2, b_1, b_2) \ll \frac{1}{\varphi(r_1 r_2)}.$$

Here the implied constants are absolute. Collecting (6.1)–(6.3), we have

$$I_{10}+I_{20}+I'_{20}=\sigma(N,r_1,r_2,b_1,b_2)\left(\frac{1}{2}N^2+O(N^2R^{-2}L^{-4B})\right). \quad (6.5)$$

Now, our Theorem 2 follows from (5.27) and (6.5).

Acknowledgments

The author would like to thank Professor Liu Jianya for constant discussions and valuable advices. He also thanks to Professor Shigeru Kanemitsu and Yoshio Tanigawa for careful reading of the manuscript.

References

[1] R. Ayoub, *On Rademacher's extension on the Goldbach-Vinogradov theorem,* Trans. Amer. Math. Soc. **74** (1953), 482–491.

[2] Z. Cui, *The ternary Goldbach problem in arithmetic progressions,* to appear.

[3] H. Davenport, *Multiplicative Number Theory,* 3rd ed., Springer, Berlin 2000.

[4] P. X. Gallagher, *A large sieve density estimate near $\sigma = 1$,* Invent. Math. **11** (1970), 329–339.

[5] D. R. Heath-Brown, *Prime numbers in short intervals and a generalized Vaughan's identity,* Canad. J. Math. **34** (1982), 1365–1377.

[6] J.-Y. Liu, *The Goldbach-Vinogradov theorem with primes in a thin subset,* Chinese Ann. Math. 19B:4 (1998), 479–488.

[7] J.-Y. Liu and T. Zhan, *The ternary Goldbach problem in arithmetic progressions,* Acta Arith. **82** (1997), 197–227.

[8] J.-Y. Liu and M.-C. Liu, *The exceptional set in the four prime squares problem,* Illinois J. Math. **44** (2000), 272–293.

[9] M.-C. Liu and K.-M. Tsang, *Small prime solutions of linear equations,* in: Théorie des Nombres, J. M. De Koninck and C. Levesque (éd. J. M. Dekoninck and C. Levesque), 595–624. Walter de Gruyter, Berlin 1989.

[10] M. C. Liu and T. Zhan, *The Goldbach problem with primes in arithmetic progressions,* in: Analytic Number Theory (Kyoto, 1996) (London Math. Soc. Lecture Note ser. 247, ed. Y. Motohashi), 227–251. Cambridge University Press, 1997.

[11] C.-D. Pan and C.-B. Pan, *Fundamentals of analytic number theory* (in Chinese), Science Press, Beijing, 1991.

[12] K. Prachar, *Primzahlverteilung,* Springer, Berlin 1957.

[13] H. Rademacher, *Über eine Erweiterung des Goldbachschen Problems,* Math. Z. **25** (1926), 627–660.

[14] E. C. Titchmarsh, *The theory of the Riemann zeta-function,* 2nd ed., University Press, Oxford 1986.

[15] R. C. Vaughan, *The Hardy-Littlewood method,* 2nd ed., Cambridge Tracts Math., vol. 125, Cambridge University Press, Cambridge, 1997.

[16] I. M. Vinogradov, *The representation of an odd number as a sum of three primes,* Dokl. Akad. Nauk. SSSR **16** (1937), 291–294.

[17] I. M. Vinogradov, *Some theorems concerning the theory of primes,* Math. Sb. N. S., 2 (1937), 179–195.

[18] D. Wolke, *Some applications to zero-density theorems for L-functions,* Acta Math. Hungar. **61** (1993), 241–258.

[19] T. Zhan and J.-Y. Liu, *A Bombieri type mean-value theorem concerning exponential sums over primes,* Chinese Sci. Bull. **41** (1990), 363–366.

[20] A. Zulauf, *Beweis einer Erweiterung des Stazes Von Goldbach-Vinogradov.* J. Reine Angew. Math. **190** (1952), 169–198.

DENSITIES OF SETS OF PRIMES RELATED TO DECIMAL EXPANSION OF RATIONAL NUMBERS

Toshihiro Hadano[1], Yoshiyuki Kitaoka[2], Tomio Kubota and Michihiro Nozaki[3]

[1] *Department of Mathematics, Meijo University*
Tenpaku, Nagoya
468-8502, Japan
hadano@ccmfs.meijo-u.ac.jp

[2] *Department of Mathematics, Meijo University*
Tenpaku, Nagoya
468-8502, Japan
kitaoka@ccmfs.meijo-u.ac.jp

[3] *Chitahigashi High School*
Dougashima, Yawata, Chita
478-0001, Japan

Abstract For a prime p and a given integer $a \neq 0, \pm 1$, not divisible by p, we suppose the order of $a \mod p$ is composite nk, $n > 1$. Let r_i be the least non-negative residue mod p of a^{ki}, $0 \le i \le n-1$ (which has a relationship to a-adic expansion of $\frac{1}{p}$), and let $\mathfrak{s}_a(p) = \frac{1}{p}\sum_{i=0}^{n-1} r_i = \sum_{i=0}^{n-1} \left\{\frac{a^{ki}}{p}\right\}$, with $\{\alpha\}$ denoting the fractional part of α. We are interested in the set of primes for which $\mathfrak{s}_a(p)$ takes a prescribed value s for a given n, and in particular in the limit of the relative frequency (density) $P_a(x, n, s)$. We propose several new conjectures on them and also on $P_{gr}(x, n, s)$ constructed in the same way from the subgroup $G_{n,p}$ of $(\mathbb{Z}/p\mathbb{Z})^\times$ of order n, where a is taken to be a primitive root mod p. We also refer to rather general probabilistic phenomena.

Keywords: decimal expansion, density, uniform distribution, normal distribution

2000 Mathematics Subject Classification: 11J71, 11K06

Number Theory: Tradition and Modernization, pp. 67–80
W. Zhang and Y. Tanigawa, eds.

1. Introduction

Among a number of numerical data which have been incentive to our research, the following is typical:

$$1/7 = 0.\dot{1}4285\dot{7} = 0.142857142857\cdots \tag{1.1}$$

where $\dot{1}4285\dot{7}$ signifies the period of the purely periodic decimal (10-adic) expansion of $1/7$. For (1.1) we have:

$$\begin{aligned} 142 + 857 &= 999 \\ 14 + 28 + 57 &= 99 \\ 1 + 4 + 2 + 8 + 5 + 7 &= 27 = 9 \times 3. \end{aligned} \tag{1.2}$$

Our starting point was the following Theorem 1 which elucidates the above phenomenon.

Before stating Theorem 1 we introduce notation which we will use throughout the paper.

Notation

p signifies a prime

$[x]$: the integral part of $x =$ the largest integer $\leq x$

$\{x\} = x - [x]$: the fractional part of x

a : an integer $\neq 0, \pm 1$, not divisible by p

$e = e_a(p)$: the order of a mod p

$e = nk$, $n > 1$ (we so suppose in anticipation of application to the group $(\mathbb{Z}/p\mathbb{Z})^\times$)

$r_i = a^{ki} - p[\frac{a^{ki}}{p}] = p\left\{\frac{a^{ki}}{p}\right\}$: the least non-negative residue mod p of a^{ki}, $a^{ki} \equiv r_i \pmod{p}$, $0 \leq r_i < p$, $0 \leq i \leq n$

$\mathfrak{s}(p) = \mathfrak{s}_a(p) = \frac{1}{p}\sum_{i=0}^{n-1} r_i$

$G_{n,p}$: the subgroup of $(\mathbb{Z}/p\mathbb{Z})^\times$ of order n.

We are now in a position to state

Theorem 1. (i) *As in Notation suppose the order e of a mod p ($p \nmid a$) is composite, $e = nk$, $n > 1$. Let r_i denote the least non-negative residue mod p of a^{ki} :*

$$r_i = a^{ki} - p\left[\frac{a^{ki}}{p}\right], \quad 0 \leq i \leq n-1. \tag{1.3}$$

Then

$$\mathfrak{s}(p) = \mathfrak{s}_a(p) = \frac{1}{p}\sum_{i=0}^{n-1} r_i \tag{1.4}$$

is an integer such that $\mathfrak{s}(p) = \frac{n}{2}$ *if* n *is even and* $1 \leq \mathfrak{s}(p) \leq n-2$ *if* n *is odd.*

(ii) *If in (i), we assume* n *further decomposes as* $n = n_1 n_2$, $n_1 \geq 3$, *then*

$$n_2 \leq \mathfrak{s}(p) \leq n - n_2 - 1$$

Before giving the proof, we state some observations which lead to plausible conjectures on the set of primes for which $\mathfrak{s}(p)$ takes the prescribed value s for given $n|e$. In tabulating the numerical data, we will restrict ourselves to the case n (≥ 5) an odd natural number, $1 \leq s \leq n-2$ and to the base 10 ($a = 10$). We introduce the relative frequency (density)

$$P_a(n,s,x) = \frac{\#\{p \mid p \leq x, n|e, \mathfrak{s}(p) = s\}}{\#\{p \mid p \leq x, n|e\}}, \tag{1.5}$$

under the notation of Theorem 1. We note that $P_a(n,s,x) = 0$ or 1 if n is even or $n = 3$.

Conjecture 1. $\lim_{x\to\infty} P_a(n,s,x)$ *exists and the value* $P(n,s)$ *is independent of* a.

Hence, in particular, $\lim_{x\to\infty} P_{10}(n,s,x) = P(n,s)$ is expected, and so we shall give tables with $a = 10$ in the following, and in §3 we shall give those for $a = 2, 5$ and $P_{gr}(n,s,x)$, which is to be introduced in §2 and is also expected to converge to the same $P(n,s)$. Table 1 is the table for $P(n,s,10^9)$, in which 0.0000 in the column for $n = 11$ means that primes which take values $s = 1, 9$ are very rare and the figure 0 for $n = 9$ means that there is no such prime (Theorem 1,(ii)).

Table 1

s	$n=5$	$n=9$	$n=11$
1	0.1666	0	0.0000
2	0.6667	0	0.0014
3	0.1667	0.2499	0.0403
4		0.5001	0.2432
5		0.2500	0.4301
6		0	0.2433
7		0	0.0403
8			0.0014
9			0.0000

Table 1 (Table 6) suggests that the values $P(n,s,x)$ are symmetrically distributed relative to $s = (n-1)/2$, and in the first instance, that

Conjecture 2.

$$P(n,s) = P(n, n-1-s) \quad \textit{for} \quad 1 \leq s \leq n-2$$

Secondly,

Conjecture 3. *$P(n,s) > 0$ holds for $1 \leq s \leq n-2$ if n is an odd prime.*

And thirdly, the distribution of $P(n,s)$ should be the normal distribution.

To make the last statement more precise, we recall standard notation of statistics. Given a table of frequency distribution

Table 2

value	x_1	x_2	$\cdots$	x_m	sum
relative frequency	r_1	r_2	$\cdots$	r_m	1

we compute the mean $\mu = \mu(x) = \mu(n,x) = \sum_{i=1}^{m} x_i r_i$ and the standard deviation $\sigma = \sigma(x) = \sigma(n,x) = \sqrt{\sum_{i=1}^{m} x_i^2 r_i - \mu^2}$ with $x_i = i$, $1 \leq i \leq n-2$, and $r_i = P(n,i,10^9)$ to obtain

Table 3

n	μ	σ
5	2.0001	0.5774
9	4.0002	0.7070
11	5.0002	0.9132

Table 3 now suggests the first half of

Conjecture 4.

$$\lim_{x\to\infty} \mu = \frac{n-1}{2}, \quad \lim_{x\to\infty} \sigma^2 = \begin{cases} (n-1)/12 & \textit{if } 3 \nmid n, \\ (n-3)/12 & \textit{if } 3 \mid n. \end{cases}$$

The last half is not apparent from Table 1, but in accord with data (Table 7) and intuitively supported by Theorem 2.

We denote the density function of the normal distribution with mean μ and standard deviation σ by $f_{\mu,\sigma}(x)$ and compare the above data with its values whose tabulation is

Table 4

n	$\max_{1\leq s\leq n-2} \lvert P(n,s,10^9) - f_{\mu,\sigma}(s)\rvert$
5	0.0243
9	0.0641
11	0.0067

Table 4 suggests

Conjecture 5.

$$\lim_{n\to\infty} \max_{1\leq s\leq n-2} \lvert P(n,s) - f_{\mu,\sigma}(s)\rvert = 0.$$

It may be more appropriate to compare distribution functions instead of density functions.

We now turn to the proof of Theorem 1.

Proof of Theorem 1.

(i) First we prove the identity

$$(a^k - 1)\mathfrak{s}(p) = \sum_{i=0}^{n-1} \left(\left[\frac{a^{k(i+1)}}{p} \right] - \left[\frac{a^{ki}}{p} \right] a^k \right) \tag{1.6}$$

Since the right-hand side of (1.6) is an integer and $p \nmid a^k - 1$, it follows that $p \mid \sum_{i=0}^{n-1} r_i$ and, *afortiori*, that $\mathfrak{s}(p) \in \mathbb{Z}$.

To prove (1.6), we substitute (1.3) in

$$(a^k - 1)\mathfrak{s}(p) = \sum_{i=0}^{n-1} \left(\frac{a^k r_i}{p} - \frac{r_i}{p} \right)$$

to write

$$(a^k - 1)\mathfrak{s}(p) = \sum_{i=0}^{n-1} \left(\frac{a^{k(i+1)}}{p} - \left[\frac{a^{ki}}{p} \right] a^k - \frac{r_i}{p} \right).$$

Then substitute (1.3) in the form $a^{k(i+1)} = r_{i+1} + p \left[\frac{a^{k(i+1)}}{p} \right]$ to obtain

$$(a^k - 1)\mathfrak{s}(p) = \sum_{i=0}^{n-1} \left(\left[\frac{a^{k(i+1)}}{p} \right] - \left[\frac{a^{ki}}{p} \right] a^k \right) + \sum_{i=0}^{n-1} \frac{r_{i+1} - r_i}{p},$$

the second term on the right being $(r_n - r_0)/p = 0$, whence (1.6) follows.

To prove the range for $\mathfrak{s}(p)$, we first note that (1.3), or its equivalent

$$a^{ki} \equiv r_i \mod p, \quad 0 \le r_i < p - 1 \tag{1.7}$$

implies $p \nmid r_i$ in view of $p \nmid a$, so that $r_i \ge 1$.

Secondly, since $ki < e$, all a^{ki} are distinct $\mod p$, and so are r_i, $0 \le i \le n-1$ with $r_0 = 1$. We conclude that r_i, $0 \le i \le n-1$ are distinct integers between 1 and p. Hence $\mathfrak{s}(p)$ is at most

$$1 + (p-1) + (p-2) + \cdots + (p - (n-1)) = (n-1)p - \frac{n(n-1)}{2} + 1,$$

which is less than $(n-1)p$ for $n \ge 3$.

Hence, we always have, for $n \geq 3$,

$$1 \leq \mathfrak{s}(p) \leq n-2.$$

Now suppose n is even, so that $\frac{n}{2}$ is an integer. Then, since

$$\left(a^{k\frac{n}{2}}-1\right)\left(a^{k\frac{n}{2}}+1\right) \equiv a^{kn}-1 \equiv 0 \mod p.$$

and $p \nmid a^{k\frac{n}{2}}-1$, we must have

$$a^{k\frac{n}{2}} \equiv -1 \mod p,$$

Hence

$$r_{\frac{n}{2}+i} \equiv a^{k(\frac{n}{2}+i)} \equiv -a^{ki} \equiv -r_i \mod p,$$

whence

$$r_{\frac{n}{2}+i}+r_i \equiv 0 \mod p.$$

But, $0 < r_{\frac{n}{2}+i}+r_i < 2p$, and so $r_{\frac{n}{2}+i}+r_i = p$.

Hence

$$\sum_{i=0}^{n-1} r_i = \sum_{i=0}^{n/2-1} (r_i + r_{\frac{n}{2}+i}) = \frac{pn}{2},$$

yielding $\mathfrak{s}(p) = n/2$.

(ii) We divide the sum $\mathfrak{s}(p)$ into n_2 equal parts and denote the ℓ-th partial sum by s_ℓ:

$$s_\ell = \sum_{i=0}^{n_1-1} \frac{r_{\ell+in_2}}{p}.$$

we then claim that $s_\ell \in \mathbb{Z}$. Indeed, by (1.7)

$$ps_\ell \equiv a^{k\ell} \sum_{i=0}^{n_1-1} (a^{kn_2})^i \equiv a^{k\ell}\frac{a^{n_1n_2k}-1}{a^{n_2k}-1} \equiv 0 \mod p,$$

and $s_\ell \in \mathbb{Z}$ follows.

In view of $1 \leq r_i \leq p-1$, we have $n_1 \leq s_\ell p \leq (p-1)n_1$ and hence $1 \leq s_\ell \leq n_1 - 1$. Noting $r_0 = 1$, we may improve the bound for s_0 as follows.

$$ps_0 \leq 1 + (n_1-1)(p-1) \leq (n_1-1)p + 2 - n_1 < (n_1-1)p,$$

whence $s_0 \leq n_1 - 2$. Therefore, summing over $\ell = 0, \cdots, n_2-1$, we conclude that

$$n_2 \leq \mathfrak{s}(p) \leq n_1 - 2 + \sum_{\ell=1}^{n_2-1} (n_1 - 1) = n - n_2 - 1,$$

and the proof is complete. □

Remark 1. Each summand $\left[\frac{a^{k(i+1)}}{p}\right] - \left[\frac{a^{ki}}{p}\right] a^k$ on the right of (1.6) admits the following familiar interpretation. We restrict to the case $a = 10$. Let $\dot{c}_1 \cdots \dot{c}_e$ be the period of the purely periodic decimal expansion of $1/p$:

$$\frac{1}{p} = 0.\dot{c}_1 \cdots \dot{c}_e = 0.c_1 \cdots c_e c_1 \cdots c_e \cdots \qquad (0 \le c_i \le 9)$$

Then, since $c_1 \cdots c_{ki} = \left[\frac{10^{ki}}{p}\right]$, it follows that

$$\left[\frac{10^{k(i+1)}}{p}\right] - \left[\frac{10^{ki}}{p}\right] 10^k = c_1 \cdots c_{k(i+1)} - c_1 \cdots c_{ki} 10^k = c_{ki+1} \cdots c_{k(i+1)},$$

viewed as a k-digit integer.

Thus $\mathfrak{s}(p)$ is the sum of these integers divided by $10^k - 1$, and Theorem 1 explain (1.2) with $e = 6$, $(k, n, \mathfrak{s}(p)) = (3, 2, 1), (2, 3, 1), (1, 6, 3)$, where for computation of r_i, formula (1.7) is most feasible.

2. Generalization And Stochastic Discussion

In the setting of Theorem 1 we suppose a is a primitive root mod p. Then the order e of a is $p-1$, which is a composite number, $e = p - 1 = nk$, $n > 1$. Let $G_{n,p}$ denote the subgroup of $(\mathbb{Z}/p\mathbb{Z})^\times$ of order n. Then it consists of a^{ki} for $i = 0, 1, \cdots, n-1$. Hence in view of (1.3) and (1.4),

$$\mathfrak{s}(p) = \sum_{g \in G_{n,p}} \{g/p\},$$

where the residue class g is identified with an integer in the class.

The relative frequency $P_{gr}(n, s, x)$ is defined in the same way as (1.5), and with the same limit density function $P(n, s)$, we propose

Conjecture 6.

$$\lim_{x \to \infty} P_{gr}(n, s, x) = P(n, s).$$

The following conjecture, also numerically supported, lays a basis for stochastic discussion on $P(n, s)$.

Conjecture 7. *For an odd natural number* $n\,(> 2)$ *and a prime number satisfying* $p \equiv 1 \bmod n$, *we set*

$$S(p) = G_{n,p} \setminus \{1/p\} = \big\{\{g/p\} \,|\, g \in G_{n,p}, g \neq 1\big\}.$$

Then the points in $\bigcup_{p<x} S(p)$ *are uniformly distributed in* $(0,1]$ *as* x *tends to infinity.*

Remark 2. Suitably generalized , this conjecture leads to the following plausible conjecture: Let $F = \mathbb{Q}(\alpha)(\neq \mathbb{Q})$ be the algebraic number field generated by an algebraic integer α, and let k be a non-negative integer. For a prime number p which decomposes fully in F and a prime ideal $\mathfrak{p}$ lying above p, we write in $F_{\mathfrak{p}} = \mathbb{Q}_p$

$$\alpha = c_{\mathfrak{p}}(0) + c_{\mathfrak{p}}(1)p + \cdots \qquad (c_{\mathfrak{p}}(i) \in \mathbb{Z}, \quad 0 \leq c_{\mathfrak{p}}(i) < p).$$

Then the points $(c_{\mathfrak{p}}(0)/p, c_{\mathfrak{p}}(1)/p, \cdots, c_{\mathfrak{p}}(k)/p)$ $(\in (0,1]^{k+1})$ are uniformly distributed as p and $\mathfrak{p}$ run through primes and prime ideals described above. This conjecture is true [DFI], [T] when F is quadratic and $k = 0$.

The next theorem with $m = n-1$ tells intuitively, under Conjecture 7, why the distribution $P(n,s)$ is approximated by the normal distribution by neglecting the small difference $1/p$ between $\mathfrak{s}(p)$ and the sum over elements of $G_{n,p}$. It also gives an intuitive evidence for the variance part of Conjecture 4 in the case $3 \nmid n$ on noting the ratio of the standard deviations of $\sqrt{m}X_m$ and X_m is $\sqrt{m}$.

Theorem 2. *Let* $x_1, x_2, \cdots, x_m$ *be random variables on* $\mathbb{R}$ *obeying the uniform distribution* $I(0,1)$*, or what amounts to the same, their distribution function are all equal to the (set-theoretic) characteristic function* $f_1(x)$ *of* $[0,1]$*. Then, putting*

$$X_m = \frac{1}{\sqrt{m}}(x_1 + x_2 + \cdots + x_m - m/2),$$

$X = \lim_{n\to\infty} X_m$ *determines the normal distribution on* $\mathbb{R}$ *with mean* $\mu = 0$ *and standard deviation* $\sigma = \frac{1}{\sqrt{12}}$.

Proof. Since the mean and standard deviation of x_i are $1/2$ and $\frac{1}{\sqrt{12}}$, X_m has the same mean and standard deviation. The theorem is then a special case of the central limit theorem.

For curiosity's sake, we shall prove below that the value $\frac{1}{\sqrt{12}}$ of σ is a natural consequence of the theory of Fourier transforms, with μ given. We may suppose $\mu = 0$ instead, i.e. we may take the characteristic function f_0 of $[-1/2, 1/2]$ instead of f_1.

We recall the standard notions of Fourier transform theory. The convolution $f * g$ of two function f and g is defined by

$$(f * g)(x) = \int_{-\infty}^{\infty} f(x-t)g(t)dt,$$

and the Fourier transform f^* of a function f is defined by

$$f^*(y) = \int_{-\infty}^{\infty} f(x)e^{2\pi ixy}dx.$$

Starting from $F_1 = f_0$, we define the m-ple convolution F_m inductively by $F_m = F_{m-1} * f_0$. It is a basic fact in probability theory that the distribution density of $x_1 + x_2 + \cdots + x_m = \sqrt{m}X_m$ is F_m, whence that of X_m is $\sqrt{m}F_m(\sqrt{m}x)$.

Now the Fourier transform $f_0{}^*$ of f_0 is $\dfrac{\sin \pi y}{\pi y}$, and so that of F_m is $\left(\dfrac{\sin \pi y}{\pi y}\right)^m$. Denote by φ_m the Fourier transform of the distribution density $\sqrt{m}F_m(\sqrt{m}x)$ of X_m. Then, it follows from the above argument that

$$\varphi_m(y) = \left(\frac{\sin \frac{\pi y}{\sqrt{m}}}{\frac{\pi y}{\sqrt{m}}}\right)^m.$$

Since the distribution density of X can be determined as the Fourier transform φ^* of φ, it suffices to find $\varphi(y) = \lim_{m\to\infty} \varphi_m(y)$. For this, we need to study $\varphi'(y)$ first.

A simple calculation shows

$$\frac{\varphi_m'(y)}{\varphi_m(y)}y = \frac{\frac{\pi y}{\sqrt{m}} \cos \frac{\pi y}{\sqrt{m}} - \sin \frac{\pi y}{\sqrt{m}}}{\sin \frac{\pi y}{\sqrt{m}}}m.$$

whence by the Taylor expansion, we deduce that

$$\begin{aligned} & y \lim_{m\to\infty} \frac{\varphi_m'(y)}{\varphi_m(y)} \\ &= \lim_{m\to\infty} \frac{\frac{\pi y}{\sqrt{m}}\left(1 - \frac{1}{2!}\left(\frac{\pi y}{\sqrt{m}}\right)^2 + \cdots\right) - \left(\frac{\pi y}{\sqrt{m}} - \frac{1}{3!}\left(\frac{\pi y}{\sqrt{m}}\right)^3 + \cdots\right)}{\frac{\pi y}{\sqrt{m}} - \frac{1}{3!}\left(\frac{\pi y}{\sqrt{m}}\right)^3 + \cdots}m \\ &= -\frac{\pi^2}{3}y^2. \end{aligned}$$

Hence

$$\lim_{m\to\infty} \frac{d}{dy} \log \varphi_m(y) = \lim_{m\to\infty} \frac{\varphi_m'(y)}{\varphi_m(y)} = -\frac{\pi^2}{3}y,$$

so that $\varphi = \lim_{m\to\infty} \varphi_m$ exists and is a solution to the differential equation

$$\frac{d\varphi}{dy} = -\frac{\pi^2}{3}y\varphi.$$

Solving the equation, we have $\varphi(y) = c_0 e^{-\frac{\pi^2}{6}y^2}$. But, since φ^* is the distribution density, we have

$$c_0 = \int_{-\infty}^{\infty} \varphi^*(x)dx = 1.$$

and $\varphi(y) = e^{-\frac{\pi^2}{6}y^2}$.

Recalling that $e^{-\pi x^2}$ is invariant under the Fourier transformation, we find that the Fourier transform of $e^{-\alpha x^2}$ is $\sqrt{\frac{\pi}{\alpha}}e^{-\frac{\pi^2}{\alpha}y^2}$. Hence with $\alpha = \pi^2/6$, we conclude that

$$\varphi^*(x) = \sqrt{\frac{6}{\pi}}e^{-6x^2},$$

which is $f_{0,\frac{1}{\sqrt{12}}}$ of the normal distribution, i.e. $\sigma = 1/\sqrt{12}$.

The proof is complete. □

Finally we state

Proposition 1. *For $n \geq 5$ an odd integer, Conjectures 1, 6 and 7 yield*

$$\lim_{x\to\infty} \mu = \frac{n-1}{2}.$$

Proof. We have, by definition,

$$\begin{aligned}\mu &= \sum_{s=1}^{n-2} P_{gr}(n,s,x)s \\ &= \frac{1}{\pi(x;n,1)} \sum_{\substack{p\leq x \\ p\equiv 1 \bmod n}} \mathfrak{s}(p) \\ &= \frac{1}{\pi(x;n,1)} \sum_{\substack{p\leq x \\ p\equiv 1 \bmod n}} \sum_{g\in G_{n,p}} \{g/p\}\end{aligned}$$

where $\pi(x;k,\ell)$ is the number of primes less than x lying in the residue class $\ell \bmod k$. Now the unity of $G_{n,p}$ gives rise to the sum

$$\frac{1}{\pi(x;n,1)} \sum_{\substack{p\leq x \\ p\equiv 1 \bmod n}} \frac{1}{p} = O\left(\frac{\log\log x}{x/\log x}\right) = o(1),$$

by Dirichlet's prime number theorem, while the sum over $g \neq 1$ is

$$(n-1)\frac{\sum_{\substack{p\le x \\ p\equiv 1 \bmod n}} \sum_{1\neq g\in G_{n,p}} \{g/p\}}{(n-1)\pi(x;n,1)},$$

which tends to $(n-1)\int_0^1 t dt = \frac{n-1}{2}$ by the defining property of the uniform distribution of points in $\bigcup_{p<x} S(p)$, whence the result follows. □

3. Numerical data

To see the dependence of numerical data on x, we give the following table of densities $P_{gr}(n, s, x)$ in the case of subgroups, where $x = 10^9$ in the upper row, and $x = 10^{10}$ in the lower row. We round off to five decimal places.

Table 5

$n = 5$

s	1	2	3
	0.16659	0.66672	0.16669
	0.16667	0.66668	0.16666

$n = 7$

s	1	2	3	4	5
	0.00837	0.21644	0.55012	0.21673	0.00834
	0.00834	0.21663	0.55002	0.21669	0.00832

$n = 9$

s	1	2	3	4	5	6	7
	0	0	0.24993	0.50014	0.24993	0	0
	0	0	0.25001	0.50000	0.24998	0	0

$n = 11$

s	1	2	3	4	5	6	7
	0.00000	0.00140	0.04026	0.24320	0.43019	0.24322	0.04035
	0.00000	0.00138	0.04021	0.24318	0.43043	0.24312	0.04029

8	9
0.00138	0.00000
0.00138	0.00000

Table 6 the table of $P_{gr}(n, s, 10^9)$.

$s\backslash n$	13	15	17	19	21	23	25
1	0.00000	0	0.00000	0.00000	0	0	0
2	0.00005	0	0.00000	0	0	0	0
3	0.00382	0	0.00001	0	0	0	0
4	0.05502	0	0.00065	0.00004	0	0.00000	0
5	0.24437	0.06392	0.01149	0.00132	0	0.00001	0.00000
6	0.39381	0.24454	0.07870	0.01613	0	0.00024	0.00003
7	0.24385	0.38327	0.23812	0.08755	0.01568	0.00353	0.00050
8	0.05527	0.24444	0.34247	0.23369	0.09389	0.02531	0.00500
9	0.00376	0.06383	0.23788	0.32296	0.23375	0.10040	0.02983
10	0.00005	0	0.07845	0.23373	0.31323	0.22442	0.10498
11	0	0	0.01159	0.08723	0.23381	0.29237	0.21953
12		0	0.00064	0.01601	0.09399	0.22434	0.28036
13		0	0.00001	0.00131	0.01565	0.10026	0.21946
14			0	0.00004	0	0.02534	0.10495
15			0	0.00000	0	0.00351	0.02979
16				0	0	0.00025	0.00505
17				0	0	0.00001	0.00049
18					0	0	0.00002
19					0	0	0.00000
20						0	0
21						0	0
22							0
23							0

In the following table, max of error means

$$\max_{1\leq s\leq n-2} |P_{gr}(n, s, 10^9) - f_{\mu,\sigma}(s)|,$$

and

$$N = \begin{cases} (n-1)/12 & \text{if } 3 \nmid n, \\ (n-3)/12 & \text{if } 3 \mid n. \end{cases}$$

Table 7

n	$\frac{\mu}{(n-1)/2}$	$\frac{\sigma^2}{N}$	max of error
5	1.00005	0.99985	0.02432
7	1.00008	1.00001	0.01407
9	1.00000	0.99973	0.06413
11	1.00003	1.00071	0.00667
13	0.99997	0.99930	0.00527
15	0.99996	0.99998	0.01567
17	0.99994	1.00020	0.00299
19	0.99989	0.99998	0.00278
21	1.00002	1.00068	0.01239
23	0.99997	0.99970	0.00231
25	0.99999	1.00146	0.00153
27	0.99991	1.00073	0.00913
29	1.00007	0.99960	0.00176
31	0.99994	0.99886	0.00143
33	0.99997	1.00120	0.00633
35	0.99993	0.99991	0.00161
37	1.00005	1.00044	0.00061
39	0.99984	0.99875	0.00504
41	0.99992	0.99840	0.00102
43	0.99990	1.00144	0.00075
45	0.99999	1.00064	0.00381
47	0.99989	0.99961	0.00054
49	1.00008	1.00209	0.00076
51	0.99995	0.99939	0.00295
53	0.99987	1.00089	0.00057

n	$\frac{\mu}{(n-1)/2}$	$\frac{\sigma^2}{N}$	max of error
55	0.99989	0.99965	0.00041
57	0.99994	1.00039	0.00226
59	1.00003	0.99673	0.00121
61	0.99999	1.00225	0.00108
63	1.00002	0.99859	0.00252
65	0.99998	1.00026	0.00044
67	0.99997	1.00264	0.00098
69	1.00000	1.00039	0.00199
71	0.99989	1.00018	0.00072
73	0.99987	1.00049	0.00119
75	1.00000	1.00135	0.00201
77	1.00001	1.00160	0.00062
79	0.99997	1.00204	0.00077
81	0.99995	0.99960	0.00186
83	0.99990	1.00135	0.00064
85	1.00000	0.99645	0.00067
87	0.99997	1.00063	0.00153
89	0.99993	0.99835	0.00107
91	0.99998	0.99901	0.00058
93	0.99996	0.99950	0.00124
95	0.99990	0.99960	0.00086
97	0.99988	1.00146	0.00100
99	0.99998	1.00207	0.00130
101	0.99995	1.00030	0.00069

Table 8

the table of $\max_{1\le s\le n-2} |P_{gr}(n,s,10^9) - P_a(n,s,10^9)|$ for $a = 10, 2, 5$.

n	$a=10$	$a=2$	$a=5$
5	0.00050	0.00013	0.00004
7	0.00008	0.00006	0.00002
9	0.00011	0.00010	0.00007
11	0.00013	0.00006	0.00010
13	0.00010	0.00008	0.00009
15	0.00022	0.00012	0.00016
17	0.00012	0.00009	0.00002
19	0.00007	0.00005	0.00003
21	0.00024	0.00021	0.00019
23	0.00008	0.00007	0.00008
25	0.00020	0.00014	0.00015
27	0.00021	0.00022	0.00018

n	$a=10$	$a=2$	$a=5$
29	0.00010	0.00011	0.00008
31	0.00011	0.00006	0.00009
33	0.00033	0.00023	0.00016
35	0.00018	0.00019	0.00020
37	0.00005	0.00004	0.00008
39	0.00032	0.00029	0.00018
41	0.00005	0.00009	0.00010
43	0.00010	0.00007	0.00011
45	0.00015	0.00035	0.00019
47	0.00011	0.00011	0.00016
49	0.00020	0.00016	0.00013
51	0.00036	0.00014	0.00032

Table 9

the table of $\max_{1\le s\le n-2} |P_{gr}(n,s,10^9) - P_{gr}(n,s,10^9;k,5)|$, where

$$P_{gr}(n,s,x;k,m) = \frac{\#\{p \mid p \le x, n|p-1, \mathfrak{s}(p)=s, p\equiv k \bmod m\}}{\#\{p \mid p\le x, n|p-1, p\equiv k \bmod m\}}.$$

n	$k=1$	$k=2$	$k=3$	$k=4$
7	0.00023	0.00036	0.00045	0.00028
9	0.00036	0.00030	0.00030	0.00027
11	0.00027	0.00036	0.00040	0.00035
13	0.00043	0.00043	0.00025	0.00026
17	0.00045	0.00047	0.00059	0.00066
19	0.00046	0.00076	0.00064	0.00063
21	0.00039	0.00051	0.00069	0.00057
23	0.00085	0.00106	0.00033	0.00032
27	0.00045	0.00063	0.00044	0.00052
29	0.00053	0.00112	0.00077	0.00074
31	0.00072	0.00089	0.00180	0.00076
33	0.00061	0.00039	0.00056	0.00055
37	0.00069	0.00107	0.00167	0.00113
39	0.00035	0.00081	0.00061	0.00041
41	0.00104	0.00141	0.00150	0.00098
43	0.00138	0.00032	0.00125	0.00100
47	0.00051	0.00062	0.00083	0.00055
49	0.00072	0.00096	0.00083	0.00059
51	0.00084	0.00088	0.00101	0.00080

Finally, we give expected densities by rationals.

Table 10

$n=5$

s	1	2	3
	1/6	4/6	1/6

$n=7$

s	1	2	3	4	5
	1/120	26/120	66/120	26/120	1/120

$n=9$

s	1	2	3	4	5	6	7
	0	0	1/4	2/4	1/4	0	0

References

[KN] Y. Kitaoka and M. Nozaki, *On the density of the set of primes which are related to decimal expansion of rational numbers,* RIMS Kokyuroku, to appear.

[DFI] W. Duke, J.B. Friedlander and H. Iwaniec, *Equidistribution of roots of a quadratic congruence to prime moduli,* Ann. of Math. **141** (1995), 423-441.

[T] Á. Tóth, *Roots of Quadratic congruences,* Internat. Math. Res. Notices 2000, 719-739.

SPHERICAL FUNCTIONS ON p-ADIC HOMOGENEOUS SPACES

Yumiko Hironaka
Department of Mathematics, Faculty of Education and Integrated Sciences, Weseda University, Nishi-Waseda, Tokyo 169-8050, Japan
hironaka@waseda.jp

Abstract In §1, after defining spherical functions on homogeneous spaces, we examine the case of symmetric forms as an enlightening example. In §2, we introduce a general formula of spherical functions using functional equations under suitable assumption. In §3, we study a certain mechanism of functional equations of spherical functions.

Keywords: spherical functions, p-adic homogeneous spaces, relative invariants, prehomogeneous vector spaces

2000 Mathematics Subject Classification: Primary 11F85; Secondly 11E95, 11F70, 22E50

1. Introduction

§1.1. Definition

Let $\mathbb{G}$ be a reductive linear algebraic group and $\mathbb{X}$ be a $\mathbb{G}$-homogeneous affine algebraic variety, where everything is assumed to be defined over a non-archimedian local filed k of characteristic 0.

We write the $\mathbb{G}$-action on $\mathbb{X}$ by $(g, x) \longmapsto g \cdot x$. For an algebraic set, we use the same ordinary letters for the set of k-rational points, e.g. $G = \mathbb{G}(k), X = \mathbb{X}(k)$.

Taking a maximal compact subgroup K, we denote by $\mathcal{H}(G, K)$ the Hecke algebra of G with respect to K

$$\mathcal{H}(G,K) = \left\{ \phi : G \longrightarrow \mathbb{C} \;\middle|\; \begin{array}{l} \phi(kgk') = \phi(g), {}^\forall k, k' \in K, {}^\forall g \in G \\ \textit{compactly supported} \end{array} \right\},$$

Number Theory: Tradition and Modernization, pp. 81–95
W. Zhang and Y. Tanigawa, eds.

which is generated by the characteristic functions ch_{KgK} of KgK ($g \in G$) over $\mathbb{C}$, and becomes a $\mathbb{C}$-algebra by the convolution product

$$\phi_1 * \phi_2(g_1) = \int_G \phi_1(g)\phi_2(g^{-1}g_1)dg,$$

where dg is the normalized Haar measure on G.

We consider the following $\mathcal{H}(G, K)$-modules

$$\mathcal{C}^\infty(K\backslash X) = \left\{ \Psi : X \longrightarrow \mathbb{C} \,\middle|\, \Psi(k \cdot x) = \Psi(x), {}^\forall k \in K \right\},$$
$$\mathcal{S}(K\backslash X) = \{ \Psi \in \mathcal{C}^\infty(K\backslash X) \mid \textit{compactly supported} \},$$

where the action is given by

$$\phi * \Psi(x) = \int_G \phi(g)\Psi(g^{-1}x)dg, \qquad \phi \in \mathcal{H}(G, K).$$

A nonzero function $\Psi \in \mathcal{C}^\infty(K\backslash X)$ is called **a spherical function on** X if it is a common $\mathcal{H}(G, K)$-eigen function, i.e. there exists a $\mathbb{C}$-algebra map $\lambda : \mathcal{H}(G, K) \longrightarrow \mathbb{C}$ such that $\phi * \Psi = \lambda(\phi)\Psi$ for every $\phi \in \mathcal{H}(G, K)$.

Typical spherical functions are constructed in the following manner.

We denote by $|\,|$ the absolute value on k normalized by $|\pi| = q^{-1}$, where π is a prime element of k and q is the cardinal number of the residue class field of k. Let $\mathbb{P}$ be a closed subgroup of $\mathbb{G}$ defined over k satisfying $G = KP = PK$. Let $f_i(x)$, $1 \leq i \leq n$ be nonzero regular functions on $\mathbb{X}$ defined over k which are relative $\mathbb{P}$-invariants; namely, there are rational characters ψ_i, $1 \leq i \leq n$ of $\mathbb{P}$ satisfying

$$f_i(p \cdot x) = \psi_i(p)f_i(p), \quad p \in \mathbb{P}. \tag{1.1}$$

Here n is arbitrary and the characters ψ_i become automatically k-rational.

Consider the following integral for $x \in X$ and $s = (s_1, \ldots, s_n) \in \mathbb{C}^n$

$$\omega(x; s) = \int_K \prod_{i=1}^n |f_i(k \cdot x)|^{s_i} \, dk.$$

The right hand side is absolutely convergent if $\mathrm{Re}(s_i) \geq 0$, $1 \leq i \leq n$, and has an analytic continuation to a rational function of $q^{s_1}, \ldots, q^{s_n}$.

Proposition 1.1. *If $\omega(x; s)$ is a nonzero function on X, then it is a spherical function on X. Indeed*

$$(\phi * \omega(\;; s))(x) = \lambda(\phi)\omega(x; s), \qquad \phi \in \mathcal{H}(G, K),$$

with

$$\lambda(\phi) = \int_P \phi(p) \prod_{i=1}^{n} |f_i(p)|^{-s_i} \, d_r p,$$

where $d_r p$ is the normalized right invariant Haar measure on P.

Spherical functions on homogeneous spaces are an interesting object to investigate and a basic tool to study harmonic analysis on G-space X. They have been studied also as spherical vectors of distinguished models, Shalika functions and Whittaker-Shintani functions, and have a close relation to the theory of automorphic functions and representation theory. When $\mathbb{G}$ and $\mathbb{X}$ are defined over $\mathbb{Q}$, spherical functions appear in local factor of global objects, e.g. Rankin-Selberg convolutions and Eisenstein series (e.g. [CS], [Fl], [HS3], [Jac], [KMS], [Sf1]).

To obtain explicit expressions of spherical functions is one of basic problems. For the group case, it has been done by I. G. Macdonald and afterwards by W. Casselman by a representation theoretical method (cf. [Ma], [Cas]). There are some results on homogeneous space cases mainly for the case that the space of spherical functions for each Satake parameter is of dimension one (e.g. [CS], [KMS], [Of]). On the other hand, the author has given an expression of them independent of the dimension, based on the data of the group G and functional equations of them, which we show in §2. Hence the knowledge of functional equations is important to obtain explicit expressions of them. In §3, we show a unified method to obtain them and explain they are reduced to those of p-adic local zeta functions of small prehomogeneous vector spaces of limited type.

Some specific X such as the space of symmetric forms, hermitian forms, or alternating forms, there is a close relation to classical problem in Number Theory. In the latter half of this section, we will explain this situation by taking the space of symmetric forms. Some more comment about this wil be given at the end of §1.

§1.2. The space of symmetric forms

For a symmetric matrix A of size m and a matrix $X \in M_{mn}$, we denote $A[X] = {}^tX \cdot A = {}^tXAX$, which is a symmetric matrix of size n.

One of the fundamental problems in the theory of symmetric forms is to find solutions $X \in M_{mn}(\mathbb{Z})$ of the matrix equation

$$A[X] = B, \qquad \text{for } A \in Sym_m(\mathbb{Z}),\ B \in Sym_n(\mathbb{Z}),$$

which is difficult in general. Even the problem

"Characterize prime numbers for which $x^2 + ny^2 = p$ is solvable" is not completely solved. The cases $n = 1, 2, 3$ are classically known (stated by P. Fermat, proved by L. Euler. $p \equiv 1 \pmod 4$; $p = 2$ or $p \equiv 1, 3 \pmod 8$; $p = 3$ or $p \equiv 1 \pmod 3$, in this order). In general, it is impossible to characterize p in a similar style. D. Cox's book [Cox] is an interesting modern approach to this problem, where Hilbert class field theory and complex multiplication are exploited.

For $A \in Sym_m(\mathbb{Z})$ and $B \in Sym_n(\mathbb{Z})$, it is easy to see that

$A[X] = B$ has a $\mathbb{Z}$-solution
$\Longrightarrow A[X] \equiv B \pmod N$ has a $\mathbb{Z}$-solution for any $N \in \mathbb{N}$
$\Longrightarrow A[X] = B$ has $\mathbb{Z}_p$-solution for any prime p.

The implication for the opposite directions is known as follows.

Theorem (Minkowski-Hasse). *Let $A \in Sym_m^{nd}(\mathbb{Q})$, $B \in Sym_n^{nd}(\mathbb{Q})$ with $m \geq n \geq 1$, where nd indicates nondegenerateness. Then*

$A[X] = B$ has a $\mathbb{Q}$-solution
$\Longleftrightarrow A[X] = B$ has an $\mathbb{R}$-solution and a $\mathbb{Q}_p$-solution for each prime p.

In general, this result cannot be extended to the case of $\mathbb{Z}$-solutions. There are equations which have $\mathbb{R}$-solutions and $\mathbb{Z}_p$-solutions for each p, but no $\mathbb{Z}$-solutions; for example

$$7x^2 + 6xy + 11y^2 = 3, \quad 5x^2 + 11y^2 = 1, \quad 5x^2 + 20xy + 31y^2 = 1.$$

While Hasse-Minkowski theorem is a qualitative theorem for $\mathbb{Q}$-solutions, Siegel's main theorem given below shows us a quantitative results for $\mathbb{Z}$-solutions. To describe the theorem, we need some more notations. For $A \in Sym_m^{nd}(\mathbb{Z})$ and $B \in Sym_n^{nd}(\mathbb{Z})$ with $m \geq n \geq 1$ and $R = \mathbb{Z}, \mathbb{Z}_p, \mathbb{R}, \ldots$, set

$$X_{B,A}(R) = \{ X \in M_{mn}(R) \mid A[X] = B \}.$$

We define **the local density of integral representations of** B **by** A $\mu_p(B, A)$ by the limit

$$\mu_p(B, A) = \lim_{l \to \infty} \frac{\sharp \{ X \in M_{mn}(\mathbb{Z}_p/(p^l)) \mid A[X] \equiv B \pmod{p^l} \}}{p^{l(mn - n(n+1)/2)}},$$

where the inside term of lim is independent of l if l is sufficiently large, and it is interpreted as the volume of $X_{B,A}(\mathbb{Z}_p)$. For the cases A is unimodular, the size of B is 1, or $A = B$, good explicit formulas are known. There is a formula for general A and B when $p \neq 2$, but it is too complicated (cf. [SH]).

For simplicity, assume that A and B are positive definite. Then one can consider the volume $\mu_\infty(B, A)$ of $\mathbb{R}$-solutions $X_{B,A}(\mathbb{R})$.

Now let $A_1, \ldots, A_h$ be a set of complete representatives of the $SL_m(\mathbb{Z})$-equivalent classes within the genus containing A, i. e. within

$$\left\{ A' \in Sym_m^{nd}(\mathbb{Z}) \;\middle|\; \begin{array}{l} A \text{ and } A' \text{ are } SL_m(\mathbb{Z}_p)\text{-equivalent for } {}^\forall p \text{ and} \\ SL_m(\mathbb{R})\text{-equivalent.} \end{array} \right\}.$$

Theorem (Siegel). *Under the notations above*

$$\frac{\displaystyle\sum_{i=1}^{h} \frac{\sharp(X_{B,A_i}(\mathbb{Z}))}{\sharp(SO(A_i)_{\mathbb{Z}})}}{\displaystyle\sum_{i=1}^{h} \frac{1}{\sharp(SO(A_i)_{\mathbb{Z}})}} = c \prod_{p \le \infty} \mu'_p(B, A),$$

where

$$c = \begin{cases} 1 & \text{if } m > n+1,\ m = n = 1 \\ 1/2 & \text{if } m = n+1,\ m = n \ge 2 \end{cases},$$

$$\mu'_p(B, A) = \begin{cases} \mu_p(B, A) & \text{if } m > n \\ \frac{1}{2}\mu_p(B, A) & \text{if } m = n \end{cases}.$$

Now we go back to our original theme, spherical functions on $X_m = Sym_m^{nd}(\mathbb{Q}_p)$. The group $G_m = GL_m(\mathbb{Q}_p)$ acts on X_m by $g \cdot x = gx{}^tg = x[{}^tg]$. For simplicity, we will write $\mu(\ ,\ )$ in stead of $\mu_p(\ ,\ )$. Set $K_m = GL_m(\mathbb{Z}_p)$ and consider the following integral

$$\omega(x; s) = \int_{K_m} \prod_{i=1}^{m} |d_i(k \cdot x)|_p^{s_i}\, dk, \qquad x \in X_m,\ s \in \mathbb{C}^m,$$

where dk is the Haar measure on K, $|\ |_p$ is the normalized p-adic value on $\mathbb{Q}_p$ and $d_i(y)$ is the determinant of upper left i by i block of y.

Here $d_i(x)$, $1 \le i \le m$, are relative B_m-invariant on X_m, where B_m is the Borel subgroup of G_m consisting of lower triangular matrices, and $G_m = K_m B_m = B_m K_m$ as is well known.

The above integral is absolutely convergent if $\mathrm{Re}(s_i) \ge 0$, $1 \le i \le m-1$, and analytically continued to a rational function of $p^{s_1}, \ldots, p^{s_m}$. As a function on X_m, $\omega(x; s)$ is an element of $\mathcal{C}^\infty(K_m \backslash X_m)$ and we obtain a typical example of spherical functions on X_m.

The following theorem shows that spherical functions can be regarded as generating functions of local densities.

Theorem ([H1-I]). *Let $m > n$ and $x \in X_m$. Then we have*

$$\omega(x; s_1, \dots, s_n, 0, \dots, 0)$$
$$= \frac{\prod_{i=1}^{n}(1-p^{-i}) \cdot \prod_{i=1}^{m-n}(1-p^{-i})}{\prod_{i=1}^{m}(1-p^{-i})} \cdot \prod_{i=1}^{n}(1-p^{-2(s_i+\cdots+s_n)-m+i-1})$$
$$\times \sum_{y \in K_n \backslash X_n} \frac{\mu(y,x)}{\mu(y,y)} \cdot \omega(y; s_1, \dots, s_n),$$

where y runs over the representatives of K_n-orbits in X_n.

By the above theorem, we can expect to extract the information on local densities from that of spherical functions when the latter is well analysed, and conversely. Unfortunately this approach has not been successful yet. Similar consideration is valid for alternating forms and hermitian forms, and in those cases we have obtained good results (cf. [HS1], [HS2], [H2], [H3]).

2. Expressions of spherical functions

In this section we introduce (formal) explicit expressions of spherical functions after [H2, §1], while some notations are changed from there.

For a connected linear algebraic group $\mathbb{H}$. We denote by $\mathfrak{X}(\mathbb{H})$ the group of k-rational characters of $\mathbb{H}$, which is a free abelian group of finite rank, and by $\mathfrak{X}_0(\mathbb{H})$ the subgroup consisting of characters corresponding to some relative $\mathbb{H}$-invariants. A set of relative $\mathbb{H}$-invariants is called *basic* if the set of the corresponding characters forms a basis for $\mathfrak{X}_0(\mathbb{H})$.

In §2 and §3, let $\mathbb{G}$ be a connected reductive linear algebraic group defined over k, $\mathbb{B}$ a minimal parabolic subgroup of $\mathbb{G}$ defined over k, and K a maximal compact subgroup of G for which $G = BK = KB$. The group $\mathbb{B}$ is not necessarily a Borel group. Denote by dg and dk the normalized invariant Haar measure on G and K, respectively, by dp the left invariant Haar measure on B normalized by $dg = dpdk$, and by δ the modulus character of B ($d(pq) = \delta^{-1}(q)dp$, $p, q \in B$).

In §2, let K be a special good maximal bounded subgroup of G, U the Iwahori subgroup in K which is compatible with B, and assume the following.

(A1) $\mathbb{X}$ has only a finite number of $\mathbb{B}$-orbit.

(A2) The group $\mathfrak{X}_0(\mathbb{B})$ has the same rank with $\mathfrak{X}(\mathbb{B})$, and a basic set of relative $\mathbb{B}$-invariants on $\mathbb{X}$ can be taken from regular functions on $\mathbb{X}$.

(A3) For any $x \in X$, there exists $\psi \in \mathfrak{X}_0(\mathbb{B})$ whose restriction to the identity component of the stabilizer $\mathbb{B}_x$ is nontrivial.

By (A1), there exists unique open $\mathbb{B}$-orbit $\mathbb{X}^{op}$ in $\mathbb{X}$, and X^{op} decomposes into a finite numbers of open B-orbits, which we write

$$X^{op} = \bigsqcup_{\nu \in J(X)} X_\nu.$$

Let $\{ f_i(x) \mid 1 \leq i \leq n\}$ a basic set of regular relative $\mathbb{B}$-invariants, and $\psi_i \in \mathfrak{X}_0(\mathbb{B})$ the character corresponding to $f_i(x)$ for each i, where $n = \mathrm{rank}(\mathfrak{X}(\mathbb{B}))$. For each $s \in \mathbb{C}^n$ and $\nu \in J(X)$, we set

$$|f(x)|_\nu^s = \begin{cases} \prod_{i=1}^{n} |f_i(x)|^{s_i} & \text{if } x \in X_\nu \\ 0 & \text{if } x \notin X_\nu, \end{cases}$$

$$|\psi(p)|^s = \prod_{i=1}^{n} |\psi_i(p)|^{s_i},$$

and determine $\varepsilon \in \mathbb{Q}^n$ by

$$|\psi(p)|^\varepsilon = \delta^{\frac{1}{2}}(p), \ p \in B.$$

Now we consider the integral

$$\omega_\nu(x; s) = \int_K |f(k \cdot x)|_\nu^{s+\varepsilon}\, dk. \tag{2.1}$$

The right hand side of (2.1) is absolutely convergent if $\mathrm{Re}(s_i) \geq -\varepsilon_i$, $1 \leq i \leq n$, analytically continued to a rational function of $q_1, \ldots, q_n$, and we get a spherical function on X.

Let W be the restricted Weyl group of $\mathbb{G}$ with respect to a maximal k-split torus $\mathbb{T}$ contained in $\mathbb{B}$, then $W \cong N_{\mathbb{G}}(\mathbb{T})/Z_{\mathbb{G}}(\mathbb{T})$, and $N_{\mathbb{G}}(\mathbb{T})$ is a Levi subgroup of $\mathbb{B}$. Thus W acts on $s \in \mathbb{C}^n$ through the canonical (conjugate) action on $\mathfrak{X}(\mathbb{B})$ and the identification $\mathfrak{X}(\mathbb{B}) \otimes_{\mathbb{Z}} \mathbb{C} \cong \mathbb{C}^n$.

Then we have

Theorem 2.1. *Let $x \in X$, s be generic, and $\mathcal{U} = \{\nu \in J(X) \mid X_\nu \cap G \cdot x \neq \emptyset\}$. Then*

$$(\omega_\nu(x; s))_{\nu \in \mathcal{U}} = \frac{1}{Q} \sum_{w \in W} \gamma(w(s)) \cdot A(s, w) \cdot \left(\int_U |f(u \cdot x)|_\nu^{w(s)+\varepsilon}\, du \right)_{\nu \in \mathcal{U}},$$

where

$$Q = \sum_{w \in W} [UwU : U]^{-1},$$

$$\gamma(s) = \prod_{\alpha \in \Sigma^+} \frac{(1 - q_{\alpha/2}^{-\frac{1}{2}} q_\alpha^{-1} |\psi(a_\alpha)|^{-s})(1 + q_{\alpha/2}^{-\frac{1}{2}} |\psi(a_\alpha)|^{-s})}{1 - |\psi(a_\alpha)|^{-2s}}$$

$$\left(= \prod_{\alpha \in \Sigma^+} \frac{(1 - q^{-1} |\psi(a_\alpha)|^{-s})}{1 - |\psi(a_\alpha)|^{-s}} \text{ if } G \text{ is split} \right),$$

and $A(s, w)$ is the invertible matrix determined by

$$\omega_\nu(x; s) = \sum_{u \in \mathcal{U}} A(s, w)_{\nu u} \cdot \omega_u(x; w(s)).$$

Remark 1. In the above theorem, Q and $\gamma(s)$ depend only on the groups $\mathbb{G}$ and $\mathbb{B}$. Σ^+ denotes the set of positive roots with respect to $\mathbb{B}$ in the set Σ of roots of $\mathbb{G}$ with respect to $\mathbb{T}$. For the definition of $a_\alpha \in T$ and numbers $q_\alpha, q_{\alpha/2}$ for $\alpha \in \Sigma$, we refer to [Cas, (9), Remark 1.1 and (12)]. If G is split, each $q_\alpha = q$ and $q_{\alpha/2} = 1$.

3. Functional equations of spherical functions

Under the assumption (AF) below we show the existence of a certain functional equation of spherical functions and it is reduced to those of p-adic local zeta functions of small prehomogeneous vector spaces, for details see [H5].

§3.1. Type (F)

Let $\mathbb{H}$ be a connected linear algebraic group $\mathbb{H}$ and $\mathbb{Y}$ an affine algebraic variety on which $\mathbb{H}$ acts, where everything is assumed to be defined over k. We say $(\mathbb{H}, \mathbb{Y})$ is *of type* (F) if it satisfies the following conditions:

(F1) $\mathbb{Y}$ has only a finite number of $\mathbb{H}$-orbits. (Then $\mathbb{Y}$ has only one open $\mathbb{H}$-orbit $\mathbb{Y}^{op}$.)

(F2) For $y \in \mathbb{Y} \backslash \mathbb{Y}^{op}$, there exists some ψ in $\mathfrak{X}_0(\mathbb{H})$ whose restriction to the identity component of the stabilizer $\mathbb{H}_y$ is not trivial.

(F3) The index of $\mathfrak{X}_0(\mathbb{H})$ in $\mathfrak{X}(\mathbb{H})$ is finite.

(F4) A basic set of relative $\mathbb{H}$-invariants on $\mathbb{Y}$ can be taken from regular functions on $\mathbb{Y}$.

For a simple root α, let $\mathbb{P}$ be the standard parabolic subgroup $\mathbb{P}_{\{\alpha\}}$. We consider a k-rational representation $\rho : \mathbb{P} \longrightarrow R_{k'/k}(GL_2)$ satisfying

$$\rho(\mathbb{P}) = R_{k'/k}(GL_2) \text{ or } R_{k'/k}(SL_2), \quad \rho(w_\alpha) = \begin{pmatrix} 0 & 1 \\ -1 & 0 \end{pmatrix}$$
$$\rho^{-1}(\mathbb{B}_2) \subset \mathbb{B}, \quad \rho(K \cap \mathbb{P}) \supset R_{k'/k}(SL_2)(\mathcal{O}_k), \tag{3.1}$$

where k' is a finite unramified extension of k, $R_{k'/k}$ is the restriction functor of base field, $w_\alpha \in N_G(T)$ is a representative of the reflection in W attached to α, and $\mathbb{B}_2$ is the Borel subgroup of $\rho(\mathbb{P})$ consisting of upper triangular matrices.

Now we assume that

(AF) $(\mathbb{B}, \mathbb{X})$ is of type (F) and there is a k-rational representation ρ satisfying (3.1) for a simple root α.

Remark 2. Chevalley groups are typical examples which have ρ as above for $k' = k$, so $R_{k'/k}(GL_2) = GL_2$. If $\mathbb{G}$ is k-split, then $\mathbb{B}$ is a Borel subgroup and $\mathbb{P} = \mathbb{B} \cup \mathbb{B}w_\alpha\mathbb{B}$. The assumption in §2 is the same that $(\mathbb{B}, \mathbb{X})$ is of type (F).

Set $\widetilde{\mathbb{X}} = \mathbb{X} \times \mathbb{V}$ and $\widetilde{\mathbb{P}} = \mathbb{P} \times R_{k'/k}(GL_1)$, where $\mathbb{V} = R_{k'/k}(M_{2,1})$ defined over k, and consider the following $\widetilde{\mathbb{P}}$-action on $\widetilde{\mathbb{X}}$:

$$(p, t) \cdot (x, v) = (p \cdot x, \rho(p)vt^{-1}), \quad (p, t) \in \widetilde{\mathbb{P}}, \ (x, v) \in \widetilde{\mathbb{X}}.$$

Here we identify k' with its image by the regular representation in $M_d(k)$ and realize $R_{k'/k}(GL_2)$ (resp. $\mathbb{V}$) in $GL_{2d}(\overline{k})$ (resp. $M_{2d,d}(\overline{k})$), where $d = [k' : k]$ and $\overline{k}$ is the algebraic closure of k. We note here that we may identify as $\widetilde{P} = P \times GL_2(k')$ and $V = k'^2$.

We regard $\mathbb{B}$ as a subgroup of $\widetilde{\mathbb{P}}$ by the embedding

$$\mathbb{B} \longrightarrow \widetilde{\mathbb{P}}, \ b \longmapsto (b, \rho(b)_1),$$

where $\rho(b)_1 \in R_{k'/k}(GL_1)$ is the upper left d by d block of $\rho(b) \in R_{k'/k}(GL_2)$. Then, one can identify $\mathbb{B}$ with the stabilizer of $\mathbb{P}$ at $v_0 = \begin{pmatrix} 1 \\ 0 \end{pmatrix}$ in $V = \mathbb{V}(k)$, i.e.

$$\mathbb{B} \cong \widetilde{\mathbb{P}}_{v_0} \left(= \left\{ (p, t) \in \widetilde{\mathbb{P}} \,\middle|\, \rho(p)v_0t^{-1} = v_0 \right\} \right).$$

There is an isomorphism

$$\begin{array}{rcl} \mathfrak{X}(\widetilde{\mathbb{P}}) \cong \mathfrak{X}(\mathbb{P}) \times \mathfrak{X}(R_{k'/k}(GL_1)) & \longrightarrow & \mathfrak{X}(\mathbb{B}) \\ (\psi_1, \psi_2) & \longmapsto & [b \longmapsto \psi_1(b)\psi_2(\rho(b)_1)]. \end{array} \tag{3.2}$$

And we have

Proposition 3.1. (i) *The space* $(\widetilde{\mathbb{P}}, \widetilde{\mathbb{X}})$ *is of type* (F).

(ii) *The set of open* $\mathbb{B}(k)$*-orbits in* $\mathbb{X}(k)$ *corresponds bijectively to the set of open* $\widetilde{\mathbb{P}}(k)$*-orbits in* $\widetilde{\mathbb{X}}(k)$ *by the map* $B \cdot x \longmapsto \widetilde{P} \cdot (x, v_0)$.

§3.2. Spherical functions and zeta distributions

Because of the assumption of type (F) for $(\mathbb{B}, \mathbb{X})$, there exists a basic set $\{f_i(x) \mid 1 \leq i \leq n\}$ of regular relative $\mathbb{B}$-invariants, where $n = \operatorname{rank}(\mathbb{B})$. Let $\left\{ \widetilde{f}_i(x,v) \,\middle|\, 1 \leq i \leq n \right\}$ be a basic set of regular relative $\widetilde{\mathbb{P}}$-invariants satisfying $f_i(x) = \widetilde{f}_i(x, v_0)$. We denote by $\widetilde{\psi}_i$ the character corresponding to $\widetilde{f}_i(x,v)$, then $\psi_i = \widetilde{\psi}_i \mid_{\mathbb{B}}$ is the character corresponding to $f_i(x)$ for each i. We may write

$$X^{op} = \bigsqcup_{u \in J(X)} X_u, \quad \widetilde{X}^{op} = \bigsqcup_{u \in J(X)} \widetilde{X}_u,$$

where $\widetilde{X_u}$ is the open $\widetilde{P}$-orbit corresponding to X_u by Proposition 3.1. Denote by $\mathcal{S}(X)$ and $\mathcal{S}(\widetilde{X})$ the spaces of Schwartz-Bruhat functions on X and $\widetilde{X}$, respectively.

For $s \in \mathbb{C}^n$ and $u \in J(X)$, we have spherical function $\omega_u(x; s)$ defined by (2.1), further we consider the following integrals

$$\Omega_u(\phi; s) = \int_X \phi(x) \cdot |f(x)|_u^{s+\varepsilon} \, dx, \quad \phi \in \mathcal{S}(X),$$

$$\widetilde{\Omega}_u(\widetilde{\phi}; s) = \int_{\widetilde{X}} \widetilde{\phi}(x,v) \cdot \left|\widetilde{f}(x,v)\right|_u^{s+\varepsilon} dx dv, \quad \widetilde{\phi} \in \mathcal{S}(\widetilde{X}),$$

where dx is a G-invariant measure on X and dv is a Haar measure on V. The above integrals are absolutely convergent for $\operatorname{Re}(s_i) \geq 0, \ 1 \leq i \leq n$, and continued to rational functions of $q^{s_1}, \ldots, q^{s_n}$. It is easy to see that

$$\omega(x; s) = \sum_{u \in \mathcal{U}} \omega_u(x; s), \quad \omega_u(x; s) = v(K \cdot x)^{-1} \Omega_u(ch_x; s), \tag{3.3}$$

where ch_x is the characteristic function of $K \cdot x$ and $v(K \cdot x)$ is the volume with respect to the above measure dx.

Since $\widetilde{f}_i(x,v)$ is a relative $R_{k'/k}(GL_1)$-invariant with respect to the action on v, it is homogeneous in the coordinates of v over k, and we set

$$e_i = \deg_v \widetilde{f}_i(x,v) \qquad (1 \leq i \leq n).$$

Then we have the following relation between $\Omega_u(\phi; s)$ and $\widetilde{\Omega}_u(\widetilde{\phi}; s)$.

Proposition 3.2. *Let $\widetilde{\phi} = \phi \otimes ch_{V(\mathfrak{p}^m)}$, where $\phi \in \mathcal{S}(K\backslash X)$ and $ch_V(\mathfrak{p}^m)$ is the characteristic function of $V(\mathfrak{p}^m)$ with $\mathfrak{p}^m = \pi^m \mathcal{O}_k$. Then for every $u \in J(X)$,*

$$\widetilde{\Omega}_u(\widetilde{\phi}; s) = c \cdot \frac{q^{-m(2d+\sum_i e_i(s_i+\varepsilon_i)}}{1 - q^{-2d-\sum_i e_i(s_i+\varepsilon_i)}} \times \Omega_u(\phi; s),$$

where c is a constant depending only on the normalization of measures.

By (3.3) and the above proposition, we expect to get functional equations of $\omega_u(x; s)$ from those of $\widetilde{\Omega}_u(\widetilde{\phi}; s)$.

§3.3. Functional equations

Take an additive character η on k' of conductor $\pi^l \mathcal{O}_{k'}$, and define the partial Fourier transform $\mathcal{F}(\widetilde{\phi})$ for $\widetilde{\phi} \in \mathcal{S}(\widetilde{X})$ by

$$\mathcal{F}(\widetilde{\phi})(x, v) = \int_V \eta({}^t v \sigma w) \widetilde{\phi}(x, w) dw, \quad \sigma = \begin{pmatrix} 0 & 1 \\ -1 & 0 \end{pmatrix}.$$

We consider the following distributions on $\mathcal{S}(\widetilde{X})$ for each $u \in J(X)$ and $s \in \mathbb{C}^n$

$$T_{u,s}(\widetilde{\phi}) = \widetilde{\Omega}_u(\widetilde{\phi}; s), \quad T^*_{u,s}(\widetilde{\phi}) = \widetilde{\Omega}_u(\mathcal{F}(\widetilde{\phi}); s).$$

By the action of $\widetilde{P} = P \times GL_1(k')$ on $\mathcal{S}(X)$ given by

$${}^{(p,t)}\widetilde{\phi}(x, v) = \widetilde{\phi}((p,t)^{-1} \cdot (x, v)) = \widetilde{\phi}(p^{-1} \cdot x, \rho(p)^{-1} vt)$$
$$(\widetilde{\phi} \in \mathcal{S}(\widetilde{X}),\ (p,t) \in \widetilde{P},\ (x, v) \in \widetilde{X}),$$

we get

$$T_{u,s}\left({}^{(p,t)}\widetilde{\phi}\right) = (w_\alpha \delta^{\frac{1}{2}})(p,t) \cdot \left|\widetilde{\psi}(p,t)\right|^s \times T_{u,s}(\widetilde{\phi}),$$
$$T^*_{u,s}\left({}^{(p,t)}\widetilde{\phi}\right) = (w_\alpha \delta^{\frac{1}{2}})(p,t) \cdot \left|\widetilde{\psi}(p,t)\right|^{w_\alpha(s)} \times T^*_{u,s}(\widetilde{\phi}).$$

On the other hand, if $(\mathbb{H}, \mathbb{Y})$ is of type (F), it satisfies the following property (F5) (See [Sf2, Lemma 2.3, Corollary 2.4]).

(F5) There is a finite set (L) of linear congruences of type

$$\sum_{i=1}^n m_i s_i \equiv \lambda \pmod{\frac{2\pi\sqrt{-1}}{\log q}\mathbb{Z}}, \qquad m_i \in \mathbb{Z}, \lambda \in \mathbb{C}$$

which satisfies the following: If T is a nonzero distribution whose support is contained in $Y \backslash Y^{op}$ and satisfies

$$T({}^g\phi) = |\psi(g)|^s \cdot T(\phi), \qquad \phi \in \mathcal{S}(Y),\ g \in H,$$

then s satisfies some relation in (L).

By the uniqueness of relatively invariant distributions on homogeneous spaces (cf. [Ig, Proposition 7.2.1]) and the property (F5), we obtain the following proposition. Let e be the group index $[\mathfrak{X}(\mathbb{B}) : \mathfrak{X}_0(\mathbb{B})]$, and $J_u = \{\nu \in J(X) \mid P \cdot X_\nu = P \cdot X_u\}$ for each $u \in J(X)$.

Proposition 3.3. *There exist rational functions $\gamma^\eta_{u\nu}(s)$ of $q^{\frac{s_1}{e}}, \dots, q^{\frac{s_n}{e}}$ which satisfy the following identity.*

$$T^*_{u,s}(\widetilde{\phi}) = \sum_{\nu \in J_u} \gamma^\eta_{u\nu}(s) T_{\nu, w_\alpha(s)}(\widetilde{\phi}), \qquad \widetilde{\phi} \in \mathcal{S}(\widetilde{X}).$$

Rewriting by terms of the zeta integrals, we obtain

Theorem 3.4. *There exist rational functions $\gamma^\eta_{u\nu}(s)$ of $q^{\frac{s_1}{e}}, \dots, q^{\frac{s_n}{e}}$ which satisfy the following identity.*

$$\widetilde{\Omega}_u(\mathcal{F}(\widetilde{\phi}); s) = \sum_{\nu \in J_u} \gamma^\eta_{u\nu}(s) \cdot \widetilde{\Omega}_\nu(\widetilde{\phi}; w_\alpha(s)), \qquad \widetilde{\phi} \in \mathcal{S}(\widetilde{X}).$$

Let normalize the measure dv on V to be self dual with respect to the inner product $(v, w) \longmapsto \eta({}^t v \sigma w)$. Then, by Theorem 3.4 and Proposition 3.2, we obtain

Corollary 3.5. *For any $\phi \in \mathcal{S}(K \backslash X)$, we have*

$$\Omega_u(\phi; s) = \frac{1 - q^{-2d - \sum_i e_i(s_i + \varepsilon_i)}}{1 - q^{-2d - \sum_i e_i(w_\alpha(s)_i + \varepsilon_i)}} \times \sum_{\nu \in J_u} \gamma_{u\nu}(s) \cdot \Omega_\nu(\phi; w_\alpha(s)),$$

where

$$\gamma_{u\nu}(s) = q^{l(d + \sum_i e_i(s_i + \varepsilon_i))} \cdot \gamma^\eta_{u\nu}(s),$$

which is independent of the choice of the character η on k'.

By Corollary 3.5 and the relation (3.3), we get

Theorem 3.6. *For any $x \in X$, we have*

$$\omega_u(x; s) = \frac{1 - q^{-2d - \sum_i e_i(s_i + \varepsilon_i)}}{1 - q^{-2d - \sum_i e_i(w_\alpha(s)_i + \varepsilon_i)}} \times \sum_{\nu \in J_u} \gamma_{u\nu}(s) \cdot \omega_\nu(x; w_\alpha(s)).$$

§3.4. Small prehomogeneous vector spaces

In this subsection, we look at the prehomogeneous vector space $(\rho(\mathbb{P}_x)\times R_{k'/k}(GL_1), \mathbb{V})$, where the action comes from the one of $\widetilde{\mathbb{P}}$ on $\widetilde{\mathbb{X}}$. For each $u \in J(X)$, fix an $x_u \in X_u$ and denote by $\mathbb{P}_u$ the stabilizer of x_u in $\mathbb{P}$.

Lemma 3.7. (i) *For any $u, \nu \in J(X)$, take $p_\nu \in \mathbb{B}$ satisfying $p_\nu \cdot x_\nu = x_u$. Then the map*

$$\begin{array}{ccc} \mathbb{P}_\nu \times R_{k'/k}(GL_1) \times \mathbb{V} & \longrightarrow & \mathbb{P}_u \times R_{k'/k}(GL_1) \times \mathbb{V}, \\ (p, r, v) & \longmapsto & (p_\nu p p_\nu^{-1}, r, \rho(p_\nu)v) \end{array}$$

gives an isomorphism of prehomogeneous vector spaces $(\rho(\mathbb{P}_\nu)\times R_{k'/k}(GL_1), \mathbb{V})$ and $(\rho(\mathbb{P}_u) \times R_{k'/k}(GL_1), \mathbb{V})$. If $\nu \in J_u$, we may take $p_\nu \in P$, and then the above isomorphism is defined over k.

(ii) *The set of k-rational points of the open orbit in $(\rho(\mathbb{B}_u)\times R_{k'/k}(GL_1), \mathbb{V})$ decomposes as*

$$\left(\rho(\mathbb{P}_u)v_0 R_{k'/k}(GL_1)\right)(k) = \bigsqcup_{\nu \in J_u} \rho(P_u p_\nu) v_0 k'^{\times}, \tag{3.4}$$

where $p_\nu \in P$ satisfying $p_\nu^{-1} \cdot x_u \in X_\nu$.

For $\phi' \in \mathcal{S}(X)$ and $\phi \in \mathcal{S}(V)$, we have

$$\mathcal{F}(\phi' \otimes \phi) = \phi' \otimes \mathcal{F}_V(\phi), \quad \mathcal{F}_V(\phi)(v) = \int_V \eta({}^t v \sigma w)\phi(w)dw.$$

Then, applying Theorem 3.4, we obtain

$$\begin{aligned} &\int_V \mathcal{F}_V(\phi)(v) \left|\widetilde{f}(x, v)\right|_u^{s+\varepsilon} dv \\ &= \sum_{\nu \in J_u} \gamma_{u\nu}^{\eta}(s) \int_V \phi(v) \left|\widetilde{f}(x, v)\right|_\nu^{w_\alpha(s)+\varepsilon} dv, \quad (\phi \in \mathcal{S}(V),\ x \in X^{op}), \end{aligned}$$

hence we obtain

Theorem 3.8. *The prehomogeneous vector space $(\rho(\mathbb{P}_u) \times GL_1, \mathbb{V})$ has the functional equation:*

$$\begin{aligned} &\int_V \mathcal{F}_V(\phi)(v) \left|\widetilde{f}(x_u, v)\right|_u^{s+\varepsilon} dv \\ &= \sum_{\nu \in J_u} \gamma_{u\nu}^{\eta}(s) \int_V \phi(v) \left|\widetilde{f}(x_u, v)\right|_\nu^{w_\alpha(s)+\varepsilon} dv, \quad (\phi \in \mathcal{S}(V)), \end{aligned}$$

where the gamma factors $\gamma^{\eta}_{u,\nu}(s)$ are the same as those for $\widetilde{\Omega}_u(\widetilde{\phi}; s)$ in Theorem 3.4.

The existence of the functional equation of the above type gives the following.

Theorem 3.9. *For the prehomogeneous vector space $(\rho(\mathbb{P}_u)\times R_{k'/k}(GL_1), \mathbb{V})$, the identity component of $\rho(\mathbb{P}_u) \times R_{k'/k}(GL_1)$ is isomorphic to $R_{k'/k}(GL_1 \times GL_1, \mathbb{V})$ over the algebraic closure $\overline{k}$ of k.*

Remark 3. The gamma factors $\gamma_{u\nu}(s)$'s of the functional equation of spherical functions $\omega_u(x; s)$ on X are reduced to those for the small prehomogeneous vector spaces $(\rho(\mathbb{P}_u) \times R_{k'/k}(GL_1), \mathbb{V})$, for which the identity component is isomorphic to $R_{k'/k}(GL_1 \times GL_1)$ over $\overline{k}$.

The set of isomorphism classes of k-forms of $GL_1 \times GL_1$ corresponds bijectively to $\mathrm{Hom}(Gal(\overline{k}/k), GL_2(\mathbb{Z}))$ (cf. [PR, §2.2.4]).

Some explicit examples are found in [H5, §4].

References

[Cas] W. Casselman, *The unramified principal series of* $\mathfrak{p}$*-adic groups I. The spherical functions*, Compositio Math. **40** (1980), 387–406.

[CS] W. Casselman and J. Shalika, *The unramified principal series of* $\mathfrak{p}$*-adic groups II. The Whittaker function*, Compositio Math. **41** (1980), 207–231.

[Cox] D. A. Cox, *Primes of the forms* $x^2 + ny^2$, Wiley Interscience, 1989.

[Fl] Y. Z. Flicker, *On distinguished representations*, J. reine angew. Mathematik **418** (1991), 139–172.

[H1] Y. Hironaka, *Spherical functions of hermitian and symmetric forms I, II, III*, Japan. J. Math. **14** (1988), 203–223; Japan. J. Math. **15** (1989), 15–51; Tôhoku Math. J. **40** (1988), 651–671.

[H2] Y. Hironaka, *Spherical functions and local densities on hermitian forms*, J. Math. Soc. Japan **51** (1999), 553–581.

[H3] Y. Hironaka, *Local zeta functions on hermitian forms and its application to local densities*, J. Number Theory **71** (1998), 40–64.

[H4] Y. Hironaka, *Spherical functions on* Sp_2 *as a spherical homogeneous space* $Sp_2 \times (Sp_1)^2$*-space*, J. Number Theory **112** (2005), 238–286.

[H5] Y. Hironaka, *Functional equations of spherical functions on p-adic homogeneous spaces*, to appear in Abh. Math. Sem. Univ. Hamburg.

[HS1] Y. Hironaka and F. Sato, *Spherical functions and local densities of alternating forms*, Amer. J. Math. **110** (1988), 473–512.

[HS2] Y. Hironaka and F. Sato, *Local densities of alternating forms*, J. Number Theory **33** (1989), 32–52.

[HS3] Y. Hironaka and F. Sato, *Eisenstein series on reductive symmetric spaces and representations of Hecke algebras*, J. Reine Angew. Math. **445** (1993), 45–108.

[Ig] J. Igusa, *An introduction to Theory of Local Zeta Functions*, AMS/IP Studies in Advanced Mathematics, vol.14, 2000.

[Jac] H. Jacquet, *Automorphic spectrum of symmetrc spaces*, Proc. Sym. Pure Math. **61** (1997), 431–455.

[KMS] S. Kato, A. Murase and T. Sugano, *Whittaker-Shintani functions for orthogonal groups*, Tohoku Math. J. **55** (2003), 1–64.

[Ma] I. G. Macdonald, *Spherical functions on a group of p-adic type*, Univ. Madras, 1971.

[Of] O. Offen, *Relative spherical functions on p-adic symmetric spaces*, Pacific J. Math. **215** (2004), 97–149.

[PR] V. Platonov and A. Rapinchuk, *Algebraic groups and number theory*, Academic Press, 1994.

[Sf1] F. Sato, *Eisenstein series on semisimple symmetric spaces of Chevalley groups*, Advanced Studies in Pure Math. **7** (1985), 295–332.

[Sf2] F. Sato, *On functional equations of zeta distributions*, Advanced Studies in Pure Math. **15** (1989), 465–508.

[SH] F. Sato and Y. Hironaka, *Local densities of representations of quadratic forms over p-adic integers 'the non-dyadic case*, J. Number Theory **83** (2000), 106–136.

ON MODULAR FORMS OF WEIGHT $(6n+1)/5$ SATISFYING A CERTAIN DIFFERENTIAL EQUATION

Masanobu Kaneko
Graduate School of Mathematics 33, Kyushu University, Fukuoka 812-8581, Japan
mkaneko@math.kyushu.u-ac.jp

Abstract We study solutions of a differential equation which arose in our previous study of supersingular elliptic curves. By choosing one fifth of an integer k as the parameter involved in the differential equation, we obtain modular forms of weight k as solutions. It is observed that this solution is also related to supersingular elliptic curves.

Keywords: modular forms, supersingular elliptic curves, differential equation

2000 Mathematics Subject Classification: Primary 11F11, 11G20

1. Introduction

In our previous work [5], [3], [4], we studied various solutions of the specific differential equation

$$(\sharp)_k \qquad f''(\tau) - \frac{k+1}{6}E_2(\tau)f'(\tau) + \frac{k(k+1)}{12}E_2'(\tau)f(\tau) = 0$$

$$\left(' = \frac{1}{2\pi i}\frac{d}{d\tau} = q\frac{d}{dq},\ q = e^{2\pi i\tau} \right),$$

where τ is a variable in the upper half-plane, k a fixed rational number, and $E_2(\tau)$ the "quasimodular" Eisenstein series of weight 2 for the full modular group $SL(2,\mathbf{Z})$:

$$E_2(\tau) = 1 - 24\sum_{n=1}^{\infty}\left(\sum_{d|n} d\right)q^n.$$

In [5], we showed that for even $k \geq 4$ with $k \not\equiv 2 \pmod 3$, this differential equation has a modular solution of weight k on $SL(2,\mathbf{Z})$ explicitly

Number Theory: Tradtion and Modernization, pp 97–102
W. Zhang and Y. Tanigawa, eds.

describable in terms of the Eisenstein series $E_4(\tau)$ and $E_6(\tau)$, and discussed its connection to supersingular elliptic curves in characteristic p when $k = p-1$. We studied further the modular/quasimodular solutions for other integral or half-integral values of k in [3], [4].

In this paper, we set $k = (6n+1)/5$, one fifth of an integer congruent to 1 modulo 6. We then encounter as solutions modular forms of weight k on the principal congruence subgroup $\Gamma(5)$. Also, modular forms on $\Gamma_1(5)$ arise naturally. In §2 we describe the solutions in terms of fundamental modular forms of weight 1/5 which already appeared in works of Klein and Ramanujan. The proof is in essence similar to the one given in [3]. In §3, we discuss a relation between our solutions and supersingular elliptic curves, which is quite analogous to the situation studied in [5] and [8].

The author should like to express his sincere gratitude to Atsushi Matsuo, whose suggestion that the cases $k = 7/5, 13/5$ would provide interesting modular solutions gave impetus to the present work. The author also learned from him that the differential equation $(\sharp)_k$, in its equivalent form, was already appeared in works of physicists, e.g., [6], [7], and its solutions, at least for small values of k, correspond to particular models in conformal field theory.

2. Main result

Let

$$\begin{aligned}\phi_1 = \phi_1(\tau) &= \frac{1}{\eta(\tau)^{3/5}} \sum_{n\in\mathbf{Z}} (-1)^n q^{(10n+1)^2/40} \\ &= 1 + \frac{3}{5}q + \frac{2}{25}q^2 - \frac{28}{125}q^3 + \frac{264}{625}q^4 + \frac{532}{15625}q^5 + \cdots ,\end{aligned}$$

and

$$\begin{aligned}\phi_2 = \phi_2(\tau) &= \frac{1}{\eta(\tau)^{3/5}} \sum_{n\in\mathbf{Z}} (-1)^n q^{(10n+3)^2/40} \\ &= q^{1/5}(1 - \frac{2}{5}q + \frac{12}{25}q^2 + \frac{37}{125}q^3 - \frac{171}{625}q^4 - \frac{3318}{15625}q^5 + \cdots).\end{aligned}$$

Here, $\eta(\tau)$ is the Dedekind eta function. These forms are of weight 1/5 (with a suitable multiplier system) on $\Gamma(5)$, and the ring of holomorphic modular forms of weight $\frac{1}{5}\mathbf{Z}$ on $\Gamma(5)$ with this multiplier system is the polynomial ring $\mathbf{C}[\phi_1, \phi_2]$ (a good reference for this is Ibukiyama [2]). Note that these forms are essentially the famous Rogers-Ramanujan functions;

$$\begin{aligned}\phi_1 &= \eta(\tau)^{2/5}q^{-1/60}\prod_{n=0}^{\infty}\frac{1}{(1-q^{5n+1})(1-q^{5n+4})},\\ \phi_2 &= \eta(\tau)^{2/5}q^{11/60}\prod_{n=0}^{\infty}\frac{1}{(1-q^{5n+2})(1-q^{5n+3})}.\end{aligned}$$

Theorem. *Assume* $k = (6n+1)/5$, $n = 0, 1, 2, \ldots$, $n \not\equiv 4 \pmod 5$. *Then the equation* $(\sharp)_k$ *has a two dimensional space of solutions in* $\mathbf{C}[\phi_1, \phi_2]_{wt=k}$, *the set of homogeneous polynomials of degree* $6n+1$ *in* ϕ_1 *and* ϕ_2.

Example: Here are a basis of solutions for small k:

$$\begin{aligned}
k = 1/5: &\quad \phi_1, \quad \phi_2,\\
k = 7/5: &\quad \phi_1^7 + 7\phi_1^2\phi_2^5, \quad 7\phi_1^5\phi_2^2 - \phi_2^7,\\
k = 13/5: &\quad \phi_1^{13} + 39\phi_1^8\phi_2^5 - 26\phi_1^3\phi_2^{10}, \quad 26\phi_1^{10}\phi_2^3 + 39\phi_1^5\phi_2^8 - \phi_2^{13},\\
k = 19/5: &\quad \phi_1^{19} + 171\phi_1^{14}\phi_2^5 + 247\phi_1^9\phi_2^{10} - 57\phi_1^4\phi_2^{15},\\
&\quad 57\phi_1^{15}\phi_2^4 + 247\phi_1^{10}\phi_2^9 - 171\phi_1^5\phi_2^{14} + \phi_2^{19}.
\end{aligned}$$

In general, we have a basis of the form $\phi_1^{n+1} \times$ (polynomial in ϕ_1^5 and ϕ_2^5) and $\phi_2^{n+1} \times$ (polynomial in ϕ_1^5 and ϕ_2^5). Here, we note that ϕ_1^5 and ϕ_2^5 become modular forms of weight 1 on

$$\Gamma_1(5) = \left\{ \begin{pmatrix} a & b \\ c & d \end{pmatrix} \mid a \equiv d \equiv 1,\ c \equiv 0 \pmod 5 \right\}.$$

In our previous cases treated in [5], [3], [4], all solutions were explicitly described with the aid of hypergeometric series. In the present case, however, a differential equation with four singularities emerges and we are so far unable to write down the explicit formulas for the solutions in general. We can nevertheless prove the theorem by giving the solutions recursively and by using an inductive structure of solutions of $(\sharp)_k$ revealed in [3].

To give a recursive description of the solutions, we change the variable by setting

$$f(\tau)/\phi_1^{5k} = F(t),$$

where

$$t = \phi_2^5/\phi_1^5 = q - 5q^2 + 15q^3 - 30q^4 + 40q^5 + \cdots$$

is (the reciprocal of) the "Hauptmodul" of $\Gamma_1(5)$, and we consider the equation locally as t a local variable. Then by a routine computation we see that $f(\tau)$ satisfies $(\sharp)_k$ if and only if $F(t)$ satisfies $(\flat)_k$:

$$(\flat)_k \quad t(t^2+11t-1)F''(t)+\left(\frac{7-11k}{6}t^2+11(1-k)t+\frac{k-5}{6}\right)F'(t)$$
$$+\frac{k(5k-1)}{6}(t+3)F(t)=0,$$

where $' = d/dt$. Incidentally, an amusing remark here is that the equation

$$(\flat)_{-1} \qquad t(t^2+11t-1)F''(t)+(3t^2+22t-1)F'(t)+(t+3)F(t)=0$$

obtained by setting $k=-1$ in $(\flat)_k$ is exactly the one used in [1] for reconstructing Apéry's irrationality proof of $\zeta(2)$. The original equation $(\sharp)_k$ when $k=-1$ becomes the trivial $f''=0$, but the solutions here are 1 and τ, "universal periods" of elliptic curves. Hence $(\flat)_{-1}$ is obtained from this trivial equation by rewriting it locally in terms of t-variable.

Now we are going to show that $(\flat)_k$ has a polynomial solution $P(t)$ if $k=(6n+1)/5$ ($n \not\equiv 4 \pmod 5$). If $P(t)$ is such a solution, then $\phi_1^{5k}P(\phi_2^5/\phi_1^5)$ and $\phi_2^{5k}P(-\phi_1^5/\phi_2^5)$ are the solutions to $(\sharp)_k$. The second one is a solution because $SL(2,\mathbf{Z})$ acts on the solution space and the action of $\begin{pmatrix}-3 & -5\\ 5 & 8\end{pmatrix}$ is $\phi_1 \mapsto \phi_2$, $\phi_2 \mapsto -\phi_1$ (we can check this by using the transformation formulas of ϕ_1 and ϕ_1, see [2]).

Proposition. *For* $0 \le n \le 8$, $n \ne 4$, *put*

$$\begin{aligned}
P_0(t) &= 1,\\
P_1(t) &= 1+7t,\\
P_2(t) &= 1+39t-26t^2,\\
P_3(t) &= 1+171t+247t^2-57t^3,\\
P_5(t) &= 1-465t-10385t^2-2945t^3-8370t^4+682t^5,\\
P_6(t) &= 1-333t-17390t^2-54390t^3+26640t^4-64158t^5+3774t^6,\\
P_7(t) &= 1-301t-36421t^2-310245t^3+10535t^4-422303t^5\\
&\quad +283843t^6-12857t^7,\\
P_8(t) &= 1-294t-101528t^2-1798692t^3-2747430t^4-387933t^5\\
&\quad -2086028t^6+740544t^7-26999t^8.
\end{aligned}$$

For $n \ge 10$, $n \not\equiv 4 \pmod 5$, *define* $P_n(t)$ *recursively by*

$$P_n(t)=(1+t^2)(1-522t-10006t^2+522t^3+t^4)P_{n-5}(t)$$

$$+12\frac{(6n-29)(6n-49)}{(n-4)(n-9)}t(1-11t-t^2)^5 P_{n-10}(t). \qquad (1)$$

Then $P_n(t)$ is a solution of $(\flat)_{(6n+1)/5}$ for all $n \geq 0$, $n \not\equiv 4 \pmod 5$.

Proof. We prove Proposition by induction on n. One may notice that the proof is essentially a translation of those of Proposition 1 and Lemma in [3].

It is straightforward to check that each $P_n(t)$ for $n \leq 8, n \neq 4$ satisfies $(\flat)_{(6n+1)/5}$. Suppose that $P_{n-5}(t)$ and $P_{n-10}(t)$ satisfy $(\flat)_{(6(n-5)+1)/5}$ and $(\flat)_{(6(n-10)+1)/5}$ respectively. If we compute the left-hand side of $(\flat)_{(6n+1)/5}$ for $F(t) = P_n(t)$ by substituting the definition (1) of $P_n(t)$ in terms of $P_{n-5}(t)$ and $P_{n-10}(t)$, and using the induction hypothesis $(\flat)_{(6(n-5)+1)/5}$ and $(\flat)_{(6(n-10)+1)/5}$, we see that $P_n(t)$ satisfies $(\flat)_{(6n+1)/5}$ if and only if the identity

$$\begin{aligned}&12(36n^2-468n+1421)(t^2+11t-1)^4 P_{n-10}(t)\\&+5(n-9)(t^4-228t^3+494t^2+228t+1)P'_{n-5}(t)\\&-(6n^2-83n+261)(t^3-171t^2+247t+57)P_{n-5}(t)=0 \qquad (2)\end{aligned}$$

holds. We prove (2) also by induction on n. Suppose $P_{n-5}(t)$ and $P_{n-10}(t)$ satisfy (2). We want to show the corresponding identity for n being replaced by $n+5$. By replacing $P_n(t)$ by the right-hand side of (1) and then replacing $P_{n-10}(t)$ by the right-hand side of (2) divided by the coefficient $12(36n^2-468n+1421)(t^2+11t-1)^4$ (thus expressing everything by $P_{n-5}(t)$ and its derivatives), we obtain a multiple of the left-hand side of the differential equation $(\flat)_{(6(n-5)+1)/5}$ satisfied by $P_{n-5}(t)$, which vanishes by the induction hypothesis. This concludes the proof of the proposition and hence the theorem is proved. □

3. Reduction modulo prime

In this section, we present some observation about reduction modulo a prime p of our polynomials $P_n(t)$ as a conjecture.

Let

$$j(t) = (1+228t+494t^2-228t^3+t^4)^3/t(1-11t-t^2)^5$$

be the elliptic modular j-invariant expressed in terms of $t = \phi_2^5/\phi_1^5$.

Conjecture. 1) *Let $p \neq 5$ be a prime. Then $P_{p-1}(t) \bmod p$ is a "supersingular t-polynomial", i.e., it is equal to $\prod_{t_0 \in \bar{\mathbf{F}}_p}(t-t_0)$ where t_0 runs through those values for which the corresponding elliptic curve with j-invariant $j(t_0)$ is supersingular.*

2) *For* $p \geq 7$, *the degrees of irreducible factors of* $P_{p-1}(t) \bmod p$ *are as follows:*

(i) *If* $p \equiv 1 \bmod 5$, *all irreducible factors have degree* 2.

(ii) *If* $p \equiv 3, 7 \bmod 20$, *one factor has degree* 2 *and all the others have degree* 4.

(iii) *If* $p \equiv 13, 17 \bmod 20$, *all irreducible factors have degree* 4.

(iv) *If* $p \equiv 4 \bmod 5$, *then there are* h *linear factors and* $(p-1-h)/2$ *quadratic factors, where*

$$h = \text{the class number of the imaginary quadratic field } \mathbf{Q}(\sqrt{-p}) \times \begin{cases} 2 & \text{if } p \equiv 1 \bmod 4 \\ 8 & \text{if } p \equiv 3 \bmod 8 \\ 4 & \text{if } p \equiv 7 \bmod 8. \end{cases}$$

At least the first part of the conjecture should be proven by looking at the Hasse invariant of a family of elliptic curves corresponding to $\Gamma_1(5)$, but we have not worked out this.

References

[1] F. Beukers, *Irrationality of* π^2, *periods of an elliptic curve and* $\Gamma_1(5)$, Approximations diophantiennes et nombres transcendants, Colloq. Luminy/Fr. 1982, *Prog. Math.* 31, 47–66 (1983).

[2] T. Ibukiyama, *Modular forms of rational weights and modular varieties*, Abh. Math. Sem. Univ. Hamburg **70** (2000), 315–339.

[3] M. Kaneko and M. Koike, *On modular forms arising from a differential equation of hypergeometric type*, The Ramanujan J. *7* (2003), 145–164.

[4] M. Kaneko and M. Koike, *Quasimodular forms as solutions to a differential equation of hypergeometric type*, Galois Theory and Modular Forms, (ed. K. Hashimoto, K. Miyake and H. Nakamura), Kluwer Academic Publishers, 329–336, (2003).

[5] M. Kaneko and D. Zagier, *Supersingular j-invariants, Hypergeometric series, and Atkin's orthogonal polynomials*, AMS/IP Studies in Advanced Mathematics, vol. 7 (1998) 97–126.

[6] E. B. Kiritsis, *Fuchsian differential equations for characters on the torus: a classification*, Nuclear Phys. B **324** (1989), no. 2, 475–494.

[7] S. D. Mathur, S. Mukhi and A. Sen, *On the classification of rational conformal field theories*, Phys. Lett. B **213** (1988), no. 3, 303–308.

[8] H. Tsutsumi, *The Atkin orthogonal polynomials for congruence subgroups of low levels*, The Ramanujan J. (to appear).

SOME ASPECTS OF THE MODULAR RELATION

Shigeru Kanemitsu[1], Yoshio Tanigawa[2], Haruo Tsukada[3] and Masami Yoshimoto[4]

[1] *Kinki University-School of Humanity-Oriented Science and Engineering, Iizuka, Fukuoka, 820-8555, Japan*
kanemitu@fuk.kindai.ac.jp

[2] *Graduate School of Mathematics, Nagoya University, Nagoya, 464-8602, Japan*
tanigawa@math.nagoya-u.ac.jp

[3] *Kinki University-School of Humanity-Oriented Science and Engineering, Iizuka, Fukuoka, 820-8555, Japan*
tsukada@fuk.kindai.ac.jp

[4] *Kinki Unversity-Interdiscriplinary Graduate School of Science and Technology, Higashiosaka, Japan*
myoshi@math.kindai.ac.jp

Abstract This paper is a companion to the forthcoming paper [19] and exhibits various manifestations of the modular relation, equivalent to the functional equation. We shall give a somewhat new proof of the functinal equation for the Hurwitz-Lerch Dirichlet L-functions in §1, elucidation of Chan's result relating the functional equation to the q-series (or vice versa) in §2, while §3 and §4 are devoted to elucidate the location of the partial fraction expansion of the coth (cot, respectively) in the modular relation framework.

Keywords: Modular relation, functional equation, Ramanujan identities, Ramanujan formula for zeta values

2000 Mathematics Subject Classification: Primary 11M35; Secondly 11M06, 33B20

Dedicated to Professor Yasuo Morita on his sixtieth birthday, with great respect

---

The first, second and fourth authors are supported by Grant-in-Aid for Scientific Research No. 14540051,14540021 and 14005245 respectively.

Number Theory: Tradition and Modernization, pp. 103–118
W. Zhang and Y. Tanigawa, eds.

In this paper we are concerned with some remote-looking manifestations of the modular relation, which is equivalent to the functional equation for the corresponding zeta-functions (here we confine ourselves to the Hecke type functional equation) ([3, 13, 18]).

In §1 we shall give a somewhat new proof of the functional equation for the Hurwitz-Lerch type Dirichlet L-function by looking at the global meromorphic function in two different regions where the Dirichlet series are absolutely convergent.

Then we go on to §2 to elucidate Chan's method of proof of the equivalence of two q-series identities of Ramanujan as the modular relation. Chan's method depends directly on Hecke's lemma and ours is just transforming his material in the upper half-plane into the right half-plane (or simply the positive real axis). In the upper half-plane, the complex exponential function arises naturally in the Fourier expansion (of automorphic function) and one is restricted to this; on the other hand, in the right half-plane, one has more freedom of choice of weights (as will be developed in [12]).

In §3 we use an equivalent form of the modular relation for the Riemann zeta-function as given by Koshlyakov [22] and elucidate the results of Bradley [4] and in particular point out that one always tacitly or explicitly uses the functional equation.

Notation.
$s = \sigma + it$, $\sigma, t \in \mathbb{R}$ — the complex variable,
$\Gamma(s,z) = \int_z^\infty u^s e^{-u} \frac{du}{u}$ — the incomplete gamma function of the second kind,
$\Gamma(s) = \Gamma(s,0) = \int_0^\infty u^s e^{-u} \frac{du}{u}$, $\sigma > 0$ — the gamma function,
$\gamma(s,z) = \Gamma(s) - \Gamma(s,z)$ — the incomplete gamma function of the first kind,
$\zeta(s,a) = \sum_{n=0}^\infty \frac{1}{(n+a)^s}$, $(\sigma > 1,\ 0 < a \le 1)$ — the Hurwitz zeta-function,
$\zeta(s) = \zeta(s,1) = \sum_{n=1}^\infty \frac{1}{n^s}$, $(\sigma > 1)$ — the Riemann zeta-function,
$\phi(w,s,a) = \sum_{n=0}^\infty \frac{e^{2\pi i w n}}{(n+a)^s}$, $(\sigma > 1,\ 0 < a \le 1, 0 < w \le 1)$ — the Hurwitz-Lerch zeta-function,
$l_s(a) = e^{-2\pi i w}\phi(w,s,1) = \sum_{n=1}^\infty \frac{e^{2\pi i w n}}{n^s}$, $(\sigma > 1)$— the polylogarithm function,
χ — Dirichlet character mod m,
$L(s,\chi) = \sum_{n=1}^\infty \frac{\chi(n)}{n^s}$, $(\sigma > 1)$ — the associated Dirichlet L-function,
$\tau(\chi) = \sum_{n \in (\mathbb{Z}/m\mathbb{Z})^\times} \chi(n) e^{2\pi i \frac{n}{m}}$ — the normalized Gauss' sum, where $(\mathbb{Z}/m\mathbb{Z})^\times$ signifies the group of reduced residue classes mod m.

1. The functional equation for the Hurwitz-Lerch Dirichlet L- functions

Here we use the following additional notation.
χ_0^* – the conventional trivial Dirichlet character to the modulus 1, the associate Dirichlet L-functions $L(s,\chi_0^*)$ being the Riemann zeta-function: $L(s,\chi_0^*) = \zeta(s)$. $\tau_z(\chi) = \sum_{n\in(\mathbb{Z}/m\mathbb{Z})^\times}\chi(n)e^{2\pi i\frac{n}{m}z}$ — the generalized Gauss' sum, with $\tau_1(\chi)=\tau(\chi)$.

We are now in a position to state

Theorem 1. *For $0<a<1$, $0<w<1$, the (global) meromorphic function*

$$G_\chi(w,s,a) = \sum_{n\in\mathbb{Z}}\frac{\chi(-n)e^{-2\pi i\frac{n+w}{m}a}}{(-2\pi i(n+w))^{1-s}}\Gamma\left(1-s,-2\pi i\frac{n+w}{m}\right) \tag{1.1}$$

has the representations:
(i) *in $\sigma>1$*

$$G_\chi(w,s,a) = \chi(-1)m^{s-1}\left(\sum_{k=0}^{\infty}\frac{\tau_k(\chi)e^{2\pi i\frac{k}{m}w}}{(k+a)^s} - \frac{\tau_0(\chi)}{2a^s}\right) \tag{1.2}$$

and
(ii) *in $\sigma<0$*

$$\begin{aligned}G_\chi(w,s,a) = \frac{\Gamma(1-s)}{(2\pi)^{1-s}}e^{-2\pi i\frac{w}{m}a}&\left(e^{\frac{\pi i}{2}(1-s)}\chi(-1)\sum_{n=0}^{\infty}\frac{\chi(n)e^{-2\pi i\frac{n}{m}a}}{(n+w)^{1-s}}\right.\\ &\left.\quad+e^{-\frac{\pi i}{2}(1-s)}\sum_{n=1}^{\infty}\frac{\chi(n)e^{2\pi i\frac{n}{m}a}}{(n-w)^{1-s}}\right)\\ &-\chi(-1)m^{s-1}\frac{\tau_0(\chi)}{2a^s},\end{aligned} \tag{1.3}$$

implying the functional equation

$$\begin{aligned}\chi(-1)m^{s-1}\sum_{k=0}^{\infty}\frac{\tau_k(\chi)e^{2\pi i\frac{k}{m}w}}{(k+a)^s} &= \frac{\Gamma(1-s)}{(2\pi)^{1-s}}e^{-2\pi i\frac{w}{m}a}\times\\ \times\left(e^{\frac{\pi i}{2}(1-s)}\chi(-1)\sum_{n=0}^{\infty}\frac{\chi(n)e^{-2\pi i\frac{n}{m}a}}{(n+w)^{1-s}}\right.&\left.+e^{-\frac{\pi i}{2}(1-s)}\sum_{n=1}^{\infty}\frac{\chi(n)e^{2\pi i\frac{n}{m}a}}{(n-w)^{1-s}}\right).\end{aligned} \tag{1.4}$$

If in particular, χ is a primitive character mod m *and $m>1$, then*

$$\chi(-1)\tau(\chi)m^{s-1}\sum_{k=0}^{\infty}\frac{\overline{\chi}(k)e^{2\pi i\frac{k}{m}w}}{(k+a)^s} = \frac{\Gamma(1-s)}{(2\pi)^{1-s}}e^{-2\pi i\frac{w}{m}a}\times$$

$$\times \left(e^{\frac{\pi i}{2}(1-s)}\chi(-1)\sum_{n=0}^{\infty}\frac{\chi(n)e^{-2\pi i\frac{n}{m}a}}{(n+w)^{1-s}} + e^{-\frac{\pi i}{2}(1-s)}\sum_{n=1}^{\infty}\frac{\chi(n)e^{2\pi i\frac{n}{m}a}}{(n-w)^{1-s}}\right). \tag{1.5}$$

Corollary 1. *For* $m > 1$ *and* χ *primitive* mod m,

$$\chi(-1)\tau(\chi)m^{s-1}L(s,\overline{\chi}) = \frac{\Gamma(1-s)}{(2\pi)^{1-s}}\left(e^{\frac{\pi i}{2}(1-s)}\chi(-1) + e^{-\frac{\pi i}{2}(1-s)}\right)L(1-s,\chi). \tag{1.6}$$

Also

$$\phi(w,s,a) = \frac{\Gamma(1-s)}{(2\pi)^{1-s}}\left(e^{\frac{\pi i}{2}(1-s)}\phi(-a,1-s,w) + e^{-\frac{\pi i}{2}(1-s)}\phi(a,1-s,1-w)\right), \tag{1.7}$$

implying

$$\zeta(s,a) = \frac{\Gamma(1-s)}{(2\pi)^{1-s}}\left(e^{\frac{\pi i}{2}(1-s)}l_{1-s}(1-a) + e^{-\frac{\pi i}{2}(1-s)}l_{1-s}(a)\right), \tag{1.8}$$

which in turn implies

$$\zeta(s) = \frac{\Gamma(1-s)}{(2\pi)^{1-s}}2\sin\frac{\pi s}{2}\zeta(1-s). \tag{1.9}$$

Proof of Theorem 1. (i) For $\sigma > 1$, we substitute the defining integral for $\Gamma\left(1-s,-2\pi i\frac{n+w}{m}a\right)$ and making the change of variable $z = at - a$ to get

$$G_\chi(w,s,a) = \chi(-1)m^{s-1}\sum_{n\in\mathbb{Z}}\chi(n)\int_0^\infty e^{2\pi i\frac{n+w}{m}z}(z+a)^{-s}dz, \tag{1.10}$$

where we inclined the integration path by $\frac{\pi}{2}$, which is justified by Cauchy's theorem.

If we invoke the pulse train (cf. [27, p.44]) and its Fourier series, we may rewrite (1.10) as

$$G_\chi(w,s,a) = \chi(-1)m^{s-1}\sum_{k\in\mathbb{Z}}\int_0^\infty \delta(z-k)\tau_z(\chi)e^{2\pi i\frac{w}{m}z}(z+a)^{-s}dz, \tag{1.11}$$

where $\delta(z)$ signifies the Dirac delta-function. Then each term of the sum corresponding to $k \in \mathbb{N}$ gives $\tau_k(\chi)e^{2\pi i\frac{w}{m}k}(k+a)^{-s}$ and the one with $k = 0$ is to be halved, whence we conclude that the right-side of (1.11) coincides with that of (1.2).

(ii) For $\sigma < 0$, we complete the incomplete gamma function by expressing it as $\Gamma(1-s) - \gamma(1-s, z)$ to get

$$G_\chi(w,s,a) = \Gamma(1-s) \sum_{n \in \mathbb{Z}} \frac{\chi(-n) e^{-2\pi i \frac{n+w}{m} a}}{(-2\pi i(n+w))^{1-s}} - \sum_{n\in\mathbb{Z}} \chi(-n) e^{-2\pi i \frac{n+w}{m} a} \left(\frac{a}{m}\right)^{1-s} \int_0^1 e^{2\pi i \frac{n+w}{m} a t} t^{-s} dt. \tag{1.12}$$

Making the same change of variable as above, we see that the second term on the right of (1.12) becomes

$$-\chi(-1) m^{s-1} \sum_{n\in\mathbb{Z}} \chi(n) \int_{-a}^{0} e^{2\pi i \frac{n+w}{m} z} (z+a)^{-s} dz,$$

to which we apply the pulse train again. But this time there is only one non-zero term corresponding to $n = 0$, which must be halved, giving

$$-\chi(-1) m^{s-1} \frac{\tau_0(\chi)}{2a^s}.$$

Hence, separating the first sum on the right of (1.12) into two, we conclude that the right side of (1.12) is equal to that of (1.3).

To prove (1.5) it is enough to note that χ is a primitive character mod m, $m > 1$, then $\tau_k(\chi) = \tau(\chi)\overline{\chi}(k)$ and $\chi(0) = 0$. □

Proof of Corollary 1. (1.7) is the special case of (1.4) with $m = 1$, (1.6) is obtained from (1.4) by letting $a \to 0+$, $w \to 0+$, and (1.8) follows by letting $w \to 0+$ in (1.4). □

Remark 1. (i) For the history of the proof of (1.7) (and (1.8)) we refer to [11], §4, the book [23], Oberhettinger [26] and Weil [32]. (1.7) is sometimes referred to as the Lipschitz tranformation formula (Grosswald [8, Chap. 8, Theorem 2, p.95], Rademacher [28, (37.1) p.77]). Another proof of (1.7) was given by Knopp and Robins [21], which is in the spirit of Eisenstein (cf. Weil [32]).

(ii) In Erdélyi [7], p.29, there occurs an expansion ([7, (8)]) of $\phi(w,s,a)$ into a power series with Hurwitz zeta-value coefficients, whose generalization was obtained by Johnson [9]. Although they deduced it from the functional equation (1.7) and then, after applying the binomial expansion, the functional equation (1.8), we may deduce [7, (8)] from [7, (7), p.26] and the functional equation [10, (2.10*)] and the Taylor expansion [10, (2.7*)].

(iii) A contour integral representation for $\phi(w, s, a)$ was given in Morita [24] and Naito [25]. Our global gamma series may be thought of as a concrete form of their contour integral representation dating back to Riemann.

2. Ramanujan's identities à la H. H. Chan

Let $\eta(z)$ denote the Dedekind eta-function

$$\eta(z) = q^{\frac{1}{24}} \prod_{n=1}^{\infty} (1 - q^n), \tag{2.1}$$

where $q = e^{2\pi i z}$ with $\operatorname{Im} z > 0$ and let $f(-q)$ denote the infinite product part of $\eta(z)$:

$$f(-q) = \prod_{n=1}^{\infty} (1 - q^n). \tag{2.2}$$

Raghavan [29] stated his guess about a possible relationship between two remarkable identities of Ramanujan

$$\sum_{n=1}^{\infty} \left(\frac{n}{5}\right) \frac{q^n}{(1-q^n)^2} = q\frac{f^5(-q^5)}{f(-q)} \tag{2.3}$$

and

$$1 - 5\sum_{n=1}^{\infty} \left(\frac{n}{5}\right) \frac{nq^n}{1-q^n} = \frac{f^5(-q)}{f(-q^5)}, \tag{2.4}$$

where $\left(\frac{n}{5}\right)$ signifies the Legendre symbol mod 5.

Chan [5] proves that (2.3) and (2.4) are equivalent under Hecke's theory. He uses the transformation formula for $f(-q)$

$$e^{\frac{2\pi i z}{24}} f\left(-e^{2\pi i z}\right) = \frac{1}{\sqrt{-iz}} e^{-\frac{2\pi i}{24}\frac{1}{z}} f\left(-e^{-2\pi i \frac{1}{z}}\right), \tag{2.5}$$

which is a consequence of the most famous transformation formula for the Dedekind eta-function

$$\eta(z) = \frac{1}{\sqrt{-iz}}\eta\left(-\frac{1}{z}\right), \quad \operatorname{Im} z > 0. \tag{2.6}$$

Using (2.5), Chan transforms the right-side of (2.3) in the form

$$\frac{-1}{5^2\sqrt{5}z^2} \frac{f^5(-q_1)}{f(-q_1^5)}, \quad q_1 = e^{-\frac{2\pi i}{5z}},$$

which is $\frac{-1}{5^2\sqrt{5}z}$ times the right-side of (2.4), whence he proceeds to prove the identity

$$\sum_{n=1}^{\infty}\left(\frac{n}{5}\right)\frac{q^n}{(1-q^n)^2}=\frac{-1}{5^2\sqrt{5}z^2}\left(1-5\sum_{n=1}^{\infty}\left(\frac{n}{5}\right)\frac{nq_1^n}{1-q_1^n}\right). \tag{2.7}$$

If we turn the complex plane by $\frac{\pi}{2}$ clockwise by writing $\tau=-iz$, then from the upper half-plane ($\operatorname{Im} z>0$) we move into the right half-plane ($\operatorname{Re}\tau>0$), and regarding (2.6) as a modular relation, we find that it is a special case of Theorem 2 [14]. But we recover this as an illustrating example.

For $\operatorname{Re}\tau>0$ we introduce the functions

$$g(\tau)=\sum_{n=1}^{\infty}\chi(n)\frac{e^{-2\pi n\tau}}{(1-e^{-2\pi n\tau})^2} \tag{2.8}$$

and

$$h(\tau)=\sum_{n=1}^{\infty}\chi(n)\frac{ne^{-2\pi n\tau}}{1-e^{-2\pi n\tau}}, \tag{2.9}$$

where χ may be any Dirichlet character mod m, but we restrict to the case $\chi(n)=\left(\frac{n}{5}\right)$.

Further, as in Chan [5] or more generally as in [14, 15] we form the Dirichlet series

$$\varphi(s)=\zeta(s)L(s-1,\chi)=\sum_{n=1}^{\infty}\frac{\sigma_{1,\chi}(n)}{n^s},\quad \sigma>2 \tag{2.10}$$

and

$$\psi(s)=\zeta(s-1)L(s,\overline{\chi})=\sum_{n=1}^{\infty}\frac{\sigma^*_{1,\overline{\chi}}(n)}{n^s},\quad \sigma>2, \tag{2.11}$$

where

$$\sigma_{z,\chi}(n)=\sum_{d|n}\chi(d)d^z,\ \sigma^*_{z,\chi}(n)=\sum_{d|n}\chi\left(\frac{n}{d}\right)d^z. \tag{2.12}$$

With $A=\frac{2\pi}{\sqrt{5}}$, $C_A(\chi)=-\tau(\chi)$, they satisfy the functional equation

$$A^{-s}\Gamma(s)\varphi(s)=C_A(\chi)A^{-(2-s)}\Gamma(2-s)\psi(2-s). \tag{2.13}$$

Now we note that

$$\sum_{n=1}^{\infty}\sigma_{1,\chi}(n)e^{-An\tau}=\sum_{n=1}^{\infty}\sum_{d|n}\chi(d)de^{-An\tau}$$

$$= \sum_{d=1}^{\infty} \chi(d) d \sum_{\delta=1}^{\infty} \left(e^{-A\delta\tau}\right)^{\delta}$$
$$= \sum_{d=1}^{\infty} \chi(d) \frac{de^{-Ad\tau}}{1-e^{-Ad\tau}} = h\left(\frac{\tau}{\sqrt{5}}\right)$$

and that

$$\sum_{n=1}^{\infty} \sigma_{1,\overline{\chi}}^{*}(n) e^{-An\tau^{-1}} = \sum_{\delta=1}^{\infty} \overline{\chi}(\delta) \sum_{d=1}^{\infty} d\left(e^{-A\delta\tau^{-1}}\right)^{d}$$
$$= \sum_{\delta=1}^{\infty} \overline{\chi}(\delta) \frac{e^{-A\delta\tau^{-1}}}{(1-e^{-A\delta\tau^{-1}})^2} = g\left(\frac{\tau^{-1}}{\sqrt{5}}\right),$$

whence we see that g and h are those two Lambert series appearing in the modular relation ([14])

$$\sum_{n=1}^{\infty} \sigma_{1,\chi}(n) e^{-An\tau} = C_A(\chi)\tau^{-1-1} \sum_{n=1}^{\infty} \sigma_{1,\overline{\chi}}^{*}(n) e^{-An\tau^{-1}} + P_A(\tau), \quad (2.14)$$

where $P_A(\tau)$ is the residual function given as the sum of the residues

$$P_A(\tau) = \sum_{s=0,\,-1} \operatorname{Res} A^{-s}\Gamma(s)\varphi(s)x^{-s}. \quad (2.15)$$

Now we have

$$P_A(\tau) = -\frac{1}{2}L(-1,\chi) + \frac{\sqrt{5}}{2\pi\tau}L(0,\chi)$$
$$= -\frac{1}{2}L(-1,\chi),$$

which may be evaluated as $-\frac{1}{5}$.

Hence, (2.15) reads

$$h\left(\frac{\tau}{\sqrt{5}}\right) - \frac{1}{5} = -\sqrt{5}\frac{1}{\tau^2}g\left(\frac{1}{\sqrt{5}\tau}\right). \quad (2.16)$$

Putting $\frac{1}{\sqrt{5}\tau} = iz$, we may rewrite (2.16) as

$$g(iz) = \frac{1}{5^2\sqrt{5}}\frac{1}{z^2}\left(5h\left(\frac{1}{5iz}\right) - 1\right), \quad (2.17)$$

which is (2.7).

Remark 2. (i) Needless to say, the procedure applies to other moduli, especially the modulus 7 studied by Chan. We will come to the study on this subject elsewhere.

(ii) We remark in passing that Weil's argument [31] has been interpreted as the special case $n = 0$ of Ramanujan's most famous formula for the zeta values $\zeta(2n+1)$ in [18] (cf. (2.14) above); Ramanujan missed this case because he confined himself to the Lambert series.

3. Bradley's results à la N.S. Koshlyakov

Recently, Bradley [4] adopted the partial fraction expansion of the hyperbolic cotangent function to obtain a generalization of Ramanujan's formula for $\zeta(2n+1)$, already referred to in §2, to the case of Dirichlet series with periodic coefficients.

Let $g(n)$ be a periodic function with period m and let $L(s,g) = \sum_{n=1}^{\infty} \frac{g(n)}{n^s}$ be absolutely convergent for $\sigma > \sigma_g > 1$. Then a typical case of Bradley's results is the following: for $x > 0$, $q \in \mathbb{N}$ and g odd (i.e. $g(-1) = -1$)

$$\frac{1}{2}\pi x L(2q,g) + \pi x \sum_{k=1}^{\infty} \frac{k^{-2q} g(k)}{e^{2\pi k/x} - 1} + \sum_{j=0}^{q} (-1)^j x^{2-2j} \zeta(2j) L(2q-2j+1, g)$$
$$= (-1)^q x^{2-2q} \frac{\pi i}{2q-1} \sum_{k=1}^{2q-2} g(k) \sum_{n=1}^{\infty} \frac{n^{-2q}}{e^{\frac{2\pi n x + 2\pi i k}{2q-1}} - 1}. \tag{3.1}$$

In the simplest case of the odd character χ_4 mod 4, (3.1) reads

$$L(2n+1, \chi_4) = \sum_{k=1}^{\infty} \frac{(-1)^k}{(2k-1)^{2n+1}} = \frac{1}{2}\left(\frac{\pi}{2}\right)^{2n+1} \frac{(-1)^n E_n}{(2n)!}, \tag{3.2}$$

where E_n is the n-th Euler number defined by

$$\frac{1}{\coth t} = \sum_{n=0}^{\infty} \frac{E_n}{n!} t^n.$$

Bradley claims that he deduces (3.1) by using the partial fraction expansion of coth and that his proof does not appeal to the functional equation. We shall show that he uses the functional equation as one form of the modular relation. This will defend his method against a possible claim that he appeals to a stronger result than the functional equation (cf. [30], in this regard, where the functional equation is deduced from the partial fraction expansion, but not conversely; [19] for the converse).

We recall from [18] that the functional equation (cf. (2.13))

$$A^{-s}\Gamma(s)\varphi(s) = \chi(s) = A^{-(r-s)}\Gamma(r-s)\psi(r-s) \tag{3.3}$$

is equivalent to the modular relation (cf. (2.14))

$$\sum_{n=1}^{\infty} a_n e^{-\lambda_n s} = \left(\frac{A}{x}\right)^r \sum_{n=1}^{\infty} b_n e^{-\mu_n \frac{A^2}{x}} + P\left(\frac{x}{A}\right), \tag{3.4}$$

where we write

$$\varphi(s) = \sum_{n=1}^{\infty} \frac{a_n}{\lambda_n^s}, \quad \psi(s) = \sum_{n=1}^{\infty} \frac{b_n}{\mu_n^s}$$

and $P(y)$ is the residual function given as the sum of residues, and to the K-Bessel expansion (cf. [18])

$$\begin{aligned} A^{-s}\Gamma(s)\varphi(s,\alpha) &= 2\alpha^{\frac{r-s}{2}} \sum_{n=1}^{\infty} b_n \mu_n^{-\frac{r-s}{2}} K_{r-s}\left(2A\sqrt{\alpha\mu_n}\right) \\ &\quad + \sum \operatorname{Res}\left(\Gamma(-(r-s)+w)\phi(w)z^{r-s-w}\right), \end{aligned} \tag{3.5}$$

where $\alpha > 0$ and $\varphi(s,\alpha)$ is the perturbed Dirichlet series $\sum_{n=1}^{\infty} \frac{a_n}{(\lambda_n \alpha)^s}$.

With slight change of notation we may rewrite (3.5) as

$$\begin{aligned} &2\sum_{k=1}^{\infty} \frac{a_k}{(2\pi\lambda_k)^{\frac{\nu}{2}}} \left(\frac{s^2}{8\pi}\right)^{\frac{\nu}{2}} K_\nu\left(s\sqrt{\lambda_k}\right) \\ &= \Gamma(r-\nu)\sum_{k=1}^{\infty} \frac{(8\pi)^{r-\nu}}{(s^2+16\pi^2\mu_k)^{r-\nu}} + \frac{1}{2\pi i}\int_C \Gamma(z-\nu)\chi(z)\left(\frac{s^2}{8\pi}\right)^{\nu-z} dz, \end{aligned} \tag{3.6}$$

where the integrand on the right side is the same as the residual function.

By applying the operator $\left(-\frac{1}{s}\frac{1}{ds}\right)^\rho$ and choosing $\nu - \rho = -\frac{1}{2}$ in the resulting formula, we deduced in [12]

$$\begin{aligned} \sum_{k=1}^{\infty} a_k \frac{1}{s} e^{-s\sqrt{\lambda_k}} &= 2^{3r}\Gamma\left(r+\frac{1}{2}\right)\pi^{r-\frac{1}{2}} \sum_{k=1}^{\infty} \frac{b_k}{(s^2+16\pi^2\mu_k)^{r+\frac{1}{2}}} \\ &\quad + \frac{1}{2\pi i}\int_C \chi(z)(2\pi)^z \frac{\Gamma(2z+1)}{\Gamma(z+1)} s^{-2z-1} dz, \end{aligned} \tag{3.7}$$

and that by applying the operator again we successfully deduced Lemma 6 [6]. Therefore, everything boils down to the formula

$$\left(-\frac{1}{s}\frac{d}{ds}\right)^\rho \{z^\nu K_\nu(z)\} = z^{\nu-\rho} K_{\nu-\rho}(x).$$

Details will appear e.g. in [11].

We now specialize (3.7) to the case of the Riemann zeta-function:

$$a_k = b_k = 1,\ \lambda_k = \mu_k = \frac{1}{2}k^2,\ \varphi(s) = \psi(s) = \zeta(2s),\ r = \frac{1}{2},\ s = \sqrt{8}\pi x,$$

to obtain

$$\sigma(x) = \frac{1}{e^{2\pi x} - 1} = -\frac{1}{2} + \frac{1}{2\pi x} + \frac{x}{\pi}\sum_{n=1}^{\infty} \frac{1}{n^2 + x^2}, \tag{3.8}$$

which can be proved to be equivalent to the functional equation of the Riemann zeta-function.

Since $\frac{1}{2}\coth \pi x = \sigma(x) + \frac{1}{2}$, (3.8) is the partial fraction expansion for the coth stated by Koshlyakov [22], as one of the forms of the modular relation, who considers also quadratic fields.

For $\sigma > \sigma_g$, let

$$T_g(s, y) = \sum_{k=1}^{\infty} \frac{k^{-s} g(k)}{k^2 + y^2}. \tag{3.9}$$

Then Bradley applies (3.8) (here the functional equation is used) to deduce that

$$\begin{aligned} &\frac{1}{2}\pi x L(s+1, g) + \pi x \sum_{k=1}^{\infty} \frac{k^{-s-1} g(k)}{e^{2\pi k/x} - 1} \\ &= \frac{1}{2}x^2 L(s+2, g) + x^2 \sum_{n=1}^{\infty} T_g(s, nx). \end{aligned} \tag{3.10}$$

Then he applies the recurrence

$$T_g(s, y) = y^{-2}\varphi(s) - y^{-2} T_g(s-2, y)$$

to reduce (3.9) to $T_g(-1, y)$ or $T_g(0, y)$ according to the parity of m and g. But in order to deduce Bradley's results it suffices to evaluate $T_g(m, y)$ for $m \in \mathbb{N}$, which we can do by using the partial fraction expansion

$$\begin{aligned} \frac{1}{x^m(x^2+y^2)} &= \frac{y^{-2}}{x^m} + \frac{y^{-4}}{x^{m-2}} + \cdots + \begin{cases} 0 & m \equiv 0\ (2) \\ \dfrac{(-1)^{\frac{m-1}{2}} y^{-m-1}}{x} & m \equiv 1\ (2) \end{cases} \\ &\quad + \frac{\frac{1}{2}(-iy)^{-m-1}}{x + iy} + \frac{\frac{1}{2}(iy)^{-m-1}}{x - iy}. \end{aligned} \tag{3.11}$$

Using (3.11), we arrive at ($m = 2q-1$, g odd)

$$T_g(2q-1,y) = \sum_{j=1}^{q} (-1)^{j-1} y^{-2j} L(2q-2j+1,g) + (-1)^q y^{-2q} T_g(-1,y). \tag{3.12}$$

Putting $y = nx$ in (3.12) and summing over $n = 1, 2, \ldots$, we obtain

$$x^2 \sum_{n=1}^{\infty} T_g(2q-1,nx) = \sum_{j=1}^{q} (-1)^{j-1} x^{2-2j} \zeta(2j) L(2q-2j+1,g) + (-1)^q x^{2-2q} \sum_{n=1}^{\infty} n^{-2q} T_g(-1,nx). \tag{3.13}$$

It remains to evaluate $\sum_{k\in\mathbb{Z}} \frac{g(k)}{y+ik}$, which can be done using another result of Koshlyakov ($\operatorname{Re} z \geq 0$)

$$2\pi i \left(\sigma(iz) + \frac{1}{2} \right) = \sum_{n\in\mathbb{Z}} \frac{1}{n+z}. \tag{3.14}$$

Indeed,

$$\begin{aligned} \sum_{k\in\mathbb{Z}} \frac{g(k)}{y+ik} &= \sum_{k=0}^{m-1} g(k) \sum_{r\in\mathbb{Z}} \frac{1}{y+i(k+mr)} \\ &= \sum_{k=0}^{m-1} g(k) \frac{1}{im} 2\pi i \left(\sigma\left(\frac{y+ik}{m} \right) + \frac{1}{2} \right), \end{aligned}$$

which is

$$\frac{\pi}{m} \sum_{k=0}^{m-1} g(k) \coth \frac{\pi(y+ik)}{m}.$$

Substituting this in (3.13) completes the proof of (3.1).

Remark 3. (i) Bradley's other expressions for the right-side of (3.1) follow by similar argument as in Katsurada [20] and [16]. The special case $g = \chi$ is already obtained in [17] which is a culmination of the thereto existing results.

(ii) For $\sigma > 1$,

$$L(s,g) = m^{-s} \sum_{a=1}^{m} g(a) \zeta\left(s, \frac{a}{m}\right), \tag{3.15}$$

so that we may transfer the results on $\zeta\left(s, \frac{a}{m}\right)$ onto those of $L(s, g)$.

One may wonder why one can deduce the results on $L(s, g)$ built on the Hurwitz zeta-function, from functional equation of the Riemann zeta-function. The reason is elucidated in [12] to the effect that the functional equations for the Riemann zeta-function and the Hurwitz zeta-function are equivalent.

4. The functional equation for $\phi(w, s, a)$

In this section we shall use the partial fraction expansion (that of the $\cot x$, slightly more general than (3.8)) of the function

$$\frac{e^{(1-a)x}}{e^x - e^{2\pi i w}} \quad (0 < w < 1)$$

to deduce the functional equation (1.7). Although in our general framework [12] the following procedure may be superfluous, we find it instructive to give the proof, which resembles that of the functional equation for the Riemann zeta-function [30] but is much closer to our standpoint, suggestive of the modular relation.

For $0 < w < 1$, we contend that

$$\frac{e^{(1-a)x}}{e^x - e^{2\pi i w}} = \sum_{n \in \mathbb{Z}} \frac{e^{-2\pi i a(w+n)}}{x - 2\pi i w - 2\pi i n}. \tag{4.1}$$

That (4.1) is true up to a constant, say c, is immediate. To determine c, we let $x \to 2\pi i w$ in the equality

$$\frac{e^{(1-a)x}}{e^x - e^{2\pi i w}} - \frac{e^{-2\pi i a w}}{x - 2\pi i w} = \sum_{0 \neq n \in \mathbb{Z}} \frac{e^{-2\pi i a(w+n)}}{x - 2\pi i(w+n)} + c. \tag{4.2}$$

The left-side is

$$e^{-2\pi i a w}\left(\frac{e^{(1-a)(x-2\pi i w)}}{e^{x-2\pi i w} - 1} - \frac{1}{x - 2\pi i w}\right) = \frac{e^{-2\pi i a w}}{x - 2\pi i w}$$

$$\times\left\{\left(1 - \frac{x - 2\pi i w}{2} + \cdots\right)\left(1 + (1-a)(x - 2\pi i w) + \cdots\right) - 1\right\}$$

$$\to e^{-2\pi i a w}\left(\frac{1}{2} - a\right),$$

while the right-side tends to

$$\frac{e^{-2\pi i a w}}{2\pi i} \sum_{0 \neq n \in \mathbb{Z}} \frac{e^{2\pi i a n}}{n} + c = e^{-2\pi i a w}\left(a - \frac{1}{2}\right) + c.$$

Hence we conclude the equality

$$\left(\frac{1}{2}-a\right)e^{-2\pi iaw} = -e^{-2\pi iaw}\left(a-\frac{1}{2}\right)+c,$$

whence $c = 0$ and (4.1) follows.

Now suppose $0 < \sigma < 1$ and substitute (4.1) for the integral in the Mellin transform formula

$$\Gamma(s)\phi(w,s,a) = \int_0^\infty \frac{e^{(1-a)x}}{e^x - e^{2\pi iw}} x^{s-1}dx \tag{4.3}$$

and integrate term-by-term to obtain

$$\Gamma(s)\phi(w,s,a) = \sum_{n\in\mathbb{Z}} e^{-2\pi i(w+n)} \int_0^\infty \frac{x^{s-1}}{x-2\pi i(w+n)}dx. \tag{4.4}$$

We now appeal to the well-known formula

$$\int_0^\infty \frac{x^{s-1}}{x-\omega i}dx = |\omega|^{s-1} e^{-\frac{\pi}{2}i(s-1)\mathrm{sgn}(\omega)} \frac{\pi}{\sin \pi s}, \tag{4.5}$$

where $\mathrm{sgn}(\omega) = 1$ for $\omega > 0$ and -1 for $\omega < 0$, to the integral on the right of (4.4):

$$\Gamma(s)\phi(w,s,a) = \frac{2^{s-1}\pi^s}{\sin \pi s} \sum_{n\in\mathbb{Z}} |w+n|^{s-1} e^{-2\pi ia(w+n)-\frac{\pi}{2}i(s-1)\mathrm{sgn}(w+n)} \tag{4.6}$$

for $0 < w < 1$, $0 < a < 1$ and $0 < \sigma < 1$. Separating the sum into two parts $n \geq 1$ and $n < 0$ (in which case we may write $n+1-w$ for $|n+w|$) and appealing to the reciprocity relation for the gamma function, we conclude (1.7).

Acknowledgments

The substantial part of the paper was completed while the authors save for the third were staying in China-Japan Number Theory Institute, North-West University, Xi'an at the end of August, 2004, and §§3-4 arose from the discussion between the first author and Professor H. H. Chan, NUS in February, 2005, who also kindly supplied the reference [21]. The authors would like to thank these institutes and colleagues for wonderful research environment and fruitful discussions.

References

[1] R. Balasubramanian, S. Kanemitsu and H. Tsukada, *Contributions to the theory of Lerch zeta-functions*, to appear.

[2] B. C. Berndt, *Ramanujan's notebooks. Part II*, Springer-Verlag, New York-Berlin, 1989.

[3] S. Bochner, *Some properties of modular relations*, Ann. of Math. (2) **53** (1951), 332–363.

[4] D. Bradley, *Series acceleration formulas for Dirichlet series with periodic coefficients*, preprint.

[5] H. H. Chan, *On the equivalence of Ramanujan's partition identities and a connection with the Rogers-Ramanujan continued fraction*, J. Math. Anal. Appl. **198** (1996), 111–120.

[6] K. Chandrasekharan and Raghavan Narasimhan, *Hecke's functional equation and arithmetical identities*, Ann. of Math. (2) **74** (1961), 1–23.

[7] A. Erdélyi, W. Magnus, F. Oberhettinger and F. G. Tricomi, *Tables of integral transforms. Vol. I.* McGraw-Hill Book Company, Inc., New York-Toronto-London, 1954.

[8] E. Grosswald, *Representations of integers as sums of two squares*, Springer Verlag, New York-Berlin-Heidelberg-Tokyo, 1985.

[9] B. R. Johnson, *Generalized Lerch zeta function*, Pacific J. Math. **53** (1974), 191–193.

[10] S. Kanemitsu, M. Katsurada and M. Yoshimoto, *On the Hurwitz-Lerch zeta-function*, Aequationes Math. **59** (2000), 1–19.

[11] S. Kanemitsu, Y. Tanigawa, H. Tsukada and M. Yoshimoto, *Contributions to the theory of the Hurwitz zeta-function*, (submitted for publication).

[12] S. Kanemitsu, Y. Tanigawa, H. Tsukada and M. Yoshimoto, *Contibutions to the theory of zeta-functions:The modular relation supremacy*, in preparation.

[13] S. Kanemitsu, Y. Tanigawa and M. Yoshimoto, *On rapidly convergent series for the Riemann zeta-values via the modular relation*, Abh. Math. Sem. Univ. Hamburg **72** (2002), 187–206.

[14] S. Kanemitsu, Y. Tanigawa and M. Yoshimoto, *On rapidly convergent series for Dirichlet L-function values via the modular relation*, Number Theory and Discrete Mathematics (ed. by A. K. Agarwal, B. C. Berndt et al.), Hindustan Book Agency, 2002, pp. 113–133.

[15] S. Kanemitsu, Y. Tanigawa and M. Yoshimoto, *On zeta- and L-function values at special rational arguments via the modular relation*, Proc. Int. Conf. SSFA, Vol.I 2001, pp.21–42

[16] S. Kanemitsu, Y. Tanigawa and M. Yoshimoto, *On multi-Hurwitz zeta-function values at rational arguments*, Acta Arith. **107** (2003), 45–67.

[17] S. Kanemitsu, Y. Tanigawa, M. Yoshimoto, *On the values of the Riemann zeta-function at rational arguments*, Hardy-Ramanujan J. **24** (2001), 10–18.

[18] S. Kanemitsu, Y. Tanigawa, M. Yoshimoto, *Ramanujan's formula and modular forms*, Number-theoretic methods – Future trends, Proceedings of a conference held in Iizuka (ed. by S. Kanemitsu and C. Jia) 2002, pp.159–212.

[19] S. Kanemitsu, Y. Tanigawa and H. Tsukada, *Some examples of a variant of the modular relation*, preprint.

[20] M. Katsurada, *On an asymptotic formula of Ramanujan for a certain theta-type series*, Acta Arith. **97** (2001), 157–172.

[21] M. Knopp and S. Robins, *Easy proofs of Riemann's functional equation for $\zeta(s)$ and of Lipschitz summation*, Proc. AMS, **129**, No. 7 (2001), 1915–1922.

[22] N. Koshlyakov, *Investigation of some questions of analytic theory of the rational and quadratic fields*, I-III, Izv. Akad. Nauk SSSR, Ser. Mat. **18** (1954), 113–144, 213–260, 307–326; Errata **19** (1955), 271 (in Russian).

[23] A. Laurinčikas and R. Garunkštis, *The Lerch zeta-function*, Kluwer Academic Publ., Dordrecht-Boston-London 2002.

[24] Y. Morita, *On the Hurwitz-Lerch L-functions*, J. Fac. Sci. Univ. Tokyo, Sect IA Math. **24** (1977), 29–43.

[25] H. Naito, *The p-adic Hurwitz L-functions*, Tôhoku Math. J. **34** (1982), 553–558.

[26] F. Oberhettinger, *Note on the Lerch zeta function*, Pacific J. Math. **6** (1956), 117–120.

[27] A. Papoulis, *The Fourier integral and its applications*, McGraw-Hill, 1962.

[28] H. Rademacher, *Topics in Analytic Number Theory*, Springer-Verlag, Berlin-Heidelberg-New York, 1973.

[29] S. Raghavan, *On certain identities due to Ramanujan*, Quart. J. Math. (Oxford) (2) **37** (1986), 221–229.

[30] E.C. Titchmarsh, *The theory of the Riemann zeta-function*, (Edited and with a preface by D.R. Heath-Brown), The Clarendon Press Oxford University Press, New York, 1986.

[31] A. Weil, *Sur une formule classique*, J. Math. Soc. Japan **20** (1968), 400-402 = *Scientific works. Collected Papers, Vol. III (1964–1978)*, Springer-Verlag, New York-Heidelberg, 1979, pp. 198–200.

[32] A. Weil, *On Eisenstein's copy of the Disquisitiones*, Advanced Studies in Pure Mathematics **17**, 1989 Algebraic Number Theory – in honor of K. Iwasawa, pp. 463–469.

ZEROS OF AUTOMORPHIC L-FUNCTIONS AND NONCYCLIC BASE CHANGE

Jianya Liu[1] and Yangbo Ye[2]

[1] *Department of Mathematics, Shandong University, Jinan, Shandong 250100, China.*
jyliu@sdu.edu.cn

[2] *Department of Mathematics, The University of Iowa, Iowa City, Iowa 52242-1419, USA.*
yey@math.uiowa.edu

Abstract Let π be an automorphic irreducible cuspidal representation of GL_m over a Galois (not necessarily cyclic) extension E of $\mathbb{Q}$ of degree ℓ. We compute the n-level correlation of normalized nontrivial zeros of $L(s, \pi)$. Assuming that π is invariant under the action of the Galois group Gal$(E/\mathbb{Q})$, we prove that it is equal to the n-level correlation of normalized nontrivial zeros of a product of ℓ distinct L-functions $L(s, \pi_1) \cdots L(s, \pi_\ell)$ attached to cuspidal representations $\pi_1, \ldots, \pi_\ell$ of GL_m over $\mathbb{Q}$. This is done unconditionally for $m = 1, 2$ and for $m = 3, 4$ with the degree ℓ having no prime factor $\leq (m^2 + 1)/2$. In other cases, the computation is made under a conjecture of bounds toward the Ramanujan conjecture over E, and a conjecture on convergence of certain series over prime powers (Hypothesis H over E and $\mathbb{Q}$). The results provide an evidence that π should be (noncyclic) base change of ℓ distinct cuspidal representations $\pi_1, \ldots, \pi_\ell$ of $GL_m(\mathbb{Q}_\mathbb{A})$, if it is invariant under the Galois action. A technique used in this article is a version of Selberg orthogonality for automorphic L-functions (Lemma 6.2 and Theorem 6.4), which is proved unconditionally, without assuming π and $\pi_1, \ldots, \pi_\ell$ being self-contragredient.

Keywords: automorphic L-function, nontrivial zero, n-level correlation, base change

2000 Mathematics Subject Classification: 11F70, 11F66, 11M41

[1] Supported by China NNSF Grant Number 10125101.
[2] Project sponsored by the USA NSA under Grant Number MDA904-03-1-0066. The United States Government is authorized to reproduce and distribute reprints notwithstanding any copyright notation herein.

Number Theory: Tradition and Modernization, pp 119–152
W. Zhang and Y. Tanigawa, eds.

1. Introduction.

According to Langlands' functoriality conjecture, the L-function $L(s,\pi)$ attached to an automorphic irreducible cuspidal representation π of GL_m over a number field E should equal a product of $L(s,\pi_j)$ for certain cuspidal representations π_j of GL_{m_j} over $\mathbb{Q}$. Arthur and Clozel [ArtClo] proved that this is indeed the case when E is a cyclic Galois extension of $\mathbb{Q}$ of degree ℓ and π is stable under $\mathrm{Gal}(E/\mathbb{Q})$. In fact in this case, π is the base change of exactly ℓ nonequivalent cuspidal representations $\pi_{\mathbb{Q}}, \pi_{\mathbb{Q}}\otimes\eta_{E/\mathbb{Q}}, \ldots, \pi_{\mathbb{Q}}\otimes\eta_{E/\mathbb{Q}}^{\ell-1}$ of $GL_m(\mathbb{Q}_{\mathbb{A}})$, where $\eta_{E/\mathbb{Q}}$ is the nontrivial character of $\mathbb{Q}_{\mathbb{A}}^{\times}/\mathbb{Q}^{\times}$ attached to the field extension $E/\mathbb{Q}$ according to the class field theory. Consequently (cf. [Bor] and [Lang])

$$L(s,\pi)=L(s,\pi_{\mathbb{Q}})L(s,\pi_{\mathbb{Q}}\otimes\eta_{E/\mathbb{Q}})\cdots L(s,\pi_{\mathbb{Q}}\otimes\eta_{E/\mathbb{Q}}^{\ell-1}), \tag{1.1}$$

where the L-functions on the right side are distinct.

When E is noncyclic over $\mathbb{Q}$, factorization of $L(s,\pi)$ into a product of L-functions of GL_{m_j} over $\mathbb{Q}$ is unknown. Recently, the authors [LiuYe2] proved that for any L-function such a factorization, if exists, must be unique. In particular, we proved that the L-function $L(s,\pi_{\mathbb{Q}})$, attached to an automorphic irreducible cuspidal representation $\pi_{\mathbb{Q}}$ of $GL_m(\mathbb{Q}_{\mathbb{A}})$, cannot be factored further as a product of automorphic L-functions for $GL_{m_j}(\mathbb{Q}_{\mathbb{A}})$. In other words, $L(s,\pi_{\mathbb{Q}})$ is primitive in the sense of [Sel] (see [Mur1] and [Mur2]).

On the other hand, Rudnick and Sarnak [RudSar] proved that the n-level correlation of normalized nontrivial zeros of this $L(s,\pi_{\mathbb{Q}})$ follows a GUE (Gaussian Unitary Ensemble) model, for a class of test functions whose Fourier transforms have restricted support. Here the zeros are normalized according to their density. For distinct L-functions $L(s,\pi_j)$, with π_j being cuspidal representations of $GL_m(\mathbb{Q}_{\mathbb{A}})$, $j=1,\ldots,\ell$, the authors [LiuYe1] proved that the n-level correlation of normalized nontrivial zeros of the product $L(s,\pi_1)\cdots L(s,\pi_\ell)$ follows a superposition distribution of individual GUEs from $L(s,\pi_j)$ and products of GUEs of lower ranks, under Selberg's orthogonality conjecture and under a conjecture on the convergence of a sum over prime powers (Hypothesis H: see §2) for $m\geq 5$. Recently, Liu, Wang, and Ye [LiuWangYe] proved this Selberg orthogonality conjecture for automorphic L-functions. Therefore, this n-level correlation is now known unconditionally for $m\leq 4$ and under the Hypothesis H for $m\geq 5$. Note that here the test functions also have a restricted support for their Fourier transforms.

In this paper, we will use this GUE correlation to study the factorization of $L(s,\pi)$. We will show that the nontrivial zeros of $L(s,\pi)$ behave in the same way as the nontrivial zeros of $L(s,\pi_1)\cdots L(s,\pi_\ell)$.

Theorem 1.1. *Let E be a Galois extension of $\mathbb{Q}$ of degree ℓ, and π an automorphic irreducible cuspidal representation of $GL_m(E_\mathbb{A})$ with unitary central character. Assume* (i) *Hypothesis H both over E and over $\mathbb{Q}$ when $m \geq 5$ and* (ii) *Conjecture 2.1 when $m \geq 3$. Suppose that $\pi \cong \pi^\sigma$ for all $\sigma \in \mathrm{Gal}(E/\mathbb{Q})$. Then the n-level correlation of normalized nontrivial zeros of $L(s,\pi)$ is equal to the n-level correlation of normalized nontrivial zeros of a product of ℓ distinct automorphic L-functions attached to cuspidal representations of $GL_m(\mathbb{Q}_\mathbb{A})$.*

Note that Theorem 1.1 is an unconditional result for $m = 1$ and 2. For $m = 3$ and 4, it is also unconditional when $(30, \ell) = 1$ and $(210, \ell) = 1$, respectively. See §2 for details.

Theorem 1.1 provides an evidence that $L(s,\pi)$ should factor into a product of L-functions ℓ distinct automorphic L-functions attached to cuspidal representations of $GL_m(\mathbb{Q}_\mathbb{A})$, and suggests that the base change and factorization of L-functions such as (1.1) should hold for noncyclic extension fields as well.

We remark that our results contain much less information than what was achieved in [ArtClo], as we cannot see individual representations $\pi_\mathbb{Q}, \dots, \pi_\mathbb{Q} \otimes \eta_{E/\mathbb{Q}}^{\ell-1}$ as in (1.1). Since we are seeking less information, we can determine through zero distributions the base change structure of π beyond the scope of [ArtClo].

The reason behind this is indeed the universality of the n-level correlation of zeros discussed in [RudSar]: The n-level correlation of normalized nontrivial zeros of $L(s,\pi_\mathbb{Q})$ is independent of $\pi_\mathbb{Q}$, as long as it is a cuspidal representation of $GL_m(\mathbb{Q}_\mathbb{A})$.

Our computation of n-level correlation will be carried our for test functions f whose Fourier transforms $\Phi(\xi)$ as in (2.8) below have restricted support $\sum_{1\leq j\leq n} |\xi_j| < 2/m$. This restriction of support is good enough for our purpose. We will not assume conjectures that the n-level GUE correlation holds for test functions whose Fourier transform has an arbitrary support.

One of the main techniques used in the computation is a version of Selberg orthogonality [Sel] for automorphic L-functions. First we prove a Mertens' theorem for Rankin-Selberg L-functions $L(s, \pi \times \tilde{\pi}')$ (Lemma 6.2 and Theorem 6.4). The proof is unconditional, and π and π' are cuspidal representations of $GL_m(E_\mathbb{A})$ not necessarily self-contragredient. Next we use Hypothesis H to control sums over prime powers and use Conjecture 2.1 to control sums over primes which do not split completely in E. The resulting orthogonality (Lemmas 7.1 and 7.3) is then relative to sums over completely split primes. By an argument in [Rog], information on whether $\pi \cong \pi^\sigma$ or not can be characterized by equivalence relationships among local representations π_v at places v lying above completely split primes. This information will therefore be described by orthogonality, and eventually lead to the n-level correlation of normalized nontrivial zeros.

2. Notation and assumption.

Let E be a Galois extension over $\mathbb{Q}$ of degree ℓ. Let $E_{\mathbb{A}} = \prod_v' E_v$ be its adele ring, where v runs over all places of E, and $\prod'$ denotes a restricted product. For any prime p, we have $E \otimes_{\mathbb{Q}} \mathbb{Q}_p = \bigoplus_{v|p} E_v$, where v with $v|p$ are places of E lying above p. Since E is Galois over $\mathbb{Q}$, all E_v with $v|p$ are isomorphic. Denote by ℓ_p the degree, by $e_p = \mathrm{ord}_{v_i}(p)$ the order of ramification, and by f_p the modular degree of E_v over $\mathbb{Q}$ for $v|p$. Then $\ell_p = e_p f_p$ and $q_v = p^{f_p}$ is the module of E_v. On the other hand, $E \otimes_{\mathbb{Q}} \mathbb{R}$ is either $= \bigoplus_{v|\infty} \mathbb{R}$, or $= \bigoplus_{v|\infty} \mathbb{C}$.

Let π be an automorphic irreducible cuspidal representation of $GL_m(E_{\mathbb{A}})$ with unitary central character. The (finite part) L-function attached to π is given by the Euler product $L(s,\pi) = \prod_{v<\infty} L(s,\pi_v)$ for $\sigma > 1$. The definition of the local factor $L(s,\pi_v)$ is given by

$$L(s,\pi_v) = \prod_{1\le j\le m} \left(1 - \alpha_\pi(j,v) q_v^{-s}\right)^{-1},$$

where $\alpha_\pi(j,v)$, $1 \le j \le m$, are complex numbers given by the Langlands correspondence. For any prime p, denote by $L_p(s,\pi) = \prod_{v|p} L(s,\pi_v)$ the product of local factors above p. Then

$$L_p(s,\pi) = \prod_{v|p} \prod_{1\le j\le m} \left(1 - \alpha_\pi(j,v) p^{-f_p s}\right)^{-1}.$$

Similarly, the product of Archimedean local factors is given by $L_\infty(s,\pi) = \prod_{v|\infty} L(s,\pi_v)$, where

$$L(s,\pi_v) = \prod_{1\le j\le m} \Gamma_{\mathbb{R}}\big(s + \mu_\pi(j,v)\big)$$

if v is real, and

$$L(s,\pi_v) = \prod_{1\le j\le m} \Gamma_{\mathbb{C}}\big(s + \mu_\pi(j,v)\big)$$

if v is complex. Here $\mu_\pi(j,v)$ are again complex numbers given by the Langlands correspondence, $\Gamma_{\mathbb{R}}(s) = \pi^{-s/2}\Gamma(s/2)$, and $\Gamma_{\mathbb{C}}(s) = (2\pi)^{-s}\Gamma(s)$.

We will need a bound for $\alpha_\pi(j,v)$:

$$|\alpha_\pi(j,v)| \le p^{f_p(1/2-1/(\ell m^2+1))}. \tag{2.1}$$

This bound holds for any π_v, either ramified or unramified. It was first observed by Serre [Ser] and appeared in published form in [LuoRudSar]. A complete proof is given in [RudSar] for the case of $E = \mathbb{Q}$, using an argument of Landau [Land].

When π_v is unramified, the generalized Ramanujan conjecture claims that

$$|\alpha_\pi(j,v)| = 1. \tag{2.2}$$

The best known bounds toward this Ramanujan conjecture over an arbitrary number field are

$$|\alpha_\pi(j,v)| \le q_v^{1/9} = p^{f_p/9} \tag{2.3}$$

for $m = 2$ ([KimShah]), and

$$|\alpha_\pi(j,v)| \le q_v^{1/2-1/(m^2+1)} = p^{f_p(1/2-1/(m^2+1))} \tag{2.4}$$

for general m ([LuoRudSar]), where $v|p$.

We will not assume (2.2), but assume a bound θ_p for it for any p which is unramified and does not split completely in E as stated in the following

Conjecture 2.1. *For any p which is unramified and does not split completely in E, we have, for any $v|p$,*

$$|\alpha_\pi(j,v)| \le q_v^{\theta_p} = p^{f_p\theta_p},$$

where $\theta_p = 1/2 - 1/(2f_p) - \varepsilon$ for a small $\varepsilon > 0$.

We remark that $e_p = 1$ and hence $f_p = \ell_p$ when p is unramified. Since p does not split completely in E, we know that $f_p \ge 2$. Consequently Conjecture 2.1 is known for $m = 2$, according to (2.3). Conjecture 2.1 is trivial for $m = 1$. Recall that $f_p|\ell$. Thus conjecture 2.1 is known when all prime factors of ℓ are $> (m^2+1)/2$. For $m = 3$ this means that any $p|\ell$ is ≥ 7, while for $m = 4$, Conjecture 2.1 is true when any $p|\ell$ is ≥ 11.

We also need the Hypothesis H ([RudSar]) generalized to E.

Hypothesis H. *Let π be an automorphic irreducible cuspidal representation of $GL_m(E_\mathbb{A})$ with unitary central character. Then for any fixed $k \ge 2$*

$$\sum_p \frac{\log^2 p}{p^{kf_p}} \sum_{v|p} \Big| \sum_{1\le j\le m} \alpha_\pi^k(j,v)\Big|^2 < \infty. \tag{2.5}$$

We note that Hypothesis H is an easy consequence of the generalized Ramanujan conjecture (2.2). Since there are only finitely many p which are not unramified in E, the sum in (2.5) may be taken over all unramified p. As we have assumed Conjecture 2.1, we know that

$$\sum_{\substack{p \text{ unramified,}\\ \text{not split completely}}} \frac{\log^2 p}{p^{kf_p}} \sum_{v|p} \Big| \sum_{1\le j\le m} \alpha_\pi^k(j,v)\Big|^2$$

$$\ll \sum_{\substack{p \text{ unramified,}\\ \text{not split completely}}} \frac{\log^2 p}{p^{kf_p}} p^{2kf_p\theta_p} < \infty.$$

Consequently under Conjecture 2.1, Hypothesis H claims that for any fixed $k \ge 2$

$$\sum_{p \text{ splits completely}} \frac{\log^2 p}{p^k} \sum_{v|p} \Big| \sum_{1\le j\le m} \alpha_\pi^k(j,v)\Big|^2 < \infty.$$

As in the case of $E = \mathbb{Q}$, Hypothesis H is trivial for $m = 1$. For $m = 2$, it can be proved using the bound in (2.3). In fact, (2.3) implies that $|\alpha_\pi^k(j,v_i)| \le p^{kf_p/9}$ and hence

$$\sum_p \frac{\log^2 p}{p^{kf_p}} \sum_{v|p} \Big| \sum_{1\le j\le m} \alpha_\pi^k(j,v)\Big|^2 \ll \sum_p \frac{\log^2 p}{p^{7k/9}} < \infty$$

when $k \ge 2$. In Appendix, we will prove Hypothesis H for $m = 3$. For $m = 4$, Hypothesis H is a consequence of [Kim1], Proposition 6.2, as pointed out by [KimSar] and proved in [Kim2].

Let g_j be a compactly supported smooth function on $\mathbb{R}$. Then its Fourier transform

$$h_j(r) = \int_{\mathbb{R}} g_j(u)e^{iru}du \tag{2.6}$$

is entire and rapidly decreasing on $\mathbb{R}$. We denote $\mathbf{h} = (h_1, \cdots, h_n)$ and define

$$\kappa(\mathbf{h}) = \int_{\mathbb{R}} h_1(r)\cdots h_n(r)dr. \tag{2.7}$$

Given $\Phi \in C_c^1(\mathbb{R}^n)$ (c meaning compactly supported) we define

$$f(x) = \int_{\mathbb{R}^n} \Phi(\xi)\delta(\xi_1 + \cdots + \xi_n)e(-x\cdot\xi)d\xi, \tag{2.8}$$

where $x = (x_1, \ldots, x_n)$, $\xi = (\xi_1, \ldots, \xi_n)$, $x \cdot \xi = x_1\xi_1 + \cdots + x_n\xi_n$, $\delta(t)$ is the Dirac mass at zero, and $e(t) = e^{2\pi it}$.

The n-level correlation of normalized nontrivial zeros of the L-function $L(s,\pi)$ is given by

$$\sum_{\substack{i_1,\dots,i_n\\ \text{distinct}}} h_1\Big(\frac{\gamma_{i_1}}{T}\Big)\cdots h_n\Big(\frac{\gamma_{i_n}}{T}\Big) f\Big(\frac{\gamma_{i_1} m\log T}{2\pi},\dots,\frac{\gamma_{i_n} m\log T}{2\pi}\Big), \tag{2.9}$$

where the sum is taken over distinct indices $i_1,\dots,i_n$, of nontrivial zeros $\rho_{i_\nu}=1/2+i\gamma_{i_\nu}$, $\nu=1,\dots,n$, of $L(s,\pi)$. Without assuming the Riemann Hypothesis, γ_{i_ν} are complex numbers. Here the factor $\frac{m}{2\pi}\log T$ provides the normalization for zeros ρ_{i_ν}. In the following sections, we will first compute the same sum as in (2.9) but taken over all indices of nontrivial zeros. We will denote this latter sum by

$$\sum_{\gamma_1,\cdots,\gamma_n} h_1\Big(\frac{\gamma_1}{T}\Big)\cdots h_n\Big(\frac{\gamma_n}{T}\Big) f\Big(\frac{\gamma_1 m\log T}{2\pi},\dots,\frac{\gamma_n m\log T}{2\pi}\Big). \tag{2.10}$$

By an argument similar to that in [RudSar] and [LiuYe1], we may deduce (2.9) from (2.10).

3. The main theorems.

Theorem 3.1. *Let E be a Galois extension of $\mathbb{Q}$ of degree ℓ, and π an automorphic irreducible cuspidal representation of $GL_m(E_{\mathbb{A}})$ with unitary central character. Let a, $1\le a\le \ell$, be the number of elements $\sigma\in \mathrm{Gal}(E/\mathbb{Q})$ with $\pi\cong\pi^\sigma$. Assume $m\le 4$ or Hypothesis H over E for $m\ge 5$. Also assume Conjecture 2.1 when $m\ge 3$ and that there is $p|\ell$ such that $p\le (m^2+1)/2$. Then*

$$\begin{aligned}
&\sum_{\gamma_1,\cdots,\gamma_n} h_1\Big(\frac{\gamma_1}{T}\Big)\cdots h_n\Big(\frac{\gamma_n}{T}\Big) f\Big(\frac{\gamma_1 m\log T}{2\pi},\dots,\frac{\gamma_n m\log T}{2\pi}\Big)\\
&= \frac{\kappa(\mathbf{h})}{2\pi} mT(\log T)\ell^n\Phi(0,\cdots,0)\\
&\quad+\frac{\kappa(\mathbf{h})}{2\pi} mT(\log T)\sum_{1\le r\le n/2}\frac{n!\ell^{n-2r}a^r}{2^r r!(n-2r)!}\int_{\mathbb{R}^r}|v_1|\cdots|v_r|\\
&\quad\times\Phi(v_1,\dots,v_r,-v_1,\dots,-v_r,0,\dots,0)dv_1\cdots dv_r+O(T),
\end{aligned}\tag{3.1}$$

where the sum on the left side is taken over all indices of zeros $\rho_j=1/2+i\gamma_j$, $j=1,\dots,n$, of $L(s,\pi)$. Here $g_j\in C_c^\infty(\mathbb{R})$, $\Phi\in C_c^1(\mathbb{R}^n)$ is supported in $\sum_{1\le j\le n}|\xi_j|<2/m$, while h_j, $\kappa(\mathbf{h})$, and f are given in (2.6) through (2.8), respectively.

When $a=\ell$, i.e., when π is invariant under the action of $\mathrm{Gal}(E/\mathbb{Q})$, we apply the combinatorial sieving argument in [LiuYe1], §9, directly to (3.1) to deduce the following corollary. We need the notation. A set partition $\underline{H}$ of $\underline{N} = (1, \cdots, n)$ is a decomposition of $\underline{N}$ into disjoint subsets $\underline{H} = [H_1, \cdots, H_\nu]$, where $\nu = \nu(\underline{H})$ is the number of subsets in $\underline{H}$. For a given set partition $\underline{H}$, define

$$W_n^{\underline{H}}(x_1, \ldots, x_n) = \prod_{1 \le k \le \nu(\underline{H})} \det(K(x_i - x_j))_{i,j \in H_k},$$

where $K(x) = (\sin \pi x)/(\pi x)$ if $x \neq 0$, and $K(x) = 1$ if $x = 0$.

Corollary 3.2. *With notation and assumption as in Theorem 3.1, we assume that π is invariant under the action of* $\mathrm{Gal}(E/\mathbb{Q})$. *Then the n-level correlation of normalized nontrivial zeros of $L(s, \pi)$ is given by*

$$\sum_{\substack{i_1, \ldots, i_n \\ \text{distinct}}} h_1\Big(\frac{\gamma_{i_1}}{T}\Big) \cdots h_n\Big(\frac{\gamma_{i_n}}{T}\Big) f\Big(\frac{\gamma_{i_1} m \log T}{2\pi}, \ldots, \frac{\gamma_{i_n} m \log T}{2\pi}\Big)$$
$$= \frac{\kappa(\mathbf{h})}{2\pi} mT(\log T) \sum_{\substack{\underline{H} \\ \nu(\underline{H}) \le \ell}} \binom{\ell}{\nu(\underline{H})}$$
$$\times \int_{\mathbb{R}^n} f(x) W_n^{\underline{H}}(x) \delta\Big(\frac{x_1 + \cdots + x_n}{n}\Big)\, dx_1 \cdots dx_n + O(T), \tag{3.2}$$

where the sum on the left side is taken over distinct indices of zeros $\rho_{i_j} = 1/2 + i\gamma_{i_j}$, $j = 1, \ldots, n$, of $L(s, \pi)$.

The following theorem and its corollary were first proved in [LiuYe1] on the assumption of Selberg's orthogonality conjecture, which was removed in [LiuWangYe].

Theorem 3.3. *Let $\pi_1, \ldots, \pi_\ell$ be automorphic irreducible cuspidal representations of $GL_m(\mathbb{Q}_\mathbb{A})$ with unitary central character, such that $\pi_i \not\cong \pi_j$ for any $i \neq j$. Assume $m \le 4$ or Hypothesis H over $\mathbb{Q}$ for $m \ge 5$. Then*

$$\sum_{\gamma_1, \ldots, \gamma_n} h_1\Big(\frac{\gamma_1}{T}\Big) \cdots h_n\Big(\frac{\gamma_n}{T}\Big) f\Big(\frac{\gamma_1 m \log T}{2\pi}, \ldots, \frac{\gamma_n m \log T}{2\pi}\Big)$$
$$= \frac{\kappa(\mathbf{h})}{2\pi} mT(\log T) \ell^n \Phi(0, \ldots, 0)$$
$$+ \frac{\kappa(\mathbf{h})}{2\pi} mT(\log T) \sum_{1 \le r \le n/2} \frac{n! \ell^{n-r}}{2^r r! (n-2r)!} \int_{\mathbb{R}^r} |v_1| \cdots |v_r|$$
$$\times \Phi(v_1, \ldots, v_r, -v_1, \ldots, -v_r, 0, \ldots, 0) dv_1 \cdots dv_r + O(T), \tag{3.3}$$

where the sum on the left side is taken over all indices of zeros $\rho_j = 1/2 + i\gamma_j$, $j = 1, \ldots, n$, of $L(s, \pi_1) \cdots L(s, \pi_\ell)$. Here $g_j \in C_c^\infty(\mathbb{R})$, $\Phi \in C_c^1(\mathbb{R}^n)$

is supported in $\sum_{1\le j\le n}|\xi_j| < 2/m$, *while* h_j, $\kappa(\mathbf{h})$, *and* f *are given in (2.6) through (2.8), respectively.*

Corollary 3.4. *Assume* $m \le 4$ *or Hypothesis H over* $\mathbb{Q}$ *for* $m \ge 5$. *With the same notation as in Theorem 3.3, the n-level correlation of normalized nontrivial zeros of* $L(s,\pi_1)\cdots L(s,\pi_\ell)$ *is given by*

$$\sum_{\substack{i_1,\ldots,i_n\\ \text{distinct}}} h_1\Big(\frac{\gamma_{i_1}}{T}\Big)\cdots h_n\Big(\frac{\gamma_{i_n}}{T}\Big) f\Big(\frac{\gamma_{i_1} m\log T}{2\pi},\ldots,\frac{\gamma_{i_n} m\log T}{2\pi}\Big)$$
$$= \frac{\kappa(\mathbf{h})}{2\pi} mT(\log T) \sum_{\substack{\underline{H}\\ \nu(\underline{H})\le \ell}} \binom{\ell}{\nu(\underline{H})}$$
$$\times \int_{\mathbb{R}^n} f(x) W_n^{\underline{H}}(x)\delta\Big(\frac{x_1+\cdots+x_n}{n}\Big)\,dx_1\cdots dx_n + O(T), \qquad (3.4)$$

where the sum on the left side is taken over distinct indices of zeros $\rho_{i_j} = 1/2 + i\gamma_{i_j}$, $j = 1,\ldots,n$, *of* $L(s,\pi_1)\cdots L(s,\pi_\ell)$.

Comparing the right side of (3.2) and (3.4), we conclude that the n-level correlation of normalized nontrivial zeros of $L(s,\pi)$ equals the n-level correlation of normalized nontrivial zeros of a product of ℓ distinct L-functions attached to cuspidal representations $\pi_1,\ldots,\pi_\ell$ of GL_m over $\mathbb{Q}$, when π is invariant under the Galois action. This is the proof of Theorem 1.1. In the following sections we will prove Theorem 3.1.

4. *L*-functions.

Again let E be a Galois extension over $\mathbb{Q}$ of degree ℓ. Denote $\Phi(s,\pi) = L_\infty(s,\pi)L(s,\pi)$, for $\sigma > 1$. Then by a classical result of [GodJac], $\Phi(s,\pi)$ extends to an entire function with the exception of $\zeta(s)$, which has a simple pole at $s = 1$. $\Phi(s,\pi)$ also has a functional equation

$$\Phi(s,\pi) = \varepsilon(s,\pi)\Phi(1-s,\widetilde{\pi}),$$

where the automorphic irreducible cuspidal representation $\widetilde{\pi}$ is contragredient to π, and $\varepsilon(s,\pi) = \tau(\pi)Q_\pi^{-s}$. Here $Q_\pi > 0$ is the conductor of π ([JacPSShal1]), $\tau(\pi) \in \mathbb{C}^\times$, $Q_{\widetilde{\pi}} = Q_\pi$, and $\tau(\pi)\tau(\widetilde{\pi}) = Q_\pi$.

Let

$$a_\pi(p^{kf_p}) = f_p \sum_{v|p}\sum_{1\le j\le m} \alpha_\pi^k(j,v), \qquad (4.1)$$

$a_\pi(p^k) = 0$ if $f_p \nmid k$, and $c_\pi(n) = \Lambda(n)a_\pi(n)$, where the von Mangoldt function $\Lambda(n)$ is defined by $\Lambda(n) = \log p$ if $n = p^k$ and zero otherwise.

Then $a_{\tilde{\pi}}(p^k) = \overline{a}_\pi(p^k)$. Then for $\sigma > 1$, we have

$$\frac{L'}{L}(s,\pi) = -\sum_{n=1}^{\infty} \frac{\Lambda(n)a_\pi(n)}{n^s}.$$

By the bound in (2.1) we have

$$|a_\pi(p^{kf_p})| \le m\ell p^{kf_p((1/2)-1/(\ell m^2+1))} \tag{4.2}$$

for any π_v, ramified or unramified. If π_v is unramified, we have sharper bounds based on (2.3) and (2.4):

$$|a_\pi(p^{kf_p})| \le m\ell p^{kf_p/9}$$

for $m = 2$, and

$$|a_\pi(p^{kf_p})| \le m\ell p^{kf_p(1/2-1/(m^2+1))} \tag{4.3}$$

for $m \ge 3$.

We will need an explicit formula for the L-functions of smooth type as in [RudSar]. Let g_j be a compactly supported smooth function on $\mathbb{R}$. Define $h_j(r)$ and $\kappa(\mathbf{h})$ as in (2.6) and (2.7). Let $\rho = 1/2 + i\gamma$ be a nontrivial zero of the L-function $L(s,\pi)$. By applying the same arguments step by step as in [RudSar], we prove that

$$\begin{aligned}\sum_\gamma h_j(\gamma) =& \delta(\pi)\Big(h_j\Big(-\frac{i}{2}\Big) + h_j\Big(\frac{i}{2}\Big)\Big) + \frac{1}{2\pi}\int_{\mathbb{R}} h_j(r)\Omega_\pi(r)dr \\ &- \sum_{n=1}^{\infty}\Big(\frac{c_\pi(n)}{\sqrt{n}}g_j(\log n) + \frac{\overline{c}_\pi(n)}{\sqrt{n}}g_j(-\log n)\Big),\end{aligned} \tag{4.4}$$

where the sum on the left side is taken over nontrivial zeros $\rho = 1/2 + i\gamma$ of $L(s,\pi)$, and $\delta(\pi)$ equals 1 if the L-function is $\zeta(s)$, and zero otherwise. Here

$$\begin{aligned}\Omega_\pi(t) =& \log Q_\pi + \sum_{v|\infty}\sum_{1\le j\le m}\Big(\frac{\Gamma'_{\mathbb{R}}}{\Gamma_{\mathbb{R}}}\Big(\frac{1}{2} + \mu_\pi(j,v) + it\Big) \\ &+ \frac{\Gamma'_{\mathbb{R}}}{\Gamma_{\mathbb{R}}}\Big(\frac{1}{2} + \overline{\mu}_\pi(j,v) - it\Big)\Big) \quad \text{if } E\otimes_{\mathbb{Q}}\mathbb{R} = \textstyle\bigoplus_{v|\infty}\mathbb{R}, \\ =& \log Q_\pi + \sum_{v|\infty}\sum_{1\le j\le m}\Big(\frac{\Gamma'_{\mathbb{C}}}{\Gamma_{\mathbb{C}}}\Big(\frac{1}{2} + \mu_\pi(j,v) + it\Big) \\ &+ \frac{\Gamma'_{\mathbb{C}}}{\Gamma_{\mathbb{C}}}\Big(\frac{1}{2} + \overline{\mu}_\pi(j,v) - it\Big)\Big) \quad \text{if } E\otimes_{\mathbb{Q}}\mathbb{R} = \textstyle\bigoplus_{v|\infty}\mathbb{C}.\end{aligned} \tag{4.5}$$

Denote $L = m\log T$. The explicit formula (4.4) can be rewritten as

$$\begin{aligned}\sum_{\gamma} h_j\Big(\frac{\gamma}{T}\Big)e^{-iL\gamma\xi_j} \\ = \delta(\pi)\Big(h_j\Big(-\frac{i}{2T}\Big)T^{-m\xi_j/2} + h_j\Big(\frac{i}{2T}\Big)T^{m\xi_j/2}\Big) \\ + Tg_{jT}(TL\xi_j) + TS_j^+(\xi_j) + TS_j^-(\xi_j),\end{aligned} \tag{4.6}$$

where

$$g_{jT}(x) = \frac{1}{2\pi}\int_{\mathbb{R}} h_j(r)\Omega_\pi(rT)e^{-irx}dr,$$

$$S_j^+(\xi_j) = -\sum_{n_j=1}^{\infty}\frac{\Lambda(n_j)a_\pi(n_j)}{\sqrt{n_j}}g_j(T(L\xi_j + \log n_j)), \tag{4.7}$$

$$S_j^-(\xi_j) = -\sum_{n_j=1}^{\infty}\frac{\Lambda(n_j)\overline{a}_\pi(n_j)}{\sqrt{n_j}}g_j(T(L\xi_j - \log n_j)). \tag{4.8}$$

In the following, we will not consider the term with $\delta(\pi)$ on the right side of (4.6), as it is non-zero only for the Riemann zeta function $\zeta(s)$.

5. Rankin-Selberg L-functions.

For two irreducible automorphic cuspidal representations π and π' of $GL_m(E_{\mathbb{A}})$ and $GL_{m'}(E_{\mathbb{A}})$, respectively, their Rankin-Selberg L-function is given by

$$L(s, \pi\times\tilde{\pi}') = \prod_{v<\infty} L_v(s, \pi\times\tilde{\pi}')$$

for $\sigma > 1$, where the local factor is

$$L_v(s, \pi\times\tilde{\pi}') = \prod_{\substack{1\le j_1\le m \\ 1\le j_2\le m'}}\Big(1 - \frac{\alpha_\pi(j_1, v)\overline{\alpha}_{\pi'}(j_2, v)}{q_{E_v}^s}\Big)^{-1}.$$

Denote

$$\begin{aligned}L_p(s, \pi\times\tilde{\pi}') &= \prod_{v|p} L_v(s, \pi\times\tilde{\pi}') \\ &= \prod_{v|p}\prod_{\substack{1\le j_1\le m \\ 1\le j_2\le m'}}\Big(1 - \frac{\alpha_\pi(j_1, v)\overline{\alpha}_{\pi'}(j_2, v)}{p^{f_p s}}\Big)^{-1}.\end{aligned}$$

Then for $\sigma > 1$,

$$\begin{aligned}\frac{L'}{L}(s, \pi \times \tilde{\pi}') &= -\sum_{v} \sum_{\substack{1\le j_1 \le m\\ 1\le j_2\le m'}} \sum_{k\ge 1} \frac{f_p \log p}{p^{kf_p s}} \alpha_\pi^k(j_1, v)\overline{\alpha}_{\pi'}^k(j_2, v)\\ &= -\sum_{n\ge 1} \frac{\Lambda(n) a_{\pi\times\tilde{\pi}'}(n)}{n^s}, \end{aligned} \tag{5.1}$$

where

$$\begin{aligned} a_{\pi\times\tilde{\pi}'}(p^{kf_p}) &= \sum_{v|p} f_p \sum_{\substack{1\le j_1 \le m\\ 1\le j_2\le m'}} \alpha_\pi^k(j_1, v)\overline{\alpha}_{\pi'}^k(j_2, v)\\ &= \sum_{v|p} f_p \Big(\sum_{1\le j_1\le m} \alpha_\pi^k(j_1, v)\Big)\Big(\sum_{1\le j_2\le m'} \overline{\alpha}_{\pi'}^k(j_2, v)\Big), \end{aligned} \tag{5.2}$$

and $a_{\pi\times\tilde{\pi}'}(p^k) = 0$ if $f_p \nmid k$. In particular, when $\pi \cong \pi'$,

$$a_{\pi\times\tilde{\pi}}(p^{kf_p}) = \sum_{v|p} f_p \Big| \sum_{1\le j\le m} \alpha_\pi^k(j, v)\Big|^2. \tag{5.3}$$

Obviously

$$\big|a_{\pi\times\tilde{\pi}'}(n)\big| \le \big|a_{\pi\times\tilde{\pi}}(n)\big| + \big|a_{\pi'\times\tilde{\pi}'}(n)\big|. \tag{5.4}$$

By the bound for $|\alpha_\pi(j, v)|$ in (2.1), we have

$$\big|a_{\pi\times\tilde{\pi}}(p^{kf_p})\big| \le m^2 \ell p^{kf_p(1-2/(\ell m^2+1))}.$$

Let $S_{\pi\times\tilde{\pi}'}$ be the finite set of primes p such that there is $v|p$ with either π_v or π'_v being ramified. Therefore for any $p \in S_{\pi\times\tilde{\pi}} = S_\pi$ we have

$$\sum_{k\ge 1} \frac{\Lambda(p^k)\big|a_{\pi\times\tilde{\pi}}(p^k)\big|}{p^k} \le m^2 \ell \log p \sum_{k\ge 1} \frac{1}{p^{2k/(\ell m^2+1)}} < \infty,$$

and hence

$$\sum_{p\in S_\pi} \sum_{k\ge 1} \frac{\Lambda(p^k)\big|a_{\pi\times\tilde{\pi}}(p^k)\big|}{p^k} < \infty. \tag{5.5}$$

On the other hand, absolute convergence of (5.1) for $\sigma > 1$, together with (5.5), implies that

$$\sum_{n\le X} \Lambda(n) \Big|a_{\pi\times\tilde{\pi}}(n)\Big| \ll_\varepsilon X^{1+\varepsilon}.$$

By partial summation, we also have

$$\sum_{n\le X} \left|c_\pi(n)\right|^2 \ll_\varepsilon X^{1+\varepsilon}, \tag{5.6}$$

where $c_\pi(n) = \Lambda(n)a_\pi(n)$ and $a_\pi(n)$ is given in (4.1).

Recall that when $\pi' \cong \pi \otimes |\det|^{i\tau_0}$ for some $\tau_0 \in \mathbb{R}$, $L(s, \pi \times \tilde{\pi})$ has simple poles at $s = 1 + i\tau_0$ and $i\tau_0$. ([JacPSShal2] and [MoeWal]). Otherwise $L(s, \pi \times \tilde{\pi})$ is entire. Note that the Archimedean part of the Rankin-Selberg L-function is

$$L_\infty(s, \pi \times \tilde{\pi}') = \prod_{v|\infty} \prod_{1\le j\le m} \prod_{1\le k\le m'} \Gamma_{\mathbb{R}}\big(s + \mu_{\pi\times\tilde{\pi}'}(j,k;v)\big)$$

or

$$L_\infty(s, \pi \times \tilde{\pi}') = \prod_{v|\infty} \prod_{1\le j\le m} \prod_{1\le k\le m'} \Gamma_{\mathbb{C}}\big(s + \mu_{\pi\times\tilde{\pi}'}(j,k;v)\big).$$

We will need a trivial bound $\operatorname{Re} \mu_{\pi\times\tilde{\pi}'}(j,k;v) > -1$.

Denote by $\Phi(s, \pi\times\tilde{\pi}') = L_\infty(s, \pi\times\tilde{\pi}')L(s, \pi\times\tilde{\pi}')$ the complete Rankin-Selberg L-function. Then by a classical result of Shahidi ([Shah1], [Shah2], [Shah3], and [Shah4]), $\Phi(s, \pi)$ has a functional equation

$$\Phi(s, \pi \times \tilde{\pi}') = \varepsilon(s, \pi \times \tilde{\pi}')\Phi(1 - s, \tilde{\pi} \times \pi'), \tag{5.7}$$

where $\varepsilon(s, \pi) = \tau(\pi \times \tilde{\pi}')Q_{\pi\times\tilde{\pi}'}^{-s}$. Here $Q_{\pi\times\tilde{\pi}'} > 0$ is the conductor.

6. Orthogonality.

When π and π' are cuspidal representations of $GL_m(\mathbb{Q}_{\mathbb{A}})$ and $GL_{m'}(\mathbb{Q}_{\mathbb{A}})$, respectively, with $\pi \not\cong \pi'$, Liu, Wang, and Ye [LiuWangYe] proved the following Selberg orthogonality

$$\sum_{n\le x} \frac{(\log n)\Lambda(n)a_\pi(n)\bar{a}_{\pi'}(n)}{n} \ll \log x, \tag{6.1}$$

if at least one of π and π' is self-contragredient. In [LiuWangYe], (6.1) was proved as a consequence of a stronger prime number theorem with weights for a Rankin-Selberg L-function, and hence a zero-free region of the classical type was required (cf. [Mor1], [Mor2], [GelLapSar], and [Sar]). This is the reason why we have to assume that at least one of π and π' is self-contragredient in (6.1).

In this section, we will use the approach in [LiuYe2] to get a weighted version of (6.1) for cuspidal representations π and π' over E, avoiding the use of the zero-free region and the self-contragredient assumption. Then we will apply an argument of Landau [Land] to remove the weight.

Lemma 6.1. *Let π and π' be irreducible automorphic cuspidal representations of $GL_m(E_{\mathbb{A}})$ and $GL_{m'}(E_{\mathbb{A}})$ with unitary central characters, respectively. Then*

$$\sum_{n\le x}\Big(1-\frac{n}{x}\Big)\frac{(\log n)\Lambda(n)a_{\pi\times\tilde{\pi}'}(n)}{n}$$
$$=\frac{1}{2}\log^2 x+O(\log x)\quad \text{if } \pi'\cong\pi; \tag{6.2}$$
$$=O(\log x)\quad \text{if } \pi'\not\cong\pi. \tag{6.3}$$

Proof. The proof closely follows [LiuYe2], and hence we will only give a brief sketch here and point out the difference. Let $\lambda(s)=\min_{n\le 0}|s-n|$. Denote by $\mathbb{C}(m,m')$ the region in the complex plane with the following discs removed:

$$|s-2n+\mu_{\pi\times\tilde{\pi}'}(j,k;v)|<\frac{1}{8mm'\ell},\ n\le 0,\ 1\le j\le m,\ 1\le k\le m',\ v|\infty,$$

if v is real, and

$$|s-n+\mu_{\pi\times\tilde{\pi}'}(j,k;v)|<\frac{1}{8mm'\ell},\ n\le 0,\ 1\le j\le m,\ 1\le k\le m',\ v|\infty,$$

if v is complex. Then for $s\in\mathbb{C}(m,m')$ and all j,k, and $v|\infty$,

$$\lambda\Big(\frac{s+\mu_{\pi\times\tilde{\pi}'}(j,k;v)}{2}\Big)\ge\frac{1}{16mm'\ell}$$

if v is real, and

$$\lambda\big(s+\mu_{\pi\times\tilde{\pi}'}(j,k;v)\big)\ge\frac{1}{16mm'\ell}$$

if v is complex. Let $\beta(j,k;v)$ for j,k in the above range be the fractional part of Re $\mu_{\pi\times\tilde{\pi}'}(j,k;v)$. In addition we let $\beta(0,0;v)=0$ and $\beta(m+1,m'+1;v)=1$. Then all $\beta(j,k;v)\in[0,1]$, and hence there exist $\beta(j_1,k_1;v_1),\beta(j_2,k_2;v_2)$ such that $\beta(j_2,k_2;v_2)-\beta(j_1,k_1;v_1)\ge 1/(3mm'\ell)$ and there is no $\beta(j,k;v)$ lying between $\beta(j_1,k_1;v_1)$ and $\beta(j_2,k_2;v_2)$. It follows that the strip

$$S_0=\{s\,|\,\beta(j_1,k_1;v_1)+1/(8mm'\ell)\le\sigma\le\beta(j_2,k_2;v_2)-1/(8mm'\ell)\}$$

is contained in $\mathbb{C}(m,m')$. Consequently, for all $n=0,-1,-2,\dots,$ the strips

$$S_n=\left\{s\ \middle|\ \begin{array}{l} n+\beta(j_1,k_1;v_1)+1/(8mm'\ell)\\ \quad\le\sigma\le n+\beta(j_2,k_2;v_2)-1/(8mm'\ell)\end{array}\right\} \tag{6.4}$$

are subsets of $\mathbb{C}(m, m')$.

Differentiating (5.1), we get

$$\left(\frac{L'}{L}\right)'(s, \pi \times \tilde{\pi}') = \sum_{n \geq 1} \frac{(\log n)\Lambda(n)a_{\pi \times \tilde{\pi}'}(n)}{n^s}$$

for $\sigma > 1$. By the same method of proof as in [LiuYe2], §4, using the fact that the Rankin-Selberg L-function is of order one away from its possible poles ([GelShah]), we have the following estimates: For $|T| > 2$ there exists τ with $T \leq \tau \leq T + 1$ such that when $-2 \leq \sigma \leq 2$

$$\left(\frac{L'}{L}\right)'(\sigma + i\tau, \pi \times \tilde{\pi}') \ll \log^3(Q_{\pi \times \tilde{\pi}'}|\tau|); \tag{6.5}$$

if s is in some strip S_n as in (6.4) with $n \leq -2$, then

$$K(s) := \left(\frac{L'}{L}\right)' L(s, \pi \times \tilde{\pi}') \ll_m 1. \tag{6.6}$$

As in [LiuYe2], §5,

$$\sum_{n \leq x} \left(1 - \frac{n}{x}\right) \frac{(\log n)\Lambda(n)a_{\pi \times \tilde{\pi}'}(n)}{n}$$
$$= \frac{1}{2\pi i} \int_{(1)} K(s+1) \frac{x^s}{s(s+1)}\, ds = \frac{1}{2\pi i} \left(\int_{1-iT}^{1+iT} + \int_{1-i\infty}^{1-iT} + \int_{1+iT}^{1+i\infty} \right). \tag{6.7}$$

The last two integrals are bounded by $\ll \int_T^\infty (x/t^2)\, dt \ll x/T$. To compute the first integral on the right side of (6.7), we choose σ_0 with $-2 < \sigma_0 < -1$ such that the line $\sigma = \sigma_0$ is contained in the strip $S_{-2} \subset \mathbb{C}(m, m')$. Let $T \sim x$ and choose τ with $T \leq \tau \leq T + 1$ such that (6.5) holds. Denote this τ simply by T. Now we consider the contour

$$\begin{aligned} C_1 &: \ 1 \geq \sigma \geq \sigma_0, \quad t = -T; \\ C_2 &: \ \sigma = \sigma_0, \quad -T \leq t \leq T; \\ C_3 &: \ \sigma_0 \leq \sigma \leq 1, \quad t = T. \end{aligned}$$

Note that the two possible poles, some trivial zeros, and certain nontrivial zeros of $L(s+1, \pi \times \tilde{\pi}')$, as well as $s = 0, -1$ are passed by the shift of the contour. The trivial zeros can be determined by the functional equation (5.7): $s = -1 - \mu_{\pi \times \tilde{\pi}'}(j, k; v)$. The trivial bound Re $\mu_{\pi \times \tilde{\pi}'}(j, k; v) > -1$ will give us $O(\log x)$ for the size of the residues from trivial zeros. Integrals along C_1 and C_3 are bounded by $O\{x \log^3(Q_{\pi \times \tilde{\pi}'}T)/T^2\}$ using

(6.5), while the integral on C_2 is bounded by $O(1/x)$ using (6.6). The residues at $s = 0, -1$ will contribute $K(1) + K(0)/x$. The two possible poles are $i\tau_0$ and $i\tau_0 - 1$ which can only happen when $m = m'$ and $\pi' \cong \pi \otimes |\det|^{i\tau_0}$ for some $\tau_0 \in \mathbb{R}$. If $\tau_0 \neq 0$, these are double poles and they will contribute $O(\log x)$ to (6.7). If $\tau_0 = 0$, i.e., if $\pi \cong \pi'$, then $s = 0$ becomes a triple pole. Its residue contributes $\frac{1}{2}\log^2 x + O(\log x)$ to (6.7). By the fact of $L(s, \pi \times \tilde{\pi}') \neq 0$ for $\sigma \geq 1$ ([Shah1]), the contribution from nontrivial zeros is also $O(\log x)$. This completes the proof of Lemma 6.1. □

Lemma 6.2. *Let π be an irreducible automorphic cuspidal representation of $GL_m(E_{\mathbb{A}})$ with unitary central character. Then*

$$\sum_{n \leq x} \frac{(\log n)\Lambda(n)a_{\pi\times\tilde{\pi}}(n)}{n} = \frac{1}{2}\log^2 x + O(\log x). \tag{6.8}$$

Proof. Note that this is the case when $\pi \cong \pi'$, and hence by (5.3) the series on the left side of (6.2) and (6.8) consists of non-negative terms. By the method of proof in [RudSar], p.282, we can remove the weight $1 - n/x$ from (6.2) when $\pi \cong \pi'$. □

Lemma 6.3. *Let π be an irreducible automorphic cuspidal representation of $GL_m(E_{\mathbb{A}})$ with unitary central character. Then*

$$\sum_{n \leq x} (\log n)\Lambda(n)a_{\pi\times\tilde{\pi}}(n) \ll x\log x. \tag{6.9}$$

Proof. This is deduced from (6.8) by a standard argument of partial summation. □

We remark that under the generalized Ramanujan conjecture, an asymptotic formula was proved in [LiuYe3] for cuspidal self-contragredient representation of GL_m over $\mathbb{Q}$:

$$\sum_{n \leq x} (\log n)\Lambda(n)a_{\pi\times\tilde{\pi}}(n) = x\log x - x + O(x\exp(-c\sqrt{\log x}))$$

for a positive constant c. The upper bound in (6.9) is nevertheless unconditional and valid for cuspidal representations not necessarily self-contragredient.

Theorem 6.4. *Let π and π' be irreducible automorphic cuspidal representations of $GL_m(E_{\mathbb{A}})$ and $GL_{m'}(E_{\mathbb{A}})$ with unitary central characters, respectively, such that $\pi' \ncong \pi$. Then*

$$\sum_{n \leq x} \frac{(\log n)\Lambda(n)a_{\pi\times\tilde{\pi}'}(n)}{n} \ll \log x. \tag{6.10}$$

Proof. The removal of the weight $1-n/x$ from (6.3) is by an argument of Landau [Land] using (6.9). The proof is given in [LiuWangYe]. To make the present paper self-contained, we reproduce the proof in [LiuWangYe] below.

Denote $c(n) = (\log n)\Lambda(n)a_{\pi\times\tilde{\pi}'}(n)/n$, and

$$C(x) = \sum_{n\le x} c(n), \qquad D(x) = \sum_{n\le x}\Big(1-\frac{n}{x}\Big)c(n).$$

Then

$$\int_0^x C(t)dt = \sum_{n\le x}(x-n)c(n) = xD(x).$$

We begin with an observation that

$$\begin{aligned}\int_x^{x+v} C(t)dt &= (x+v)D(x+v) - xD(x)\\ &= vD(x+v) + x(D(x+v)-D(v)),\end{aligned} \tag{6.11}$$

where $v=\sqrt{x}$.

By (6.3), we have

$$vD(x+v) \ll v\log x \tag{6.12}$$

as $\pi' \not\cong \pi$. The last term in (6.11) is

$$\begin{aligned}&x(D(x+v)-D(v))\\ &= x\sum_{n\le x}\Big(\frac{n}{x}-\frac{n}{x+v}\Big)c(n) + x\sum_{x<n\le x+v}\Big(1-\frac{n}{x+v}\Big)c(n)\\ &=: E_1+E_2,\end{aligned}$$

say. By (5.4) and Lemma 6.3 we have

$$\begin{aligned}|E_1| &\le \frac{v}{x+v}\sum_{n\le x} n|c(n)|\\ &\ll \frac{v}{x+v}\sum_{n\le x}(\log n)\Lambda(n)\big(|a_{\pi\times\tilde{\pi}}(n)| + |a_{\pi'\times\tilde{\pi}'}(n)|\big)\\ &\ll v\log x,\end{aligned} \tag{6.13}$$

and

$$\begin{aligned}|E_2| &\le \frac{xv}{x+v}\sum_{x<n\le x+v}|c(n)| \ll \frac{v}{x+v}\sum_{x<n\le x+v} n|c(n)|\\ &\ll \frac{v}{x+v}\sum_{x<n\le x+v}(\log n)\Lambda(n)\big(|a_{\pi\times\tilde{\pi}}(n)| + |a_{\pi'\times\tilde{\pi}'}(n)|\big)\\ &\ll v\log x.\end{aligned} \tag{6.14}$$

Putting (6.12) through (6.14) into (6.11), we get

$$\int_x^{x+v} C(t)dt \ll v \log x. \tag{6.15}$$

Now consider the difference

$$\begin{aligned}\Big|\int_x^{x+v} C(t)dt - vC(x)\Big| &= \Big|\int_x^{x+v} (C(t) - C(x))dt\Big| \\ &\ll v \sum_{x<n\le x+v} |c(n)| \ll \frac{v}{x} \sum_{x<n\le x+v} n|c(n)| \\ &\ll \frac{v}{x}\sum_{n\le x}(\log n)\Lambda(n)\big(|a_{\pi\times\tilde{\pi}}(n)| + |a_{\pi'\times\tilde{\pi}'}(n)|\big) \ll v\log x,\end{aligned}$$

by the same argument as above. The desired result (6.10) now follows from this and (6.15). □

Under the generalized Ramanujan conjecture, a much precise version of (6.8) and (6.10) was proved in [LiuYe3] for self-contragredient cuspidal representations over $\mathbb{Q}$:

$$\begin{aligned}&\sum_{n\le x} \frac{(\log n)\Lambda(n)a_{\pi\times\tilde{\pi}'}(n)}{n} \\ &= \frac{1}{2}\log^2 x + c_1 + O\{\exp(-c\sqrt{\log x})\} \quad \text{if } \pi' \cong \pi; \\ &= \frac{x^{i\tau_0}}{i\tau_0}\log x + \frac{x^{i\tau_0}-1}{\tau_0^2} + c_2 + O\{\exp(-c\sqrt{\log x})\} \\ &\qquad\qquad \text{if } \pi' \cong \pi\otimes\alpha^{i\tau_0} \text{ for some } \tau_0 \in \mathbb{R}^\times; \\ &= c_3 + O\{\exp(-c\sqrt{\log x})\} \quad \text{if } \pi' \not\cong \pi\otimes\alpha^{it} \text{ for any } t\in\mathbb{R},\end{aligned}$$

where $c, c_1, ..., c_3$ are positive constants.

7. Orthogonality over primes.

In this section we will rewrite Lemmas 6.2 and 6.4 as orthogonality relation over primes using Conjecture 2.1 and Hypothesis H. First, according to (6.8) we have

$$\sum_{\substack{p,k\\ p^{kf_p}\le x}} \frac{k\log^2 p}{p^{kf_p}} \sum_{v_i|p} f_p^2 \Big|\sum_{1\le j\le m} \alpha_\pi^k(j, v_i)\Big|^2 = \frac{1}{2}\log^2 x + O(\log x). \tag{7.1}$$

Using the bound for $\alpha_\pi(j, v_i)$ in (2.1), we have

$$\sum_{k>(\ell m^2+1)/2} \sum_p \frac{k \log^2 p}{p^{k f_p}} \sum_{v|p} f_p^2 \Big| \sum_{1\le j\le m} \alpha_\pi^k(j,v)\Big|^2$$

$$\ll \sum_{k>(\ell m^2+1)/2} \sum_p \frac{k \log^2 p}{p^{2k/(\ell m^2+1)}} < \infty.$$

Therefore Hypothesis H implies

$$\sum_{k\ge 2} \sum_p \frac{k \log^2 p}{p^{k f_p}} \sum_{v|p} f_p^2 \Big| \sum_{1\le j\le m} \alpha_\pi^k(j,v)\Big|^2 < \infty. \tag{7.2}$$

From (7.1) and (7.2) we conclude that

$$\sum_{p\le x} \frac{\log^2 p}{p^{f_p}} \sum_{v|p} f_p^2 \Big| \sum_{1\le j\le m} \alpha_\pi(j,v)\Big|^2 = \frac{1}{2}\log^2 x + O(\log x). \tag{7.3}$$

What we will show below is that the outer sum in (7.3) can be taken over all primes $p \le x$ which split completely in E.

Theorem 7.1. *Let π be an automorphic irreducible cuspidal representation π of $GL_m(E_{\mathbb{A}})$ with unitary central character. Assume Hypothesis H over E when $m \ge 5$. Assume Conjecture 2.1 when $m \ge 3$ and that there is $p|\ell$ such that $p \le (m^2+1)/2$. Then*

$$\sum_{\substack{p\le x \\ \text{splits completely}}} \frac{\log^2 p}{p} \sum_{v|p} \Big| \sum_{1\le j\le m} \alpha_\pi(j,v)\Big|^2 = \frac{1}{2}\log^2 x + O(\log x). \tag{7.4}$$

Proof. We know that there are only finitely many p with its order of ramification $e_p > 1$. Thus we can ignore these primes in (7.3) and consider only unramified primes p with $f_p = \ell_p$. If such a prime p does not split completely in E, then $f_p \ge 2$. Under Conjecture 2.1, we then have

$$\sum_{\substack{p \text{ unramified,} \\ \text{not split completely}}} \frac{\log^2 p}{p^{f_p}} \sum_{v|p} f_p^2 \Big| \sum_{1\le j\le m} \alpha_\pi(j,v)\Big|^2$$

$$\ll \sum_{\substack{p \text{ unramified,} \\ \text{not split completely}}} \frac{\log^2 p}{p^{f_p}} p^{2f_p(1/2-1/(2f_p)-\varepsilon)}$$

$$\ll \sum_p \frac{\log^2 p}{p^{1+2\varepsilon f_p}} \ll \sum_p \frac{\log^2 p}{p^{1+2\varepsilon}} < \infty.$$

Consequently (7.4) holds. □

As in [RudSar], p.300, we can apply partial summation to (7.4) and prove an asymptotic formula for a weighted sum.

Lemma 7.2. *Let $\phi(v)$ be a C^1 function supported on $|v| \le (1-\delta)/m$ for some positive δ. Let π be an irreducible automorphic cuspidal representations of $GL_m(E_{\mathbb{A}})$ with unitary central character. Assume Hypothesis H over E when $m \ge 5$. Assume Conjecture 2.1 when $m \ge 3$ and that there is $p|\ell$ such that $p \le (m^2+1)/2$. Then we have*

$$\sum_{\substack{p\le x \\ \text{splits completely}}} \frac{\log^2 p}{p}\phi\Big(\frac{\log p}{m\log x}\Big)\sum_{v_i|p}\Big|\sum_{1\le j\le m}\alpha_\pi(j,v_i)\Big|^2$$

$$= m^2(\log x)^2\int_0^{1/m} v\phi(v)\ dv + O(\log x),$$

where the implied constant depends on m and δ.

Now let us turn to the case of $\pi \not\cong \pi'$. By Cauchy's inequality, we may remove terms on prime powers from the left side of (6.10). Using the same proof as for Lemma 7.1, we can further remove terms on those primes which do not split completely in E.

Lemma 7.3. *Let π and π' be irreducible automorphic cuspidal representations of $GL_m(E_{\mathbb{A}})$ and $GL_{m'}(E_{\mathbb{A}})$ with unitary central characters, respectively, such that $\pi' \not\cong \pi$. Assume Hypothesis H over E when $m \ge 5$ or $m' \ge 5$. Also assume Conjecture 2.1 when $m \ge 3$ and that there is $p|\ell$ such that $p \le (m^2+1)/2$, or when $m' \ge 3$ and that there is $p|\ell$ such that $p \le (m'^2+1)/2$. Then*

$$\sum_{\substack{p\le x \\ \text{splits completely}}} \frac{\log^2 p}{p}\sum_{v_i|p}\Big(\sum_{1\le j_1\le m}\alpha_\pi(j_1,v_i)\Big)\Big(\sum_{1\le j_2\le m'}\overline{\alpha}_{\pi'}(j_2,v_i)\Big)$$

$$\ll \log x. \tag{7.5}$$

Partial summation now can be applied to (7.5) and the following weighted sum can be obtained as in the proof of Proposition 4.5 in [LiuYe1], p.436.

Lemma 7.4. *Let $\phi(v)$ be a C^1 function supported on $|v| \le (1-\delta)/m$ for some positive δ. Let π and π' be irreducible automorphic cuspidal* *representations of $GL_m(E_{\mathbb{A}})$ and $GL_{m'}(E_{\mathbb{A}})$ with unitary central characters, respectively, such that $\pi' \not\cong \pi$. Then under the same assumption as in Lemma 7.3, we have*

$$\sum_{\substack{p\le x\\ \text{splits completely}}} \frac{\log^2 p}{p}\phi\Big(\frac{\log p}{m\log x}\Big)\sum_{v_i|p}\Big(\sum_{1\le j_1\le m}\alpha_\pi(j_1,v_i)\Big)\Big(\sum_{1\le j_2\le m}\overline{\alpha}_{\pi'}(j_2,v_i)\Big)$$

$$\ll (\log x)\int_0^{1/m}\big(|\phi(v)|+|\phi'(v)|\big)dv,$$

where the implied constant depends on m and δ.

8. The n-level correlation.

Let $\Phi\in C^1(\mathbb{R}^n)$ be supported in the region $\sum_{1\le j\le n}|\xi_j|<2/m$. Define $f(x)$ as in (2.8). We will compute the left side of (3.1):

$$C_n(f,\mathbf{h},T)=\sum_{\gamma_1,\cdots,\gamma_n} h_1\Big(\frac{\gamma_1}{T}\Big)\cdots h_n\Big(\frac{\gamma_n}{T}\Big)f\Big(\frac{L}{2\pi}\gamma_1,\dots,\frac{L}{2\pi}\gamma_n\Big),$$

where $L=m\log T$ and the sum over each γ_j $(1\le j\le n)$ is taken over all non-trivial zeros $\rho_j=1/2+i\gamma_j$ of $L(s,\pi)$. Now we use the Fourier transform and get

$$\begin{aligned}C_n(f,\mathbf{h},T)=&\int_{\mathbb{R}^n}\prod_{j=1}^n\Big(\sum_{\gamma_j}h_j\Big(\frac{\gamma_j}{T}\Big)e^{-iL\gamma_j\xi_j}\Big)\Phi(\xi)\delta(\xi_1+\cdots+\xi_n)d\xi\\ =&T^n\int_{\mathbb{R}^n}\prod_{j=1}^n\Big(g_{jT}(TL\xi_j)+S_j^+(\xi_j)+S_j^-(\xi_j)\Big)\\ &\times\Phi(\xi)\delta(\xi_1+\cdots+\xi_n)d\xi.\end{aligned}$$

To compute the product, we set $i_\mu=0$ or ±1 for $\mu=1,\dots,n$, and use i_μ to indicate which one of $g_{\mu T}, S_\mu^+, S_\mu^-$ appears in the term:

$$\begin{aligned}C_n(f,\mathbf{h},T)=&T^n\sum_{-1\le i_1,\dots,i_n\le1}\int_{\mathbb{R}^n}\prod_{i_\mu=0}g_{\mu T}(TL\xi_\mu)\prod_{i_\mu=1}S_\mu^+(\xi_\mu)\prod_{i_\mu=-1}S_\mu^-(\xi_\mu)\\ &\times\Phi(\xi)\delta(\xi_1+\cdots+\xi_n)d\xi.\end{aligned}$$

Now we use (4.7) and (4.8) to expand $S_\mu^\pm(\xi_\mu)$. Recall that $\Lambda(n_\mu)a_\pi(n_\mu)$ $=c_\pi(n_\mu)$. We have

$$C_n(f,\mathbf{h},T)=\sum_{-1\le i_1,\dots,i_n\le1}(-1)^{i_1+\cdots+i_n}C_{i_1\cdots i_n}(T) \tag{8.1}$$

and

$$C_{i_1\cdots i_n}(T) = \sum_{\substack{n_\mu\geq 1\\ \text{for } i_\mu\neq 0}} \Big(\prod_{i_\mu=1} \frac{c_\pi(n_\mu)}{n_\mu^{1/2}}\Big)\Big(\prod_{i_\mu=-1} \frac{\overline{c_\pi}(n_\mu)}{n_\mu^{1/2}}\Big) A_{i_1\cdots i_n}(\mathbf{n},T).$$

Here $\mathbf{n} = (n_1, \cdots, n_n)$ where $n_\mu = 1$ if $i_\mu = 0$, and

$$\begin{aligned} &A_{i_1\cdots i_n}(\mathbf{n},T)\\ &= T^n \int_{\mathbb{R}^n} \prod_{i_\mu=0} g_{\mu T}(TL\xi_\mu) \prod_{i_\mu\neq 0} g_\mu\Big(T(L\xi_\mu + i_\mu \log n_\mu)\Big)\\ &\quad\times \Phi(\xi)\delta(\xi_1+\cdots+\xi_n)d\xi. \end{aligned} \tag{8.2}$$

We will follow [RudSar] and [LiuYe1] closely to estimate $C_n(f,\mathbf{h},T)$ and point out differences between the situation at present and those discussed there. First, in the definition of $\Omega_\pi(r)$ and hence of $g_T(x)$, there are sums of $\Gamma'_{\mathbb{C}}/\Gamma_{\mathbb{C}}$ or possibly more terms of $\Gamma'_{\mathbb{R}}/\Gamma_{\mathbb{R}}$. These will cause no problem, since either over $\mathbb{C}$ or over $\mathbb{R}$, Re $\mu_\pi(j,v) > -1/2$ by [RudSar]. By Stirling's formula, we thus have

$$\frac{\Gamma'_{\mathbb{C}}}{\Gamma_{\mathbb{C}}}(s) = 2\frac{\Gamma'_{\mathbb{R}}}{\Gamma_{\mathbb{R}}}(s) + O\Big(\frac{1}{s}\Big). \tag{8.3}$$

Therefore, Lemma 4.1 in [RudSar] is still true here. That is,

$$g_T(x) \ll \frac{e^{-\sigma x/T}}{x}(1-e^{-\sigma x/2}) + e^{-\sigma x}\log T,$$

$$g_T(x) \ll \begin{cases} \log T & \text{if } |x| \ll \log\log T,\\ 1/|x| & \text{if } |x| \gg \log\log T, \end{cases}$$

and

$$\int_{\mathbb{R}} |g_T(x)|dx \ll \log T.$$

Consequently, the integral in (8.2) converges absolutely.

We are now going to prove Proposition 4.1 of [RudSar] in our case. That is, we will show that

$$C_{i_1,\ldots,i_n}(T) = \widetilde{C}_{i_1,\ldots,i_n}(T) + O(T^{1-\delta/3}), \tag{8.4}$$

if $\Phi(\xi)$ in (2.8) is supported in $|\xi_1| + \cdots + |\xi_n| \leq (2-\delta)/m$, where

$$\tilde{C}_{i_1\cdots i_n}(T) = \sum_{\substack{1\leq n_\mu \ll T \text{ for } i_\mu\neq 0,\\ MN\ll T^{2-\delta},\\ M=N}} \Big(\prod_{i_\mu=1} \frac{c_\pi(n_\mu)}{n_\mu^{1/2}}\Big)\Big(\prod_{i_\mu=-1} \frac{\overline{c_\pi}(n_\mu)}{n_\mu^{1/2}}\Big)\tilde{A}_{i_1\cdots i_n}(\mathbf{n},T), \tag{8.5}$$

and

$$\begin{aligned}
&\tilde{A}_{i_1\cdots i_n}(\mathbf{n},T) \\
&= T^n \int_{\substack{\mathbb{R}^n \\ |TL\xi_\mu|\ll T^{\delta/3} \\ \text{for } i_\mu=0}} \prod_{i_\mu=0} g_{\mu T}(TL\xi_\mu) \\
&\quad \times \prod_{i_\mu\neq 0} g_\mu\Big(T(L\xi_\mu + i_\mu \log n_\mu)\Big)\Phi(\xi)\delta(\xi_1+\cdots+\xi_n)d\xi. \qquad (8.6)
\end{aligned}$$

To do this, we first point out that Lemmas 4.2 and 4.3 of [RudSar], i.e., Lemmas 7.1 and 7.2 of [LiuYe1], hold in the present case.

Lemma 8.1. *$A_{i_1\cdots i_n}(\mathbf{n},T)=0$ unless $|n_\mu|\ll T$ for $i_\mu\neq 0$ and $\prod_{i_\mu\neq 0} n_\mu \ll T^{2-\delta}$.*

Lemma 8.2. *$\tilde{A}_{i_1\cdots i_n}(\mathbf{n},T)=0$ unless $|n_\mu|\ll T$ for $i_\mu\neq 0$, $\prod_{i_\mu\neq 0} n_\mu \ll T^{2-\delta}$, and $\prod_{i_\mu=1} n_\mu = \prod_{i_\mu=-1} n_\mu$.*

Lemma 4.4 of [RudSar] then holds in our case, since it is based on the bounds for g_T and Lemmas 8.1 and 8.2 above.

Lemma 8.3. *If $MN\ll T^{2-\delta}$, then*

$$A_{i_1\cdots i_n}(\mathbf{n},T) - \tilde{A}_{i_1\cdots i_n}(\mathbf{n},T) \ll \begin{cases} T^{1-\delta/3}L^{-r-s} & \text{if } \left|\log \frac{M}{N}\right| \ll T^{\delta/3-1}, \\ \dfrac{L^{-r-s}}{|\log M/N|} & \text{if } \left|\log \frac{M}{N}\right| \gg T^{\delta/3-1}, \end{cases}$$

where $r=\sum_{i_\mu=1} 1$ and $s=\sum_{i_\mu=-1} 1$.

Now we can write

$$C_{i_1\cdots i_n}(T) = \tilde{C}_{i_1\cdots i_n}(T) + \sum\nolimits_{\text{diag}} + \sum\nolimits_{\text{off}}$$

where

$$\begin{aligned}
\sum\nolimits_{\text{diag}} &= \sum_{\substack{1\le n_\mu \ll T \text{ for } i_\mu\neq 0, \\ MN\ll T^{2-\delta}, \\ |\log M/N|\ll T^{\delta/3-1}}} \Big(\prod_{i_\mu=1} \frac{c_\pi(n_\mu)}{n_\mu^{1/2}}\Big)\Big(\prod_{i_\mu=-1} \frac{\overline{c_\pi}(n_\mu)}{n_\mu^{1/2}}\Big) \\
&\quad \times \Big(A_{i_1\cdots i_n}(\mathbf{n},T) - \tilde{A}_{i_1\cdots i_n}(\mathbf{n},T)\Big) \\
&\ll \frac{T^{1-\delta/3}}{L^{r+s}} \sum_{\substack{M=N, \\ \prod_{i_\mu\neq 0} n_\mu \ll T^{2-\delta}}} \prod_{i_\mu\neq 0} \frac{|c_\pi(n_\mu)|}{\sqrt{n_\mu}},
\end{aligned}$$

and

$$\sum\nolimits_{\text{off}} = \sum_{\substack{1 \le n_\mu \ll T \text{ for } i_\mu \ne 0, \\ MN \ll T^{2-\delta}, \\ |\log M/N| \gg T^{\delta/3-1}}} \Big(\prod_{i_\mu = 1} \frac{c_\pi(n_\mu)}{n_\mu^{1/2}} \Big) \Big(\prod_{i_\mu = -1} \frac{\overline{c_\pi}(n_\mu)}{n_\mu^{1/2}} \Big)$$

$$\times \left(A_{i_1 \cdots i_n}(\mathbf{n}, T) - \tilde{A}_{i_1 \cdots i_n}(\mathbf{n}, T) \right)$$

$$\ll \frac{1}{L^{r+s}} \sum_{\substack{MN \ll T^{2-\delta}, \\ M \ne N}} \frac{1}{\sqrt{MN} |\log M/N|}$$

$$\times \sum_{\prod_{i_\mu = 1} n_\mu = M} \prod_{i_\mu = 1} |c_\pi(n_\mu)| \sum_{\prod_{i_\mu = -1} n_\mu = N} \prod_{i_\mu = -1} |c_\pi(n_\mu)|$$

by Lemma 8.3. We will prove later that

$$\sum_{\substack{M = N, \\ \prod_{i_\mu \ne 0} n_\mu \ll T^{2-\delta}}} \prod_{i_\mu \ne 0} \frac{|c_\pi(n_\mu)|}{n_\mu^{1/2}} \ll \left(\log T\right)^{r+s} \tag{8.7}$$

from which we can deduce $\sum_{\text{diag}} \ll T^{1-\delta/3}$. Denote

$$a_k(M) = \sum_{n_1 \cdots n_k = M} \prod_{1 \le j \le k} |c_\pi(n_j)|.$$

Then using (5.6) we can still prove Lemma 4.5 of [RudSar]:

$$\sum_{m \le X} a_k(m)^2 \ll_\varepsilon X^{1+\varepsilon}.$$

Using the same arguments as in [RudSar], pp.293-294, we get

$$\sum\nolimits_{\text{off}} \ll T^{1+\varepsilon-\delta/2}$$

for any $\varepsilon > 0$. Therefore, (8.4) is valid if we prove (8.7).

Back to (8.6), we change variables to

$$y_\mu = \begin{cases} TL\xi_\mu & \text{if } i_\mu = 0, \\ T(L\xi_\mu + i_\mu \log n_\mu) & \text{if } i_\mu \ne 0 \end{cases}$$

and get

$$\begin{aligned}\tilde{A}_{i_1\cdots i_n}(\mathbf{n},T) =&\frac{T}{L^{n-1}}\int_V \prod_{i_\mu=0} g_{\mu T}(y_\mu)\prod_{i_\mu\neq 0} g_\mu(y_\mu)\\ &\times \Phi\Big(\frac{y_1}{TL}-\frac{i_1\log n_1}{L},\dots,\frac{y_n}{TL}-\frac{i_n\log n_n}{L}\Big)dy\\ &+\frac{T}{L^{n-1}}\int_V \prod_{i_\mu=0} g_{\mu T}(y_\mu)\prod_{i_\mu\neq 0} g_\mu(y_\mu)dy\\ &\times\Big(\Phi\Big(-\frac{i_1\log n_1}{L},\dots,-\frac{i_n\log n_n}{L}\Big)+O\big(T^{-1+\delta/3}\big)\Big),\end{aligned}$$

where V is defined by $\sum_j y_j = 0$, $|y_\mu| \ll 1$ for $i_\mu \neq 0$, and $|y_\mu| \ll T^{\delta/3}$ for $i_\mu = 0$.

By Stirling's approximation formula and (8.3), we get from (4.5) that

$$\Omega_\pi(Tr) = m\ell\log(Tr) + O(1)$$

for $r \geq 1$. Consequently

$$\frac{1}{2\pi}\int_{\mathbb{R}} h_1(r)\cdots h_n(r)\Omega_\pi(Tr)^k dr = \frac{\kappa(\mathbf{h})}{2\pi}L^k\ell^k + O(L^{k-1}),$$

and hence

$$\int_V \prod_{i_\mu=0} g_{\mu T}(y_\mu)\prod_{i_\mu\neq 0} g_\mu(y_\mu)dy = \frac{\kappa(\mathbf{h})}{2\pi}L^k\ell^k + O(L^{k-1}),$$

where $k = n - r - s$ is the number of μ with $i_\mu = 0$. This is Lemma 4.6 of [RudSar]. Back to (8.6), we now have for $r + s > 0$ that

$$\tilde{A}_{i_1\cdots i_n}(\mathbf{n},T) = \frac{T\ell^{n-r-s}}{L^{r+s-1}}\frac{\kappa(\mathbf{h})}{2\pi}\Phi\Big(-\frac{i_1\log n_1}{L},\dots,-\frac{i_n\log n_n}{L}\Big)+O\Big(\frac{T}{L^{r+s}}\Big).$$

From (8.4) and (8.5) we thus obtain

$$\begin{aligned}C_{i_1\cdots i_n}(T) =&\frac{\kappa(\mathbf{h})}{2\pi}\frac{T\ell^{n-r-s}}{L^{r+s-1}}\Big(1+O\Big(\frac{1}{L}\Big)\Big)\\ &\times\sum_{\substack{1\leq n_\mu\ll T \text{ for } i_\mu\neq 0,\\ \prod_{i_\mu=1} n_\mu=\prod_{i_\mu=-1} n_\mu\\ \prod_{i_\mu\neq 0} n_\mu\ll T^2}}\Big(\prod_{i_\mu=1}\frac{c_\pi(n_\mu)}{n_\mu^{1/2}}\Big)\Big(\prod_{i_\mu=-1}\frac{\overline{c_\pi}(n_\mu)}{n_\mu^{1/2}}\Big)\\ &\times\Phi\Big(-\frac{i_1\log n_1}{L},\dots,-\frac{i_n\log n_n}{L}\Big)+O(T^{1-\delta/3}) \qquad (8.8)\end{aligned}$$

when $i_1, \ldots, i_n$ are not all zero, and

$$C_{0\cdots 0}(T) = \frac{\kappa(\mathbf{h})}{2\pi} T L \ell^n \Phi(0, \ldots, 0) + O(T). \tag{8.9}$$

If we assume Lemma 8.4 below, then (8.8) can be rewritten as

$$\begin{aligned} &C_{i_1\cdots i_n}(T) \\ &= \frac{\kappa(\mathbf{h})}{2\pi} \frac{T\ell^{n-r-s}}{L^{r+s-1}} \sum_{\substack{1\le n_\mu \ll T \text{ for } i_\mu \neq 0,\\ \prod_{i_\mu=1} n_\mu = \prod_{i_\mu=-1} n_\mu \\ \prod_{i_\mu\neq 0} n_\mu \ll T^2}} \Big(\prod_{i_\mu=1} \frac{c_\pi(n_\mu)}{n_\mu^{1/2}} \Big) \\ &\quad \times \Big(\prod_{i_\mu=-1} \frac{\overline{c_\pi}(n_\mu)}{n_\mu^{1/2}} \Big) \Phi\Big(-\frac{i_1 \log n_1}{L}, \ldots, -\frac{i_n \log n_n}{L} \Big) + O(T) \end{aligned} \tag{8.10}$$

when $r + s > 0$. This is indeed Lemma 4.7 of [RudSar].

Now let us turn to Lemma 3.8 of [RudSar]. From the bound for $a_\pi(p^k)$ in (4.2) we get

$$\left| c_\pi(p^k) \right| \le m\ell(\log p) p^{k(1/2-\beta)}$$

with $\beta = 1/(\ell m^2 + 1) > 0$, which is essentially the same bound as in [RudSar] and [LiuYe1]. Consequently for any $K > 1/\beta$ we have (4.54) of [RudSar]:

$$\sum_p \sum_{k_1+\cdots+k_t \ge K} \frac{\left| c_\pi(p^{k_1}) \cdots c_\pi(p^{k_t}) \right|}{p^{(k_1+\cdots+k_t)/2}} \ll \sum_p \frac{\log^t p}{p^{K\beta}} < \infty. \tag{8.11}$$

From (5.5) we deduce that

$$\sum_p \frac{\left| c_\pi(p^k) \right|^2}{p^{k+\delta}} < \infty$$

for fixed k and $\delta > 0$. By an argument in [RudSar], we can prove that

$$\sum_p \sum_{k_1+\cdots+k_t = s} \frac{\left| c_\pi(p^{k_1}) \cdots c_\pi(p^{k_t}) \right|}{p^{s/2}}$$

is bounded for $s < K$ fixed. Together with (8.11), this implies that Lemma 3.8 of [RudSar] is valid:

$$\sum_p \sum_{\substack{k_i, l_j \ge 1, \\ k_1+\cdots+k_r = l_1+\cdots+l_s}} \frac{\left| c_\pi(p^{k_1}) \cdots c_\pi(p^{k_r}) \bar{c}_\pi(p^{l_1}) \cdots \bar{c}_\pi(p^{l_s}) \right|}{p^{k_1+\cdots+k_r}} < \infty$$

when $r + s \geq 3$.

Now we prove the following lemma which is Lemma 3.9 of [RudSar].

Lemma 8.4. *Assume $m = 2$ or Hypothesis H. If $1 \leq r \leq s$, then*

$$\sum_{\substack{p_\mu^{k_\mu} \leq x \text{ for } i_\mu \neq 0, \\ \prod_{i_\mu=1} p_\mu^{k_\mu} = \prod_{i_\mu=-1} p_\mu^{k_\mu}}} \Big(\prod_{i_\mu=1} \frac{c_\pi(p_\mu^{k_\mu})}{p_\mu^{k_\mu/2}} \Big) \Big(\prod_{i_\mu=-1} \frac{\overline{c_\pi}(p_\mu^{k_\mu})}{p_\mu^{k_\mu/2}} \Big) \ll \begin{cases} (\log x)^{2r} & \text{if } r = s, \\ (\log x)^{2r-2} & \text{if } r < s. \end{cases}$$

Proof. Using the argument in [RudSar] after (4.59), the Lemma is reduced to estimation of

$$\begin{aligned}
\sum_{\substack{p,k \\ p^k \leq x}} \frac{|c_\pi(p^k)|^2}{p^k} &= \sum_{\substack{p,k \\ p^k \leq x}} \frac{\log^2 p}{p^k} \sum_{\substack{v_1, v_2 | p, \\ f_1, f_2 | k}} f_1 f_2 \Big(\sum_{1 \leq j_1 \leq m} \alpha_\pi^{k/f_1}(j_1, v_1) \Big) \\
&\quad \times \Big(\sum_{1 \leq j_2 \leq m} \alpha_\pi^{k/f_2}(j_2, v_2) \Big) \\
&\ll \Big(\sum_{\substack{p,k \\ p^k \leq x}} \frac{\log^2 p}{p^k} \sum_{\substack{v_1 | p, \\ f_1 | k}} f_1^2 \Big| \sum_{1 \leq j_1 \leq m} \alpha_\pi^{k/f_1}(j_1, v_1) \Big|^2 \Big)^{1/2} \\
&\quad \times \Big(\sum_{\substack{p,k \\ p^k \leq x}} \frac{\log^2 p}{p^k} \sum_{\substack{v_2 | p, \\ f_2 | k}} f_2^2 \Big| \sum_{1 \leq j_2 \leq m} \alpha_\pi^{k/f_2}(j_2, v_2) \Big|^2 \Big)^{1/2}
\end{aligned}$$

by Cauchy's inequality. Using our results in (7.1) from the Rankin-Selberg L-functions, we conclude that

$$\sum_{\substack{p,k \\ p^k \leq x}} \frac{|c_\pi(p^k)|^2}{p^k} \ll \log^2 x.$$

This proves the Lemma. □

Lemma 8.4 implies (8.7) and hence (8.4). It also completed the proof of (8.10). Let $S_{i_1 \cdots i_n}$ be the set of bijective maps from the set $\{\mu : i_\mu = 1\}$ onto the set $\{\mu : i_\mu = -1\}$ when $r = s$. Then (8.10) can be further reduced to

$$C_{i_1\cdots i_n}(T) = \frac{\kappa(\mathbf{h})}{2\pi}\frac{TL}{L^{2r}} \sum_{p_\mu \le T \text{ for } i_\mu=1} \sum_{\sigma \in S_{i_1\cdots i_n}} \prod_{i_\mu=1} \frac{|c_\pi(p_\mu)|^2}{p_\mu} \times \Phi(z_1, \cdots, z_n) + O(T)$$

where $z_\mu = -(\log p_\mu)/L$ and $z_{\sigma(\mu)} = (\log p_\mu)/L$ if $i_\mu = 1$, and $z_\mu = 0$ if $i_\mu = 0$, when $r = s > 0$. When $r \neq s$, $C_{i_1\cdots i_n}(T) = O(T)$ by Lemma 8.4. Consequently we can go back to (8.1) and (8.9) to write

$$\begin{aligned}
&\sum_{\gamma_1,\cdots,\gamma_n} h_1\Big(\frac{\gamma_1}{T}\Big)\cdots h_n\Big(\frac{\gamma_n}{T}\Big) f\Big(\frac{L\gamma_1}{2\pi},\cdots,\frac{L\gamma_n}{2\pi}\Big)\\
&= \frac{\kappa(\mathbf{h})}{2\pi} TL\ell^n \Phi(0,\cdots,0)\\
&+ \sum_{\substack{-1\le i_1,\cdots,i_n \le 1\\ r=s\ge 1}} \frac{\kappa(\mathbf{h})}{2\pi}\frac{T\ell^{n-2r}}{L^{2r-1}} \sum_{\sigma\in S_{i_1\cdots i_n}} \prod_{i_\mu=1} \sum_{p_\mu \le T} \frac{|c_\pi(p_\mu)|^2}{p_\mu}\\
&\times \Phi\Big(\frac{\log p_1}{L},\dots,\frac{\log p_r}{L},-\frac{\log p_1}{L},\dots,-\frac{\log p_r}{L},0,\dots,0\Big)\\
&+ O(T),
\end{aligned} \tag{8.12}$$

where on the left side, for each $j = 1, \dots, n$, $\rho_j = 1/2 + i\gamma_j$ is taken over all nontrivial zeros of $L(s, \pi)$.

Estimation of (8.12) is thus reduced to determining the asymptotic behavior of

$$\sum_{p\le x} \frac{|c_\pi(p)|^2}{p} \phi\Big(\frac{\log p}{m\log x}\Big),$$

where $\phi(v)$ is a C^1 function supported in $|v| \le (1-\delta)/m$ for some positive δ. By the definition of $c_\pi(n)$ in (4.1), we get

$$\begin{aligned}
&\sum_{p\le x} \frac{|c_\pi(p)|^2}{p} \phi\Big(\frac{\log p}{m\log x}\Big)\\
&= \sum_{p\le x} \frac{\log^2 p}{p} \phi\Big(\frac{\log p}{m\log x}\Big) \Big| \sum_{\substack{v|p\\ f_p=1}} \sum_{1\le j\le m} \alpha_\pi(j,v)\Big|^2\\
&= \sum_{p\le x} \frac{\log^2 p}{p} \phi\Big(\frac{\log p}{m\log x}\Big)
\end{aligned}$$

$$\times \sum_{\substack{v_1|p,\ v_2|p \\ f_p=1}} \Big(\sum_{1\le j_1\le m} \alpha_\pi(j_1, v_1)\Big)\Big(\sum_{1\le j_2\le m} \overline{\alpha}_\pi(j_2, v_2)\Big). \tag{8.13}$$

9. Proof of Theorem 3.1.

In this section we will assume that E is a Galois (not necessarily cyclic) extension of $\mathbb{Q}$ of degree ℓ. We are going to compute (8.12) and deduce the n-level correlation of normalized nontrivial zeros of $L(s,\pi)$ from (8.12). The estimation (8.12) is based on the asymptotic behavior of (8.13).

Note that only those primes p with $f_p = 1$ contribute to (8.13). Consequently, the outermost sum on the right side of (8.13) is actually taken over (i) primes $p \le x$ which split completely in E, and (ii) primes $p \le x$ above which each local field E_v is fully ramified ($f_p = 1$). Note that there are only finitely many primes p which satisfy (ii) but not (i). We can thus rewrite (8.13) as

$$\begin{aligned}
&\sum_{p\le x} \frac{|c_\pi(p)|^2}{p}\phi\Big(\frac{\log p}{m\log x}\Big) \\
&= \sum_{\substack{p\le x \\ \text{splits completely}}} \frac{\log^2 p}{p}\phi\Big(\frac{\log p}{m\log x}\Big) \\
&\quad\times \sum_{v_1|p,\ v_2|p} \Big(\sum_{1\le j_1\le m} \alpha_\pi(j_1, v_1)\Big)\Big(\sum_{1\le j_2\le m} \overline{\alpha}_\pi(j_2, v_2)\Big) + O(1).
\end{aligned} \tag{9.1}$$

Now we distinguish two cases. First let us assume that for any $\sigma \in \mathrm{Gal}(E/\mathbb{Q})$, π and π^σ are equivalent. Under this condition, we have $\pi_{v_1} \cong \pi_{v_2}$ and

$$\{\alpha_\pi(j_1, v_1) \,|\, 1 \le j_1 \le m\} = \{\alpha_\pi(j_2, v_2) \,|\, 1 \le j_2 \le m\}$$

for any v_1 and v_2 lying above a completely splitting p. Consequently (9.1) becomes

$$\begin{aligned}
&\sum_{p\le x} \frac{|c_\pi(p)|^2}{p}\phi\Big(\frac{\log p}{m\log x}\Big) \\
&= \ell \sum_{\substack{p\le x \\ \text{splits completely}}} \frac{\log^2 p}{p}\phi\Big(\frac{\log p}{m\log x}\Big) \sum_{v|p} \Big| \sum_{1\le j\le m} \alpha_\pi(j_1, v)\Big|^2 + O(1).
\end{aligned}$$

By Lemma 7.2, we get

$$\sum_{p\le T}\frac{|c_\pi(p)|^2}{p}\phi\Big(\frac{\log p}{L}\Big)=\ell m^2\log^2 T\int_0^{1/m}v\phi(v)dv+O(\log T). \tag{9.2}$$

This implies the asymptotic formula

$$\begin{aligned}
&\sum_{\gamma_1,\cdots,\gamma_n} h_1\Big(\frac{\gamma_1}{T}\Big)\cdots h_n\Big(\frac{\gamma_n}{T}\Big)f\Big(\frac{L\gamma_1}{2\pi},\cdots,\frac{L\gamma_n}{2\pi}\Big)\\
&=\frac{\kappa(\mathbf{h})}{2\pi}TL\ell^n\Phi(0,\cdots,0)\\
&\quad+\frac{\kappa(\mathbf{h})}{2\pi}TL\sum_{\substack{-1\le i_1,\dots,i_n\le 1\\ 1\le r=s\le n/2}}\ell^{n-r}\sum_{\sigma\in S_{i_1\cdots i_n}}\int_0^{1/m}\cdots\int_0^{1/m}v_1\cdots v_r\\
&\quad\times\Phi\big(v_1,\dots,v_r,-v_{\sigma(1)},\dots,-v_{\sigma(r)},0,\dots,0\big)dv_1\cdots dv_r+O(T)\\
&=\frac{\kappa(\mathbf{h})}{2\pi}TL\ell^n\Phi(0,\cdots,0)\\
&\quad+\frac{\kappa(\mathbf{h})}{2\pi}TL\sum_{1\le r\le n/2}\frac{n!\ell^{n-r}}{2^r r!(n-2r)!}\int_{\mathbb{R}^r}|v_1|\cdots|v_r|\\
&\quad\times\Phi\big(v_1,\dots,v_r,-v_1,\dots,-v_r,0,\dots,0\big)dv_1\cdots dv_r+O(T).
\end{aligned}\tag{9.3}$$

Next let us consider the case that π is not stable under the whole $\mathrm{Gal}(E/\mathbb{Q})$. Let $\mathrm{G}_\pi=\{\sigma\in\mathrm{Gal}(E/\mathbb{Q})\mid\pi^\sigma\cong\pi\}$ be the subgroup of $\mathrm{Gal}(E/\mathbb{Q})$ fixing π. Denote by a the number of elements in G_π, so that $a|\ell$ and $1\le a<\ell$. Then according to [Rog], for any two places v_1 and v_2 lying above a completely splitting p, $\pi_{v_1}\cong\pi_{v_2}$ if and only if E_{v_1} is mapped onto E_{v_2} by some $\sigma\in\mathrm{G}_\pi$. Consequently (9.1) becomes

$$\begin{aligned}
&\sum_{p\le x}\frac{|c_\pi(p)|^2}{p}\phi\Big(\frac{\log p}{m\log x}\Big)\\
&=\sum_{\substack{\sigma\in\mathrm{Gal}(E/\mathbb{Q})\\ \pi\cong\pi^\sigma}}\ \sum_{\substack{p\le x\\ \text{splits}\\ \text{completely}}}\frac{\log^2 p}{p}\phi\Big(\frac{\log p}{m\log x}\Big)\sum_{v|p}\Big|\sum_{1\le j\le m}\alpha_\pi(j,v)\Big|^2\\
&\quad+\sum_{\substack{\sigma\in\mathrm{Gal}(E/\mathbb{Q})\\ \pi\not\cong\pi^\sigma}}\ \sum_{\substack{p\le x\\ \text{splits}\\ \text{completely}}}\frac{\log^2 p}{p}\phi\Big(\frac{\log p}{m\log x}\Big)\\
&\quad\times\sum_{v|p}\Big(\sum_{1\le j_1\le m}\alpha_\pi(j_1,v)\Big)\Big(\sum_{1\le j_2\le m}\overline{\alpha}_{\pi^\sigma}(j_2,v)\Big)+O(1).
\end{aligned}\tag{9.4}$$

The estimation of the first term on the right of (9.4) is the same as in the first case, following (9.2) and (9.3):

$$= am^2(\log T)^2 \int_0^{1/m} v\phi(v)dv + O(\log T).$$

On the other hand, the estimation of the second term on the right of (9.4) is done by Lemma 7.4; we get $O(\log T)$. Consequently,

$$\sum_{p\le T} \frac{|c_\pi(p)|^2}{p}\phi\Big(\frac{\log p}{L}\Big) = am^2(\log T)^2 \int_0^{1/m} v\phi(v)dv + O(\log T),$$

and

$$\begin{aligned}
&\sum_{\gamma_1,\cdots,\gamma_n} h_1\Big(\frac{\gamma_1}{T}\Big)\cdots h_n\Big(\frac{\gamma_n}{T}\Big) f\Big(\frac{L\gamma_1}{2\pi},\cdots,\frac{L\gamma_n}{2\pi}\Big)\\
&= \frac{\kappa(\mathbf{h})}{2\pi} TL\ell^n \Phi(0,\cdots,0)\\
&\quad + \frac{\kappa(\mathbf{h})}{2\pi} TL \sum_{1\le r\le n/2} \frac{n!\ell^{n-2r}a^r}{2^r r!(n-2r)!}\int_{\mathbb{R}^r} |v_1|\cdots|v_r|\\
&\quad \times \Phi(v_1,\ldots,v_r,-v_1,\ldots,-v_r,0,\ldots,0)dv_1\cdots dv_r + O(T).
\end{aligned}$$

This proves Theorem 3.1. □

Appendix. Proof of Hypothesis H for GL_3 over E.

Now we prove Hypothesis H over E for $m = 3$. Over $\mathbb{Q}$, this was proved in [RudSar]. We follow their approach closely. Since the series on the right side of (5.1) converges absolutely for $\sigma > 1$, we know that for any $\delta > 0$

$$\sum_{p,k}\sum_{v|p} kf_p^2 \frac{\log^2 p}{p^{kf_p(1+\delta)}}\Big|\sum_{j=1}^{3}\alpha_\pi^k(j,v)\Big|^2 < \infty.$$

Taking $k = 1$, we get

$$\sum_{p}\sum_{v|p} \frac{\log^2 p}{p^{f_p(1+\delta)}}\Big|\sum_{j=1}^{3}\alpha_\pi(j,v)\Big|^2 < \infty. \tag{A.1}$$

On the other hand, as in [RudSar], p.283, we can have $|\alpha_\pi(2,v)| = 1$ and $|\alpha_\pi(1,v)| = 1/|\alpha_\pi(3,v)| \ge 1$, because the central character of π is assumed to be unitary. Therefore

$$\Big|\sum_{j=1}^{3}\alpha_\pi^k(j,v)\Big|^2 \ll \sum_{j=1}^{3}\Big|\alpha_\pi(j,v)\Big|^{2k}$$

$$\ll \sum_{j=1}^{3}\Big|\alpha_\pi(j,v)\Big|^{2k} \le \Big(\Big|\sum_{j=1}^{3}\alpha_\pi(j,v)\Big| + 4\Big)^{2k} \ll 1 + \Big|\sum_{j=1}^{3}\alpha_\pi(j,v)\Big|^{2k}.$$

Consequently

$$\sum_p \frac{\log^2 p}{p^{kf_p}} \sum_{v|p}\Big|\sum_{1\le j\le 3}\alpha_\pi^k(j,v)\Big|^2$$

$$\ll \sum_p\sum_{v|p}\frac{\log^2 p}{p^{kf_p}} + \sum_p\sum_{v|p}\frac{\log^2 p}{p^{kf_p}}\Big|\sum_{j=1}^{3}\alpha_\pi(j,v)\Big|^{2k}. \tag{A.2}$$

The first sum on the right side of (A.2) is $\le \ell\sum_p(\log p)^2/p^k < \infty$ when $k \ge 2$. Applying the bound in (4.3) to $|\sum_{j=1}^{3}\alpha_\pi(j,v)|^{2k-2}$, we see that the last sum in (A.2) is

$$\ll \sum_p\sum_{v|p}\frac{\log^2 p}{p^{f_p(1+(k-1)/5)}}\Big|\sum_{j=1}^{3}\alpha_\pi(j,v)\Big|^2 < \infty,$$

by (A.1). Therefore Hypothesis H is true for $m = 3$. □

Acknowledgments

The second author wishes to acknowledge support from the Obermann Center for Advanced Studies, the University of Iowa.

References

[ArtClo] J. Arthur and L. Clozel, *Simple Algebras, Base Change, and the Advanced Theory of the Trace Formula*, Annals of Math. Studies, no. 120, Princeton Univ. Press, Princeton, 1989.

[Bor] A. Borel, *Automorphic L-functions*, Proc. Symp. Pure Math. **33** (1979), part 2, 27–61.

[GelLapSar] S.S. Gelbart, E.M. Lapid, and P. Sarnak, *A new method for lower bounds of L-functions*, C.R. Acad. Sci. Paris, Ser. I **339** (2004), 91–94.

[GelShah] S. Gelbart and F. Shahidi, *Boundedness of automorphic L-functions in vertical strips*, J. Amer. Math. Soc. **14** (2001), 79–107.

[GodJac] R. Godement and H. Jacquet, *Zeta functions of simple algebras*, Lecture Notes in Math. **260**, Springer-Verlag, Berlin, 1972.

[JacPSShal1] H. Jacquet, I.I. Piatetski-Shapiro, and J. Shalika, *Conducteur des réprésentations du groupe linéaire*, Math. Ann. **256** (1981), 199–214.

[JacPSShal2] H. Jacquet, I.I. Piatetski-Shapiro, and J. Shalika, *Rankin-Selberg convolutions*, Amer. J. Math. **105** (1983), 367–464.

[Kim1] H. Kim, *Functoriality for the exterior square of GL_4 and the symmetric fourth of GL_2*, J. Amer. Math. Soc. **16** (2003), 139–183.

[Kim2] H. Kim, *A note on Fourier coefficients of cusp forms on GL_n*, to appear in Forum Math.

[KimSar] H. Kim and P. Sarnak, *Appendix: Refined estimates towards the Ramanujan and Selberg conjectures*, Appendix to [Kim1].

[KimShah] H. Kim and F. Shahidi, *Cuspidality of symmetric powers with applications*, Duke Math. J. **112** (2002), 177–197.

[Land] E. Landau, *Über die Anzahl der Gitterpunkte in gewissen Bereichen*, Gött. Nachr. (1915), 209–243.

[Lang] R.P. Langlands, *Base Change for GL_2*, Annals of Math. Studies, no. 96, Princeton Univ. Press, Princeton, 1980.

[LiuWangYe] Jianya Liu, Yonghui Wang, and Yangbo Ye, *A proof of Selberg's orthogonality for automorphic L-functions*, to appera in Manuscripta Math.

[LiuYe1] Jianya Liu and Yangbo Ye, *Superposition of zeros of distinct L-functions*, Forum Math. **14** (2002), 419–455.

[LiuYe2] Jianya Liu and Yangbo Ye, *Weighted Selberg orthogonality and uniqueness of factorization of automorphic L-functions*, Forum Math. **17** (2005), 493–512.

[LiuYe3] Jianya Liu and Yangbo Ye, *Selberg's orthogonality conjecture for automorphic L-functions*, Amer. J. Math. **127** (2005), 837–849.

[LuoRudSar] W. Luo, Z. Rudnick, and P. Sarnak, On the generalized Ramanujan conjecture for $GL(n)$, Proc. Symp. Pure Math. **66** (1999), part 2, 301–310.

[MoeWal] C. Moeglin and J.-L. Waldspurger, *Le spectre résiduel de $GL(n)$*, Ann. Sci. École Norm. Sup. (4) **22** (1989), 605–674.

[Mor1] C.J. Moreno, *Explicit formulas in the theory of automorphic forms*, Lecture Notes Math. vol. **626**, Springer, Berlin, 1977, 73–216.

[Mor2] C.J. Moreno, *Analytic proof of the strong multiplicity one theorem*, Amer. J. Math. **107** (1985), 163–206.

[Mur1] M. Ram Murty, *Selberg conjectures and Artin L-functions*, Bull. Amer. Math. Soc. **31** (1994), 1–14.

[Mur2] M. Ram Murty, *Selberg's conjectures and Artin L-functions II*, Current trends in mathematics and physics, Narosa, New Delhi, 1995, 154–168.

[Rog] J.D. Rogawski, *Functoriality and the Artin conjecture*, Proc. Symp. Pure Math. **61** (1997), 331–353.

[RudSar] Z. Rudnick and P. Sarnak, *Zeros of principal L-functions and random matrix theory*, Duke Math. J. **81** (1996), 269–322.

[Sar] P. Sarnak, *Nonvanishing of L-functions on $\Re(s) = 1$*, Contributions to Automorphic Forms, Geometry & Number Theory, ed. by Hida,

Ramakrishnan and Shahidi, Johns Hopkins Univ. Press, Baltimore, 2004, 719–732.

[Sel] A. Selberg, *Old and new conjectures and results about a class of Dirichlet series*, in Collected Papers, vol. 2, Springer, Berlin 1991, 47–65.

[Ser] J.-P. Serre, Letter to J.-M. Deshouillers (1981).

[Shah1] F. Shahidi, *On certain L-functions*, Amer. J. Math. **103** (1981), 297–355.

[Shah2] F. Shahidi, *Fourier transforms of intertwining operators and Plancherel measures for $GL(n)$*, Amer. J. Math. **106** (1984), 67–111.

[Shah3] F. Shahidi, *Local coefficients as Artin factors for real groups*, Duke Math. J. **52** (1985), 973–1007.

[Shah4] F. Shahidi, *A proof of Langlands' conjecture on Plancherel measures; Complementary series for p-adic groups*, Ann. Math. **132** (1990), 273–330.

ANALYTIC PROPERTIES OF MULTIPLE ZETA-FUNCTIONS IN SEVERAL VARIABLES

Kohji Matsumoto
Graduate School of Mathematics, Nagoya University, Chikusa-ku, Nagoya 464-8602, Japan
kohjimat@math.nagoya-u.ac.jp

Abstract We report several recent results on analytic properties of multiple zeta-functions, mainly in several variables, such as the analytic continuation, the asymptotic behaviour, the location of singularities, and the recursive structure. Some results presented in this paper have never been published before.

Keywords: multiple zeta-function, analytic continuation, Mellin-Barnes integral

2000 Mathematics Subject Classification: 11M41

1. Euler-Zagier sums

Let r be a positive integer. We begin with the discussion on the Euler-Zagier r-fold sum

$$\begin{aligned}\zeta_{EZ,r}(s_1,\ldots,s_r) &= \sum\cdots\sum_{1\leq m_1<\cdots<m_r<\infty} m_1^{-s_1}m_2^{-s_2}\cdots m_r^{-s_r} \\ &= \sum_{m_1=1}^{\infty}\sum_{m_2=1}^{\infty}\cdots\sum_{m_r=1}^{\infty} m_1^{-s_1}(m_1+m_2)^{-s_2}\cdots(m_1+\cdots+m_r)^{-s_r}, \quad (1.1)\end{aligned}$$

where $s_1,\ldots,s_r$ are complex variables. This multiple series is convergent absolutely in the region

$$\{(s_1,\ldots,s_r)\in \mathbf{C}^r \mid \Re(s_{r-k+1}+\cdots+s_r)>k \quad (1\leq k\leq r)\}.$$

The case $r=2$ of (1.1) was already investigated by L. Euler in the eighteenth century, and the general r-fold case has recently been studied by Zagier [48] and others. In particular, their research on special values

Number Theory: Tradition and Modernization, pp. 153–173
W. Zhang and Y. Tanigawa, eds.

of $\zeta_{EZ,r}$ at positive integers shows the great importance of this function in various fields of mathematics and mathematical physics.

The meromorphic continuation of (1.1) to $\mathbf{C}^r$ has been achieved by various methods; see Arakawa and Kaneko [5], Zhao [49], Akiyama, Egami and Tanigawa [1], and the author [25]. It can also be regarded as a special case of Essouabri's general result [9]; see Section 3.

We shall describe the method of [1] which is based on the Euler-Maclaurin summation formula. Let n_1 be a positive integer, η be a real number, $\Re s > 1$, $\alpha \geq 0$, $B_j(x)$ the j-th Bernoulli polynomial, and $\tilde{B}_j(x) = B_j(x-[x])$. Akiyama and Ishikawa [2] proved a modified version of the Euler-Maclaurin formula, including a parameter η, which implies

$$\sum_{n>n_1+\eta} \frac{1}{(n+\alpha)^s} = \sum_{j=-1}^{J} \frac{\tilde{B}_{j+1}(\eta)}{(j+1)!} \frac{(s)_j}{(n_1+\eta+\alpha)^{s+j}} - \frac{(s)_{J+1}}{(J+1)!} \int_{n_1+\eta}^{\infty} \frac{\tilde{B}_{J+1}(u)}{(u+\alpha)^{s+J+1}} du \qquad (1.2)$$

for any positive integer J, where $(s)_j = \Gamma(s+j)/\Gamma(s)$ (see Lemma 1 of [2]). In [1], formula (1.2) (with $\alpha = \eta = 0$) is applied to the sum with respect to m_r on the second member of (1.1), and an expression of $\zeta_{EZ,r}(s_1, \ldots, s_r)$ as a sum involving $\zeta_{EZ,r-1}(s_1, \ldots, s_{r-2}, s_{r-1}+s_r+j)$ $(-1 \leq j \leq J)$ is obtained. Hence the analytic continuation can be shown by induction on r, because the integral on the right-hand side of (1.2) converges in a wider region of s when J becomes larger.

On the other hand, the basic tool of the author's method is the Mellin-Barnes integral formula

$$(1+\lambda)^{-s} = \frac{1}{2\pi i} \int_{(c)} \frac{\Gamma(s+z)\Gamma(-z)}{\Gamma(s)} \lambda^z dz, \qquad (1.3)$$

where $s, \lambda \in \mathbf{C}$, $\Re s > 0$, $|\arg \lambda| < \pi$, $\lambda \neq 0$, and $c \in \mathbf{R}$, $-\Re s < c < 0$. The path of integration is the vertical line from $c - i\infty$ to $c + i\infty$. The key point is to apply (1.3) (with $\lambda = m_r/(m_1 + \cdots + m_{r-1})$) to the factor

$$(m_1 + \cdots + m_r)^{-s_r} = (m_1 + \cdots + m_{r-1})^{-s_r} \left(1 + \frac{m_r}{m_1 + \cdots + m_{r-1}}\right)^{-s_r}$$

on the last member of (1.1), and express $\zeta_{EZ,r}(s_1, \ldots, s_r)$ as an integral whose integrand includes $\zeta_{EZ,r-1}(s_1, \ldots, s_{r-2}, s_{r-1}+s_r+z)$ as a factor. Then the analytic continuation can be shown by shifting the path of integration suitably.

We can also see the location of possible singularities by both of the above methods. Akiyama, Egami and Tanigawa [1] considered this matter more carefully, and proved

Theorem 1. (Akiyama, Egami and Tanigawa [1]) *Singularities of* $\zeta_{EZ,r}(s_1,\ldots,s_r)$ *are located only on*

$$s_r = 1, \quad s_{r-1}+s_r = 2,1,0,-2,-4,-6,\ldots$$

and

$$s_{r-k+1}+s_{r-k+2}+\cdots+s_r = k-n \quad (3\leq k\leq r,\ n\in\mathbf{N}_0),$$

where $\mathbf{N}_0$ *denotes the set of non-negative integers. All of the above sets are indeed singularity sets.*

In the same paper [1], they also studied the values of $\zeta_{EZ,r}(s_1,\ldots,s_r)$ at non-positive integers. This direction of research has been continued by Akiyama and Tanigawa [3], and Kamano [16].

It is an important problem to generalize the analytic theory of Euler-Zagier sums to a more general situation. Akiyama and Ishikawa [2] studied the series

$$\begin{aligned}&\zeta_{EZ,r}(s_1,\ldots,s_r;\alpha_1,\ldots,\alpha_r)\\ &=\sum_{1\leq m_1<\cdots<m_r<\infty}\cdots\sum(m_1+\alpha_1)^{-s_1}(m_2+\alpha_2)^{-s_2}\cdots(m_r+\alpha_r)^{-s_r}\end{aligned} \tag{1.4}$$

and

$$\begin{aligned}&L_{EZ,r}(s_1,\ldots,s_r;\chi_1,\ldots,\chi_r)\\ &=\sum_{1\leq m_1<\cdots<m_r<\infty}\cdots\sum\frac{\chi_1(m_1)}{m_1^{s_1}}\frac{\chi_2(m_2)}{m_2^{s_2}}\cdots\frac{\chi_r(m_r)}{m_r^{s_r}},\end{aligned} \tag{1.5}$$

where $0\leq\alpha_k<1$ $(1\leq k\leq r)$ and χ_k $(1\leq k\leq r)$ are Dirichlet characters of the same conductor. It is clear that (1.5) can be expressed as a linear combination of several series of the form (1.4). Akiyama and Ishikawa [2] applied (1.2) to the right-hand side of (1.4) to obtain an expression involving

$$\zeta_{EZ,r-1}(s_1,\ldots,s_{r-2},s_{r-1}+s_r+j;\alpha_1,\ldots,\alpha_{r-1}) \quad (-1\leq j\leq J).$$

This expression gives the analytic continuation of (1.4). Akiyama and Ishikawa also discussed the location of singularities of (1.4) and (1.5). Ishikawa [13] further studied the location of poles of (1.5) in the special case $s_1=\cdots=s_r=s$, and applied the result to the evaluation of certain multiple character sums (Ishikawa [14]).

The author [26], [27] considered a further generalization of (1.4), that is the series

$$\sum_{m_1=0}^{\infty}\cdots\sum_{m_r=0}^{\infty}(\alpha_1+m_1w_1)^{-s_1}(\alpha_2+m_1w_1+m_2w_2)^{-s_2}\\ \times\cdots\times(\alpha_r+m_1w_1+\cdots+m_rw_r)^{-s_r}, \tag{1.6}$$

where α_k, w_k $(1\le k\le r)$ are complex parameters. Let $-\pi<\theta\le\pi$ and

$$H(\theta)=\{w\in\mathbf{C}\mid w\neq 0,\ \theta-\pi/2<\arg w<\theta+\pi/2\}.$$

If we assume that $w_k\in H(\theta)$ $(1\le k\le r)$, then the series (1.6) is convergent absolutely when $\Re s_k$ $(1\le k\le r)$ are sufficiently large. Under the same assumption, the author proved the meromorphic continuation of (1.6) to $\mathbf{C}^r$ by using the Mellin-Barnes formula (1.3), discussed the asymptotic behaviour with respect to w_r and the order estimate with respect to $\Im s_r$.

The aim of introducing the above generalized form (1.6) is to treat the Barnes multiple zeta-function

$$\zeta_{B,r}(s;\alpha;w_1,\ldots,w_r)=\sum_{m_1=0}^{\infty}\cdots\sum_{m_r=0}^{\infty}(\alpha_r+m_1w_1+\cdots+m_rw_r)^{-s} \tag{1.7}$$

as a special case $s_1=\cdots=s_{r-1}=0$ and $s_r=s$. The author first considered the asymptotic behaviour of $\zeta_{B,2}(s;\alpha;w_1,w_2)$ in [22] by a different method (contour integration), and then by using (1.3) in [25]. These studies have applications to Hecke's zeta and L-functions attached to real quadratic fields; see Corrigendum and addendum of [22], and [23].

Multiple Dirichlet series of the Euler-Zagier type with general coefficients, of the form

$$\sum_{m_1=1}^{\infty}\cdots\sum_{m_r=1}^{\infty}\frac{a_1(m_1)}{m_1^{s_1}}\frac{a_2(m_2)}{(m_1+m_2)^{s_2}}\cdots\frac{a_r(m_r)}{(m_1+\cdots+m_r)^{s_r}}, \tag{1.8}$$

have been introduced and studied by Matsumoto and Tanigawa [30], under the assumption that the series $\sum_{m=1}^{\infty}a_k(m)m^{-s}$ $(1\le k\le r)$ have nice properties. The above (1.8) includes the multiple L-series of Arakawa and Kaneko [6]. In [30], the analytic continuation and a certain order estimate of (1.8) have been obtained by using the Mellin-Barnes integral (1.3).

It seems that the method of using the Mellin-Barnes integral is suitable to consider upper bound estimates of multiple zeta-functions. The case of the Euler-Zagier sum $\zeta_{EZ,r}(s_1,\ldots,s_r)$ was studied by Ishikawa and Matsumoto [15]; especially, non-trivial estimates in the cases $r = 2$ and $r = 3$ have been obtained. However it is still not clear how is the real order of magnitude of $\zeta_{EZ,r}(s_1, \ldots, s_r)$.

Recently, Matsumoto and Tsumura [31] introduced further generalized series

$$\sum_{m_1=0}^{\infty} \cdots \sum_{m_r=0}^{\infty} \frac{a_1(m_1) \cdots a_r(m_r) u^{-(m_1+\cdots+m_r)}}{(\alpha_1 + m_1 w_1)^{s_1} \cdots (\alpha_r + m_1 w_1 + \cdots + m_r w_r)^{s_r}}, \tag{1.9}$$

where $u \geq 1$, $\alpha_k, w_k \in \mathbf{R}$, $0 < \alpha_k - \alpha_{k-1} \leq w_k$, in connection with a study of certain generalized multiple polylogarithms. (As for multiple polylogarithms, see, for example, [7].)

2. Multiple series defined by linear forms

Let $A_{Nr} = (a_{nj})_{1 \leq n \leq N, 1 \leq j \leq r}$ be an (N, r)-matrix, where a_{nj} are non-negative real numbers. Assume that all rows and all columns of A_{Nr} include at least one non-zero element. Let

$$\zeta_r(s_1, \ldots, s_N; A_{Nr}) = \sum_{m_1=1}^{\infty} \cdots \sum_{m_r=1}^{\infty} (a_{11}m_1 + \cdots + a_{1r}m_r)^{-s_1} \times \cdots \times (a_{N1}m_1 + \cdots + a_{Nr}m_r)^{-s_N}. \tag{2.1}$$

The Euler-Zagier sum (1.1) is a special case of (2.1). Shintani [35], [36] considered the situation when all a_{nj} are positive (with characters and additional constant terms). Shintani actually treated the case $s_1 = \cdots = s_N$, but Hida [11] introduced multi-variable Shintani zeta-functions.

Other typical examples of (2.1) are the Mordell-Tornheim multiple series

$$\zeta_{MT,r}(s_1, \ldots, s_r; s_{r+1}) = \sum_{m_1=1}^{\infty} \cdots \sum_{m_r=1}^{\infty} m_1^{-s_1} \cdots m_r^{-s_r} (m_1 + \cdots + m_r)^{-s_{r+1}} \tag{2.2}$$

and the Apostol-Vu multiple series

$$\zeta_{AV,r}(s_1, \ldots, s_r; s_{r+1}) = \sum \cdots \sum_{1 \leq m_1 < \cdots < m_r < \infty} m_1^{-s_1} \cdots m_r^{-s_r} (m_1 + \cdots + m_r)^{-s_{r+1}}. \tag{2.3}$$

Both of the above series (2.2) and (2.3) were introduced in the author's paper [28], though the history of some special cases goes back to Tornheim [38], Mordell [34], and Apostol and Vu [4]. The following theorem has been proved by the author in [24] for the case $r = 2$, and in [28] for general r.

Theorem 2. *The series (2.2) and (2.3) can be continued meromorphically to* $\mathbf{C}^{r+1}$*. The possible singularities of (2.2) are located only on the subset of* $\mathbf{C}^{r+1}$ *defined by one of the following equations:*

$$\sum_{a=1}^{h} s_{j_a} + s_{r+1} = h - \ell\left(1 - \left[\frac{h}{r}\right]\right)$$

$$(1 \leq h \leq r,\ 1 \leq j_1 < \cdots < j_h \leq r,\ \ell \in \mathbf{N}_0).$$

Also, the possible singularities of (2.3) are located only on the subset of $\mathbf{C}^{r+1}$ *defined by one of the following equations:*

$$s_i + \cdots + s_{r+1} = r + 1 - i - \ell \quad (1 \leq i \leq r,\ \ell \in \mathbf{N}_0).$$

The proof of Theorem 2 in [24] [28] is again based on (1.3). As in the case of the Euler-Zagier sum, using (1.3) we obtain

$$\zeta_{MT,r}(s_1, \ldots, s_r; s_{r+1}) = \frac{1}{2\pi i} \int_{(c)} \frac{\Gamma(s_{r+1} + z)\Gamma(-z)}{\Gamma(s_{r+1})} \times \zeta_{MT,r-1}(s_1, \ldots, s_{r-1}; s_{r+1} + z)\zeta(s_r - z)dz, \tag{2.4}$$

where $\zeta(\cdot)$ is the Riemann zeta-function. In the case of the Apostol-Vu series, we have

$$\varphi_{j,r}(s_1, \ldots, s_j; s_{j+1}, \ldots, s_r; s_{r+1}) = \frac{1}{2\pi i} \int_{(c)} \frac{\Gamma(s_{r+1} + z)\Gamma(-z)}{\Gamma(s_{r+1})} \times \varphi_{j-1,r}(s_1, \ldots, s_{j-1}; s_j - z, s_{j+1}, \ldots, s_r; s_{r+1} + z)dz, \tag{2.5}$$

where

$$\varphi_{j,r}(s_1, \ldots,, s_j; s_{j+1}, \ldots, s_r; s_{r+1}) = \sum \cdots \sum_{1 \leq m_1 < \cdots < m_r < \infty} m_1^{-s_1} \cdots m_r^{-s_r} (m_1 + \cdots + m_j)^{-s_{r+1}}. \tag{2.6}$$

Note that $\varphi_{r,r} = \zeta_{AV,r}$ and

$$\varphi_{1,r}(s_1; s_2, \ldots, s_r; s_{r+1}) = \zeta_{EZ,r}(s_1 + s_{r+1}, s_2, \ldots, s_r).$$

The relations (2.4) and (2.5) imply the recursive sequences

$$\zeta_{MT,r} \to \zeta_{MT,r-1} \to \cdots \to \zeta_{MT,1} = \zeta$$

and

$$\zeta_{AV,r} = \varphi_{r,r} \to \varphi_{r-1,r} \to \cdots \to \varphi_{1,r} = \zeta_{EZ,r},$$

along which the proof of Theorem 2 goes inductively. The discussion in Section 1 implies another recursive sequence

$$\zeta_{EZ,r} \to \zeta_{EZ,r-1} \to \cdots \to \zeta_{EZ,1} = \zeta.$$

Thus we find a recursive structure in the family of multiple zeta-functions. This viewpoint is discussed in the last section of [28].

In [47], Maoxiang Wu introduced the χ-analogues of (2.2) and (2.3). Let $\chi_1, \ldots, \chi_r$ be Dirichlet characters of the same modulus q (≥ 2), and define

$$\begin{aligned} &L_{MT,r}(s_1, \ldots, s_r; s_{r+1}; \chi_1, \ldots, \chi_r) \\ &= \sum_{m_1=1}^{\infty} \cdots \sum_{m_r=1}^{\infty} \frac{\chi_1(m_1) \cdots \chi_r(m_r)}{m_1^{s_1} \cdots m_r^{s_r} (m_1 + \cdots + m_r)^{s_{r+1}}}, \end{aligned} \tag{2.7}$$

$$\begin{aligned} &L_{AV,r}(s_1, \ldots, s_r; s_{r+1}; \chi_1, \ldots, \chi_r) \\ &= \sum \cdots \sum_{1 \leq m_1 < \cdots < m_r < \infty} \frac{\chi_1(m_1) \cdots \chi_r(m_r)}{m_1^{s_1} \cdots m_r^{s_r} (m_1 + \cdots + m_r)^{s_{r+1}}}. \end{aligned} \tag{2.8}$$

These series are convergent absolutely for $\Re s_k > 1$ $(1 \leq k \leq r)$, $\Re s_{r+1} > 0$. Wu proved the following two theorems.

Theorem 3. (Wu [47]) *The series (2.7) can be continued meromorphically to* $\mathbf{C}^{r+1}$. *If none of the characters* $\chi_1, \ldots, \chi_r$ *are principal, then* $L_{MT,r}$ *is entire. If there are* k *principal characters* $\chi_{j_1}, \ldots, \chi_{j_k}$ *among them, then possible singularities are located only on the subsets of* $\mathbf{C}^{r+1}$ *defined by one of the following equations:*

$$\sum_{a=1}^{h} s_{j_{i(a)}} + s_{r+1} = h - \ell \left(1 - \left[\frac{h}{r}\right]\right),$$

where $1 \leq h \leq k$, $1 \leq i(1) < \cdots < i(h) \leq k$, $\ell \in \mathbf{N}_0$.

Theorem 4. (Wu [47]) *The series (2.8) can be continued meromorphically to* $\mathbf{C}^{r+1}$, *and possible singularities are located only on the subsets of* $\mathbf{C}^{r+1}$ *defined by one of the following equations:*

$$\sum_{a=1}^{h} s_{r-a+1} + s_{r+1} = h - \ell,$$

where $1 \leq h \leq r$, $\ell \in \mathbf{N}_0$.

Since Wu [47] is unpublished, we briefly outline his proof of these two theorems here.

The proof of Theorem 3 is just a direct generalization of the argument developed in [28]. We omit the details, only noting that the basic formula, corresponding to (2.4), is

$$L_{MT,r}(s_1,\ldots,s_r;s_{r+1};\chi_1,\ldots,\chi_r)=\frac{1}{2\pi i}\int_{(c)}\frac{\Gamma(s_{r+1}+z)\Gamma(-z)}{\Gamma(s_{r+1})}$$
$$\times L_{MT,r-1}(s_1,\ldots,s_{r-1};s_{r+1}+z;\chi_1,\ldots,\chi_{r-1})L(s_r-z,\chi_r)dz, \tag{2.9}$$

where $L(\cdot,\chi_r)$ is the Dirichlet L-function attached to χ_r.

To prove Theorem 4, we define

$$\Phi_{j,r}(s_1,\ldots,s_j;s_{j+1},\cdots,s_r;s_{r+1};\chi_1,\ldots,\chi_r)$$
$$=\sum_{1\leq m_1<\cdots<m_r<\infty}\cdots\sum\frac{\chi_1(m_1)\cdots\chi_r(m_r)}{m_1^{s_1}\cdots m_r^{s_r}(m_1+\cdots+m_j)^{s_{r+1}}}. \tag{2.10}$$

Then corresponding to (2.5), we have

$$\Phi_{j,r}(s_1,\ldots,s_j;s_{j+1},\cdots,s_r;s_{r+1};\chi_1,\ldots,\chi_r)=\frac{1}{2\pi i}\int_{(c)}\frac{\Gamma(s_{r+1}+z)\Gamma(-z)}{\Gamma(s_{r+1})}$$
$$\times\Phi_{j-1,r}(s_1,\ldots,s_{j-1};s_j-z,s_{j+1},\cdots,s_r;s_{r+1}+z;\chi_1,\ldots,\chi_r)dz. \tag{2.11}$$

Hence the induction argument goes along the sequence

$$L_{AV,r}=\Phi_{r,r}\to\Phi_{r-1,r}\to\cdots\to\Phi_{1,r},$$

but

$$\Phi_{1,r}(s_1;s_2,\ldots,s_r;\chi_1,\ldots,\chi_r)=L_{EZ,r}(s_1+s_{r+1},s_2,\ldots,s_r;\chi_1,\ldots,\chi_r),$$

whose basic analytic properties has already been discussed by Akiyama and Ishikawa [2].

As explained in Section 1, $L_{EZ,r}(s_1,\ldots,s_r;\chi_1,\ldots,\chi_r)$ can be expressed in terms of $\zeta_{EZ,r}(s_1,\ldots,s_r;\alpha_1,\ldots,\alpha_r)$, and the latter can be expressed as a sum involving

$$\zeta_{EZ,r-1}(s_1,\ldots,s_{r-2},s_{r-1}+s_r+j;\alpha_1,\ldots,\alpha_{r-1})\quad(-1\leq j\leq J).$$

Using these expressions, Wu [47] proved (by induction) that both of the functions $\zeta_{EZ,r}(s_1,\ldots,s_r;\alpha_1,\ldots,\alpha_r)$ and $L_{EZ,r}(s_1,\ldots,s_r;\chi_1,\ldots,\chi_r)$ are of polynomial order with respect to $|\Im s_1|, \ldots, |\Im s_r|$. Hence, using (2.11), we can show that $\Phi_{j,r}$ $(1 \leq j \leq r)$ is also of polynomial order. Therefore it is possible to shift the path of integration on the right-hand side of (2.11) freely. The remaining part of the proof is the same as in [28].

Special values of $\zeta_{MT,r}$, $\zeta_{AV,r}$ and their relatives have been studied by several mathematicians, including Tornheim, Mordell and Apostol and Vu themselves. Special values in the domain of absolute convergence have been further studied by Huard, Williams and Zhang [12], Subbarao and Sitaramachandrarao [37], and Tsumura in his recent series of papers [39], [40], [41], [42], [43] and [44]. In those papers, various relations among special values at integer arguments have been obtained. From an analytic point of view, however, it is important to reveal *whether those relations are valid only at integer points, or valid also at other values.* Tsumura [45], [46] discovered that some relations at integer points, proved in his previous articles [39], [40], [42], are actually valid continuously at other values. These relations of Tsumura may be regarded as functional relations among multiple zeta-functions.

Another functional relation has been found by the author [29], which implies, as a special case, a certain relation between $\zeta_{EZ,2}(s_1, s_2)$ and $\zeta_{EZ,2}(1 - s_2, 1 - s_1)$. More generally, in [29] the author defined the double Hurwitz-Lerch zeta-function

$$\zeta_2(s_1, s_2; \alpha, \beta, w) = \sum_{m=0}^{\infty} (\alpha + m)^{-s_1} \sum_{n=1}^{\infty} e^{2\pi i n\beta} (\alpha + m + nw)^{-s_2},$$

where $0 < \alpha \leq 1$, $0 \leq \beta \leq 1$, $w > 0$, and proved a certain relation between $\zeta_2(s_1, s_2; \alpha, \beta, w)$ and $\zeta_2(1 - s_2, 1 - s_1; 1 - \beta, 1 - \alpha, w)$. Note that the case $w = 1$ of this function was already introduced by Katsurada [17] in his study on the mean square of Lerch zeta-functions.

It is also possible to regard Proposition 1 of [29] as a double analogue of the functinal equation for Hurwitz-Lerch zeta-functions.

3. Multiple series defined by polynomials

In [28], it has been shown that any multiple series of the form (2.1) can be continued meromorphically to the whole space $\mathbf{C}^N$, by the method of Mellin-Barnes integrals. It is in fact possible to prove a much more general result by the same method. Let

$$P_n(X_1, \ldots, X_r) = \sum_{k=1}^{K(n)} a_k(n) X_1^{p_1(k,n)} \cdots X_r^{p_r(k,n)} \qquad (1 \leq n \leq N)$$

be polynomials, where $a_k(n) \in \mathbf{C}$, $p_j(k,n) \in \mathbf{N}_0$, and for any fixed j, at least one of $p_j(k,n)$ $(1 \le n \le N,\ 1 \le k \le K(n))$ is positive. We assume that $\Re a_k(n) > 0$ for all k and n. Hence

$$\theta_n = \max\{|\arg a_k(n)|\ ;\ 1 \le k \le K(n)\}$$

is smaller than $\pi/2$. Define

$$\zeta_r(s_1,\ldots,s_N;P_1,\ldots,P_N) = \sum_{m_1=1}^{\infty}\cdots\sum_{m_r=1}^{\infty} P_1(\mathbf{m})^{-s_1}\cdots P_N(\mathbf{m})^{-s_N}, \tag{3.1}$$

where $\mathbf{m} = (m_1,\ldots,m_r)$ and $s_n = \sigma_n + it_n \in \mathbf{C}$ $(1 \le n \le N)$. It is clear that there exists a positive constant $\sigma_a = \sigma_a(P_1,\ldots,P_N)$ such that the series (3.1) is absolutely convergent when $\sigma_n > \sigma_a$ for $1 \le n \le N$.

By a *multiple strip* we mean a set of the form

$$\{(s_1,\ldots,s_N) \in \mathbf{C}^N \mid \sigma_{n1} \le \sigma_n \le \sigma_{n2} \quad (1 \le n \le N)\}, \tag{3.2}$$

where σ_{n1}, σ_{n2} $(1 \le n \le N)$ are any fixed real numbers with $\sigma_{n1} < \sigma_{n2}$. By $\mathcal{F}(\cdot)$ we denote a quantity, not necessarily the same at each occurrence, which is of polynomial order with respect to the indicated variables.

Theorem 5. *When $\theta_n < \pi/2$ $(1 \le n \le N)$, the multiple zeta-function (3.1) can be continued meromorphically to the whole space $\mathbf{C}^N$. The possible singularities of it are located only on hyperplanes of the form*

$$c_1 s_1 + \cdots + c_N s_N = u(c_1,\ldots,c_N) - \ell \qquad (\ell \in \mathbf{N}_0), \tag{3.3}$$

where $c_1,\ldots,c_N \in \mathbf{N}_0$ and $u(c_1,\ldots,c_N)$ is an integer determined by $c_1,\ldots,c_N$. Moreover, the estimate

$$\zeta_r(s_1,\ldots,s_N;P_1,\ldots,P_N) = O\left(\mathcal{F}(t_1,\ldots,t_N)\prod_{n=1}^{N} e^{\theta_n|t_n|}\right) \tag{3.4}$$

holds uniformly in any multiple strip (3.2), except in neighbourhoods of possible polar sets (3.3).

The case $N = 1$ of (3.1) was first studied by Mellin [32], [33]. The Mellin-Barnes integral (1.3) already appeared in those papers. After Mellin, many people including K. Mahler, P. Cassou-Noguès, and P. Sargos continued his research. The multi-variable form (3.1) was first discussed by Lichtin [18], [19], [20], [21], and he proved the continuation of (3.1) when polynomials are hypoelliptic. Then Essouabri [9], [10] introduced the condition $\mathrm{H_0S}$, under which he proved the continuation. Here we do not give the exact definition of $\mathrm{H_0S}$, but only state that it is satisfied if all coefficients of polynomials have positive real parts. Moreover, though only the case $N = 1$ is discussed in [10], Essouabri mentioned in his thesis [9] that his result can be generalized to the multi-variable case. See also de Crisenoy [8], in which a twisted version of (3.1) (for general N) was studied.

Therefore, the meromorphic continuation of (3.1) in the above theorem is included, as a special case, in Essouabri's theorem. Nevertheless we give a proof of the above theorem here, because of several reasons. First, our method is quite different from Essouabri's and rather simple. Secondly, formula (3.5) below, which is the key of our proof, implies the recursive structure similar to those discussed in the preceding section. Thirdly, our method is suitable to obtain various explicit information, such as location of poles and order estimates, inductively. And finally, our method can be generalized to the case with general coefficients (similar to (1.8) and (1.9)).

Remark 1. When we write the (possible) polar sets of ζ_r in the form (3.3), we can choose $c_1, \ldots, c_N$ whose common greatest divisor is as small as possible. We call such tuples $(c_1, \ldots, c_N)$ *primitive*. Then, in the proof of Theorem 5 it will be shown that, for any fixed ζ_r, there are only finitely many primitive tuples $(c_1, \ldots, c_N)$ such that the (possible) polar sets of ζ_r are of the form (3.3).

Remark 2. For any fixed $c_1, \ldots, c_N$, there exists a positive integer $v(c_1, \ldots, c_N)$, by which the order of the singularity (3.3) is bounded uniformly for any ℓ.

Now we start the proof. We prove Theorem 5 with Remarks 1 and 2 by induction on

$$K(P_1, \ldots, P_N) = \prod_{n=1}^{N} K(n).$$

The argument is a generalization of the proof of Theorem 3 in [28].

First consider the case $K(P_1, \ldots, P_N) = 1$. Then $K(n) = 1$ for $1 \leq n \leq N$, so all the P_n's are monomials and

$$\begin{aligned}
&\zeta_r(s_1, \ldots, s_N; P_1, \ldots, P_N) \\
&= \sum_{m_1=1}^{\infty} \cdots \sum_{m_r=1}^{\infty} \prod_{n=1}^{N} \left(a_1(n) m_1^{p_1(1,n)} \cdots m_r^{p_r(1,n)} \right)^{-s_n} \\
&= a_1(1)^{-s_1} \cdots a_1(N)^{-s_N} \prod_{j=1}^{r} \zeta(p_j(1,1)s_1 + \cdots + p_j(1,N)s_N).
\end{aligned}$$

Hence all the assertions of Theorem 5, Remarks 1 and 2 clearly hold.

Now consider the case $K(P_1,\ldots,P_N)\geq 2$. Let $\sigma_a^*\geq\sigma_a$, and at first assume that $(s_1,\ldots,s_N)$ is in the region

$$\mathcal{B}^*=\{(s_1,\ldots,s_N)\mid \sigma_n>2\sigma_a^*(1\leq n\leq N)\}.$$

Since at least one $K(n)\geq 2$, changing the parameters if necessary, we may assume that $K(N)\geq 2$. Then

$$\begin{aligned}P_N(\mathbf{m})^{-s_N}&=\left(a_1(N)M_1(N)+\sum_{k=2}^{K(N)}a_k(N)M_k(N)\right)^{-s_N}\\&=\left(\sum_{k=2}^{K(N)}a_k(N)M_k(N)\right)^{-s_N}\left(1+\frac{a_1(N)M_1(N)}{\sum_{k=2}^{K(N)}a_k(N)M_k(N)}\right)^{-s_N},\end{aligned}$$

where $M_k(N)=m_1^{p_1(k,N)}\cdots m_r^{p_r(k,N)}$. Hence, applying (1.3), we obtain

$$\zeta_r(s_1,\ldots,s_N;P_1,\ldots,P_N)=\frac{1}{2\pi i}\int_{(\gamma)}\frac{\Gamma(s_N+z)\Gamma(-z)}{\Gamma(s_N)}$$
$$\times\sum_{m_1=1}^{\infty}\cdots\sum_{m_r=1}^{\infty}P_1(\mathbf{m})^{-s_1}\cdots P_{N-1}(\mathbf{m})^{-s_{N-1}}P_N^*(\mathbf{m})^{-s_N-z}P_N^{**}(\mathbf{m})^z dz, \tag{3.5}$$

where

$$P_N^*(\mathbf{m})=\sum_{k=2}^{K(N)}a_k(N)M_k(N),\qquad P_N^{**}(\mathbf{m})=a_1(N)M_1(N),$$

and we can choose γ as

$$-\sigma_N+\sigma_a<\gamma<-\sigma_a. \tag{3.6}$$

Then the multiple series on the right-hand side of (3.5) is absolutely convergent and is the zeta-function

$$\zeta_r(s_1,\ldots,s_{N-1},s_N+z,-z;P_1,\ldots,P_{N-1},P_N^*,P_N^{**}). \tag{3.7}$$

Since

$$\begin{aligned}K(P_1,\ldots,P_{N-1},P_N^*,P_N^{**})&=K(1)\times\cdots\times K(N-1)\times(K(N)-1)\times 1\\&<\prod_{n=1}^{N}K(n)=K(P_1,\ldots,P_N),\end{aligned}$$

by the induction assumption we see that (3.7) can be continued meromorphically to the whole space $\mathbf{C}^{N+1}$, and possible singularities are of the form

$$c_1 s_1 + \cdots + c_N s_N + (c_N - c_{N+1})z = u(c_1, \ldots, c_{N+1}) - \ell,$$

where $c_1, \ldots, c_{N+1}, \ell \in \mathbf{N}_0$ and $u(c_1, \ldots, c_{N+1}) \in \mathbf{Z}$. If $c_N = c_{N+1}$, then this is

$$c_1 s_1 + \cdots + c_N s_N = u(c_1, \ldots, c_N) - \ell \quad (\ell \in \mathbf{N}_0), \tag{3.8}$$

which is irrelevant to z. If $c_N - c_{N+1} = d_0 > 0$, then

$$z = d_0^{-1}\{-c_1 s_1 - \cdots - c_N s_N + u(c_1, \ldots, c_N, d_0)\} - d_0^{-1}\ell \quad (\ell \in \mathbf{N}_0), \tag{3.9}$$

and if $c_N - c_{N+1} = -e_0 < 0$, then

$$z = e_0^{-1}\{c_1 s_1 + \cdots + c_N s_N - u(c_1, \ldots, c_N, e_0)\} + e_0^{-1}\ell \quad (\ell \in \mathbf{N}_0). \tag{3.10}$$

We write the first term on the right-hand side of (3.9) (resp. (3.10)) as $D(s_1, \ldots, s_N; \mathbf{c})$ (resp. $E(s_1, \ldots, s_N; \mathbf{c})$) for brevity, where $\mathbf{c} = (c_1, \ldots, c_N)$. Denote the set of all primitive tuples $\mathbf{c} = (c_1, \ldots, c_N)$ appearing in (3.8) (resp. (3.9), (3.10)) by T_0 (resp. T_D, T_E). These sets are finite because of Remark 1. The above (3.9) and (3.10) can be poles, with respect to z, of the integrand on the right-hand side of (3.5). The other poles of the integrand are

$$z = -s_N - \ell \quad (\ell \in \mathbf{N}_0) \tag{3.11}$$

and

$$z = \ell \quad (\ell \in \mathbf{N}_0). \tag{3.12}$$

We can assume that σ_a^* is so large that all the poles (3.9) and (3.11) are on the left of the line $\Re z = \gamma$, while all the poles (3.10) and (3.12) are on the right of $\Re z = \gamma$.

Now, let $(s_1^0, \ldots, s_N^0)$ be any point in the space $\mathbf{C}^N$, and we show that the right-hand side of (3.5) can be continued meromorphically to $(s_1^0, \ldots, s_N^0)$.

First, remove the singularities of the form (3.8) from the integrand. These singularities are cancelled by the factor

$$(c_1 s_1 + \cdots + c_N s_N - u(c_1, \ldots, c_N) + \ell)^{v(c_1, \ldots, c_N)}$$

(by Remark 2 as a part of the induction assumption). Let L be a sufficiently large positive integer such that, if $\sigma_n \geq \Re s_n^0$ $(1 \leq n \leq N)$,

$$c_1 s_1 + \cdots + c_N s_N = u(c_1, \ldots, c_N) - L$$

does not hold for any $\mathbf{c}=(c_1,\ldots,c_N)\in T_0$. Define

$$\Phi(s_1,\ldots,s_N)=\prod_{\mathbf{c}\in T_0}\prod_{\ell=0}^{L-1}(c_1s_1+\cdots+c_Ns_N-u(c_1,\ldots,c_N)+\ell)^{v(c_1,\ldots,c_N)},$$

and rewrite (3.5) as

$$\zeta_r(s_1,\ldots,s_N;P_1,\ldots,P_N)=\Phi(s_1,\ldots,s_N)^{-1}J(s_1,\ldots,s_N), \tag{3.13}$$

where

$$J(s_1,\ldots,s_N)=\frac{1}{2\pi i}\int_{(\gamma)}\frac{\Gamma(s_N+z)\Gamma(-z)}{\Gamma(s_N)}\Phi(s_1,\ldots,s_N)$$
$$\times\zeta_r(s_1,\ldots,s_{N-1},s_N+z,-z;P_1,\ldots,P_{N-1},P_N^*,P_N^{**})dz. \tag{3.14}$$

Then the integrand on the right-hand side of (3.14) does not have singularities of the form (3.8) in the region $\sigma_n\geq\Re s_n^0$ $(1\leq n\leq N)$.

Since $\Phi(s_1,\ldots,s_N)^{-1}$ is meromorphic in the whole space, in order to complete the proof of the continuation, our remaining task is to show the continuation of $J(s_1,\ldots,s_N)$. Let M be a positive integer, and $s_n^*=s_n^0+M$ $(1\leq n\leq N)$. We may choose M so large that $(s_1^*,\ldots,s_N^*)\in\mathcal{B}^*$. Let $\mathcal{I}_1$ be the set of all imaginary parts of the poles (3.9) and (3.11), and $\mathcal{I}_2$ be the set of all imaginary parts of the poles (3.10) and (3.12), for $(s_1,\ldots,s_N)=(s_1^*,\ldots,s_N^*)$.

Case 1. In the case $\mathcal{I}_1\cap\mathcal{I}_2=\emptyset$, we join $D(s_1^*,\ldots,s_N^*;\mathbf{c})$ and $D(s_1^0,\ldots,s_N^0;\mathbf{c})$ by the segment $S(D;\mathbf{c})$ which is parallel to the real axis. Similarly join $E(s_1^*,\ldots,s_N^*;\mathbf{c})$ and $E(s_1^0,\ldots,s_N^0;\mathbf{c})$ by the segment $S(E;\mathbf{c})$, and join $-s_N^*$ and $-s_N^0$ by the segment $S(N)$. Since $\mathcal{I}_1\cap\mathcal{I}_2=\emptyset$, we can deform the path $\Re z=\gamma$ to obtain a new path $\mathcal{C}$ from $\gamma-i\infty$ to $\gamma+i\infty$, such that all the segments $S(D;\mathbf{c})$ and $S(N)$ are on the left of $\mathcal{C}$, while all the segments $S(E;\mathbf{c})$ and the poles (3.12) are on the right of $\mathcal{C}$ (see Fig.1). Then we have

$$J(s_1,\ldots,s_N)=\frac{1}{2\pi i}\int_{\mathcal{C}}\frac{\Gamma(s_N+z)\Gamma(-z)}{\Gamma(s_N)}\Phi(s_1,\ldots,s_N)$$
$$\times\zeta_r(s_1,\ldots,s_{N-1},s_N+z,-z;P_1,\ldots,P_{N-1},P_N^*,P_N^{**})dz \tag{3.15}$$

in a sufficiently small neighbourhood of $(s_1^*,\ldots,s_N^*)$. Next, on the right-hand side of (3.15), we move $(s_1,\ldots,s_N)$ from $(s_1^*,\ldots,s_N^*)$ to $(s_1^0,\ldots,s_N^0)$ with keeping the values of imaginary parts of each s_n. Since ζ_r in the integrand satisfies an estimate of the form (3.4) by the induction assumption, this procedure is possible; and, in the course of this procedure, the path $\mathcal{C}$ does not cross any poles of the integrand. Hence the

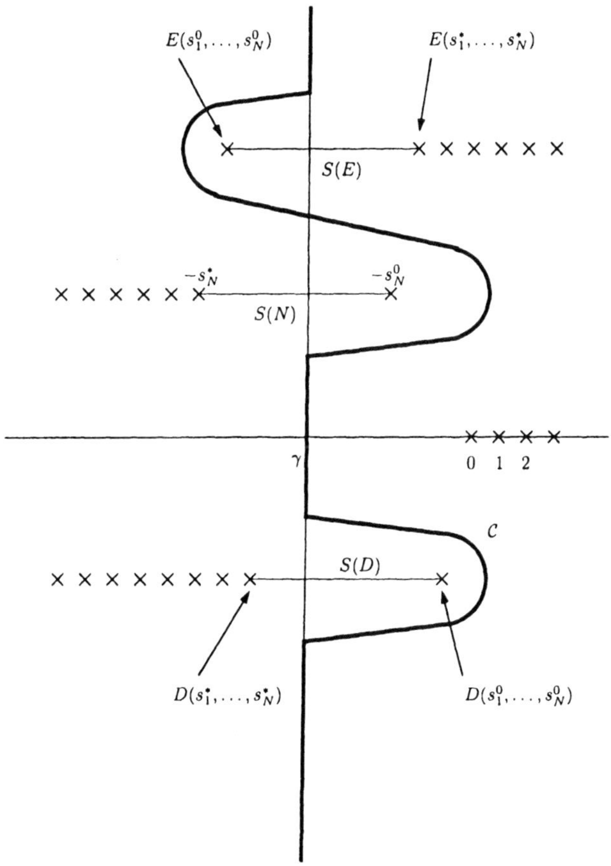

Fig.1

expression (3.15) gives the holomorphic continuation of $J(s_1,\ldots,s_N)$ to a neighbourhood of $(s_1^0,\ldots,s_N^0)$.

Case 2. Next consider the case $\mathcal{I}_1 \cap \mathcal{I}_2 \neq \emptyset$. Then the imaginary part of some member of $\{D(s_1^*,\ldots,s_N^*;\mathbf{c}),\ -s_N^* \mid \mathbf{c} \in T_D\}$ coincides with the imaginary part of some member of $\{E(s_1^*,\ldots,s_N^*;\mathbf{c}),\ 0 \mid \mathbf{c} \in T_E\}$. We consider the case

$$\Im D(s_1^*,\ldots,s_N^*;\mathbf{c}_1) = \Im E(s_1^*,\ldots,s_N^*;\mathbf{c}_2) \tag{3.16}$$

for some $\mathbf{c}_1$ and $\mathbf{c}_2$, because other cases can be treated similarly. The associated poles are $D(s_1^*,\ldots,s_N^*;\mathbf{c}_1) - d_0^{-1}\ell_1$ and $E(s_1^*,\ldots,s_N^*;\mathbf{c}_1) + e_0^{-1}\ell_2$ $(\ell_1,\ell_2 \in \mathbf{N}_0)$. When $(s_1^*,\ldots,s_N^*)$ is moved to $(s_1^0,\ldots,s_N^0)$, these poles are moved to $D(s_1^0,\ldots,s_N^0;\mathbf{c}_1) - d_0^{-1}\ell_1$ and $E(s_1^0,\ldots,s_N^0;\mathbf{c}_1) + e_0^{-1}\ell_2$, respectively. In the case

$$\Re D(s_1^0,\ldots,s_N^0;\mathbf{c}_1) - d_0^{-1}\ell_1 \neq \Re E(s_1^0,\ldots,s_N^0;\mathbf{c}_1) + e_0^{-1}\ell_2 \tag{3.17}$$

for any ℓ_1 and ℓ_2, we modify the argument in Case 1 as follows. Let η be a small positive number, and consider the oriented polygonal path $S'(D;\mathbf{c}_1)$ joining the points $D(s_1^*,\ldots,s_N^*;\mathbf{c}_1)$, $D(s_1^*+i\eta,\ldots,s_N^*+i\eta;\mathbf{c}_1)$, $D(s_1^0+i\eta,\ldots,s_N^0+i\eta;\mathbf{c}_1)$, and then $D(s_1^0,\ldots,s_N^0;\mathbf{c}_1)$ in that order. Similarly define the path $S'(E;\mathbf{c}_2)$ which joins $E(s_1^*,\ldots,s_N^*;\mathbf{c}_2)$, $E(s_1^*+i\eta,\ldots,s_N^*+i\eta;\mathbf{c}_2)$, $E(s_1^0+i\eta,\ldots,s_N^0+i\eta;\mathbf{c}_2)$, and then $E(s_1^0,\ldots,s_n^0;\mathbf{c}_2)$. Then $S'(D;\mathbf{c}_1)$ lies on the lower side of the line

$$\mathcal{L}=\{z\mid \Im z=\Im D(s_1^*,\ldots,s_N^*;\mathbf{c}_1)=\Im E(s_1^*,\ldots,s_N^*;\mathbf{c}_2)\},$$

while $S'(E;\mathbf{c}_2)$ lies on the upper side of $\mathcal{L}$. Because of (3.17), we can define the path $\mathcal{C}'$, which is almost the same as $\mathcal{C}$, but near the line $\mathcal{L}$ we draw $\mathcal{C}'$ such that it separates

$$\bigcup_{\ell_1\in\mathbf{N}_0}(S'(D;\mathbf{c}_1)-d_0^{-1}\ell_1)\qquad\text{and}\qquad\bigcup_{\ell_2\in\mathbf{N}_0}(S'(E;\mathbf{c}_2)+e_0^{-1}\ell_2)$$

(see Fig.2). Then the expression (3.15), with replacing $\mathcal{C}$ by $\mathcal{C}'$, is valid in a sufficiently small neighbourhood of $(s_1^*,\ldots,s_N^*)$. When $(s_1,\ldots,s_N)$ moves along the polygonal path joining $(s_1^*,\ldots,s_N^*)$, $(s_1^*+i\eta,\ldots,s_N^*+i\eta)$, $(s_1^0+i\eta,\ldots,s_N^0+i\eta)$, and then $(s_1^0,\ldots,s_N^0)$ in that order, the path $\mathcal{C}'$ encounters no pole, hence we obtain the holomorphic continuation.

Case 3. The remaining case is that

$$D(s_1^0,\ldots,s_N^0;\mathbf{c}_1)-d_0^{-1}\ell_1=E(s_1^0,\ldots,s_N^0;\mathbf{c}_2)+e_0^{-1}\ell_2 \tag{3.18}$$

holds for some ℓ_1 and ℓ_2. Then this might hold for some other pairs of (ℓ_1,ℓ_2). In this case we consider the path $\mathcal{C}''$ which is almost the same as $\mathcal{C}$, but near the line $\mathcal{L}$ we only require that $S(D;\mathbf{c}_1)$ is on the left of $\mathcal{C}''$, and that the points $E(s_1^*,\ldots,s_N^*;\mathbf{c}_2)+e_0^{-1}\ell_2$, $E(s_1^0,\ldots,s_N^0;\mathbf{c}_2)+e_0^{-1}\ell_2$ are not on $\mathcal{C}''$ for any ℓ_2. When we deform the path $\Re z=\gamma$ on the right-hand side of (3.14) to $\mathcal{C}''$, we might encounter several poles of the form (3.10). Then we move $(s_1,\ldots,s_N)$ from $(s_1^*,\ldots,s_N^*)$ to $(s_1^0,\ldots,s_N^0)$; again the path might encounter several poles of the same type. Hence, in a sufficiently small neighbourhood U of $(s_1^0,\ldots,s_N^0)$, the integral $J(s_1,\ldots,s_N)$ has the expression

$$\begin{aligned}&-R(s_1,\ldots,s_N)+\frac{1}{2\pi i}\int_{\mathcal{C}''}\frac{\Gamma(s_N+z)\Gamma(-z)}{\Gamma(s_N)}\Phi(s_1,\ldots,s_N)\\&\times\zeta_r(s_1,\ldots,s_{N-1},s_N+z,-z;P_1,\ldots,P_{N-1},P_N^*,P_N^{**})dz,\end{aligned} \tag{3.19}$$

where $R(s_1,\ldots,s_N)$ is the sum of residues of the above poles. Hence we see that $R(s_1,\ldots,s_N)$ is a (finite) sum of residues of the form

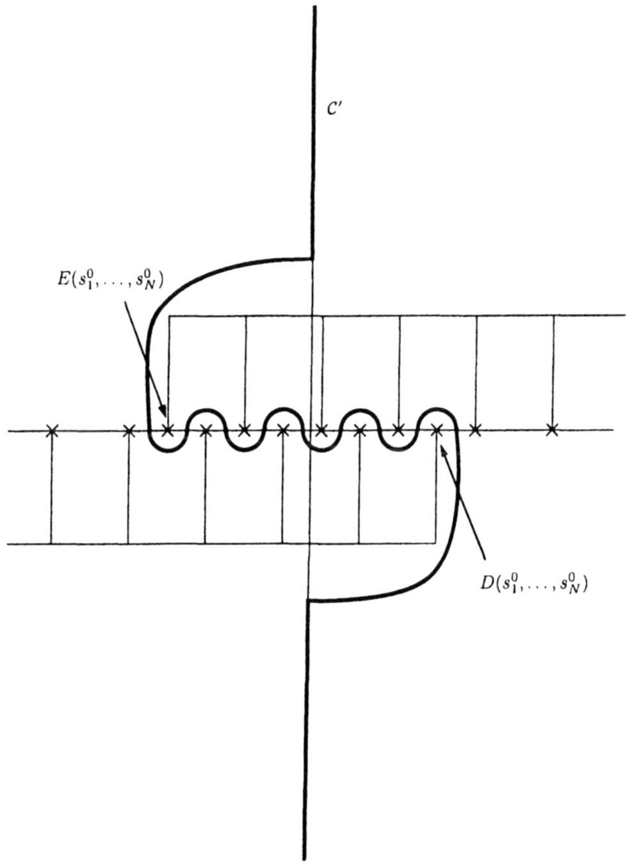

Fig.2

$\Gamma(s_N)^{-1}\Phi(s_1,\ldots,s_N)R(\ell_2)$, where

$$R(\ell_2)=\frac{1}{(h-1)!}\frac{d^{h-1}}{dz^{h-1}}\Big\{(z-z(\ell_2))^h\Gamma(s_N+z)\Gamma(-z)$$

$$\times\zeta_r(s_1,\ldots,s_{N-1},s_N+z,-z;P_1,\ldots,P_{N-1},P_N^*,P_N^{**})\Big\}\Big|_{z=z(\ell_2)} \qquad (3.20)$$

with $z(\ell_2)=E(s_1,\ldots,s_N;\mathbf{c}_2)+e_0^{-1}\ell_2$, if the order of the pole is h. This implies that all possible singularities of $R(s_1,\ldots,s_N)$ are polar sets. Therefore expression (3.19) gives the meromorphic continuation of $J(s_1,\ldots,s_N)$ to U.

Now the meromorphic continuation of $\zeta_r(s_1,\ldots,s_N;P_1,\ldots,P_N)$ has been proved. Next we show that all the possible polar sets of $\zeta_r(s_1,\ldots,s_N;P_1,\ldots,P_N)$ are of the form (3.3). This is clear for the polar sets of $\Phi(s_1,\ldots,s_N)^{-1}$. The polar sets of $J(s_1,\ldots,s_N)$ only appear in Case 3. Hence from condition (3.18) we see that those polar sets are also of the form (3.3). The assertions of Remarks 1 and 2 are easily verified from the above argument.

Lastly we show the assertion on the order of $\zeta_r(s_1,\ldots,s_N;P_1,\ldots,P_N)$. Since (3.7) satisfies an estimate of the form (3.4) by the induction assumption, the integral on the right-hand side of (3.15) clearly satisfies the same type of estimate with respect to $t_1,\ldots,t_{N-1}$. As for t_N, using Stirling's formula we find that the integral is

$$\ll \int_{-\infty}^{\infty} \exp\left(\frac{\pi}{2}(|t_N|-|t_N+y|-|y|)+\theta_N(|t_N+y|+|y|)\right)\mathcal{F}(t_N,y)dy, \tag{3.21}$$

which is $O(e^{\theta_N|t_N|}\mathcal{F}(t_N))$ by Lemma 4 of [26]. Hence we obtain the desired assertion in Case 1, and the treatment of Case 2 is similar.

In Case 3, we have to estimate $R(s_1,\ldots,s_N)$. Since

$$R(\ell_2)=\frac{1}{2\pi i}\int_{\mathcal{K}}\Gamma(s_N+z)\Gamma(-z) \\ \times\zeta_r(s_1,\ldots,s_{N-1},s_N+z,-z;P_1,\ldots,P_{N-1},P_N^*,P_N^{**})dz,$$

where $\mathcal{K}$ is a small circle around the point $z(\ell_2)$, we see that $R(s_1,\ldots,s_N)$ satisfies an estimate of the form (3.4) with respect to $t_1,\ldots,t_{N-1}$. As for t_N, the relevant exponential factor is the same as the exponential factor in (3.21), and hence we can obtain the desired estimate as above. The proof of Theorem 5 is now complete.

Acknowledgments

The author expresses his gratitude to Dr. Yasushi Komori for helping him to draw the figures, and to Professor Hirofumi Tsumura for pointing out several misprints in the original version of the manuscript.

References

[1] S. Akiyama, S. Egami and Y. Tanigawa, *Analytic continuation of multiple zeta functions and their values at non-positive integers*, Acta Arith. **98** (2001), 107-116.

[2] S. Akiyama and H. Ishikawa, *On analytic continuation of multiple L-functions and related zeta-functions*, in "Analytic Number Theory", C. Jia and K. Matsumoto (eds.), Developments in Math. Vol. 6, Kluwer, 2002, pp. 1-16.

[3] S. Akiyama and Y. Tanigawa, *Multiple zeta values at non-positive integers*, Ramanujan J. **5** (2001), 327-351.

[4] T. M. Apostol and T. H. Vu, *Dirichlet series related to the Riemann zeta function*, J. Number Theory **19** (1984), 85-102.

[5] T. Arakawa and M. Kaneko, *Multiple zeta values, poly-Bernoulli numbers, and related zeta functions*, Nagoya Math. J. **153** (1999), 189-209.

[6] T. Arakawa and M. Kaneko, *On multiple L-values*, J. Math. Soc. Japan, **56** (2004), 967-991.

[7] D. Bowman and D. M. Bradley, *Multiple polylogarithms: a brief survey*, in "q-series with Applications to Combinatorics, Number Theory and Physics", B. C. Berndt and K. Ono (eds.), Contem. Math. vol 291, Amer. Math. Soc., 2001, pp. 71-92.

[8] M. de Crisenoy, *Values at T-tuples of negative integers of twisted multivariable zeta series associated to polynomials of several variables*, Compositio Math., to appear.

[9] D. Essouabri, *Singularités des séries de Dirichlet associées à des polynômes de plusieurs variables et applications à la théorie analytique des nombres*, Thése, Univ. Henri Poincaré - Nancy I, 1995.

[10] D. Essouabri, *Singularités des séries de Dirichlet associées à des polynômes de plusieurs variables et applications en théorie analytique des nombres*, Ann. Inst. Fourier **47** (1997), 429-483.

[11] H. Hida, Elementary Theory of L-functions and Eisenstein Series, London Math. Soc. Student Text **26**, Cambridge Univ. Press, 1993.

[12] J. G. Huard, K. S. Williams and N. Y. Zhang, *On Tornheim's double series*, Acta Arith. **75** (1996), 105-117.

[13] H. Ishikawa, *On analytic properties of a multiple L-function*, in "Analytic Extension Formulas and their Applications", S. Saitoh et al. (eds.), Soc. Anal. Appl. Comput. Vol. 9, Kluwer, 2001, pp. 105-122.

[14] H. Ishikawa, *A multiple character sum and a multiple L-function*, Arch. Math. **79** (2002), 439-448.

[15] H. Ishikawa and K. Matsumoto, *On the estimation of the order of Euler-Zagier multiple zeta-functions*, Illinois J. Math. **47** (2003), 1151-1166.

[16] K. Kamano, *Multiple zeta values at non-positive integers and a generalization of Lerch's formula*, Tokyo J. Math. to appear.

[17] M. Katsurada, *An application of Mellin-Barnes' type integrals to the mean square of Lerch zeta-functions*, Collect. Math. **48** (1997), 137-153.

[18] B. Lichtin, *Poles of Dirichlet series and D-modules*, in "Théorie des Nombres/Number Theory", Proc. Intern. Number Theory Conf. (Laval 1987), J.-M. De Koninck and C. Levesque (eds.), Walter de Gruyter, 1989, pp. 579-594.

[19] B. Lichtin, *The asymptotics of a lattice point problem associated to a finite number of polynomials I*, Duke math. J. **63** (1991), 139-192; *II*, ibid. **77** (1995), 699-751.

[20] B. Lichtin, *Volumes and lattice points – proof of a conjecture of L. Ehrenpreis*, in "Singularities, Lille 1991", London Math. Soc. Lect. Note Vol. 201, J.-P. Brasselet (ed.), Cambridge Univ. Press, 1994, pp. 211-250.

[21] B. Lichtin, *Asymptotics determined by pairs of additive polynomials*, Compos. Math. **107** (1997), 233-267.

[22] K. Matsumoto, *Asymptotic series for double zeta, double gamma, and Hecke L-functions*, Math. Proc. Cambridge Phil. Soc. **123** (1998), 385-405; *Corrigendum and addendum*, ibid, **132** (2002), 377-384.

[23] K. Matsumoto, *Asymptotic expansions of double gamma-functions and related remarks*, in "Analytic Number Theory", C. Jia and K. Matsumoto (eds.), Developments in Math. Vol. 6, Kluwer, 2002, pp. 243-268.

[24] K. Matsumoto, *On the analytic continuation of various multiple zeta-functions*, in "Number Theory for the Millennium II, Proc. of the Millennial Conference on Number Theory" M. A. Bennett et al. (eds.), A K Peters, 2002, pp. 417-440.

[25] K. Matsumoto, *Asymptotic expansions of double zeta-functions of Barnes, of Shintani, and Eisenstein series*, Nagoya Math. J. **172** (2003), 59-102.

[26] K. Matsumoto, *The analytic continuation and the asymptotic behaviour of certain multiple zeta-functions I*, J. Number Theory **101** (2003), 223-243.

[27] K. Matsumoto, *The analytic continuation and the asymptotic behaviour of certain multiple zeta-functions II*, in "Analytic and Probabilistic Methods in Number Theory", Proc. 3rd Intern. Conf. in Honour of J. Kubilius, (Palanga, Lithuania, 2001), A. Dubickas et al. (eds.), TEV, Vilnius, 2002, pp. 188-194.

[28] K. Matsumoto, *On Mordell-Tornheim and other multiple zeta-functions*, in "Proceedings of the Session in Analytic Number Theory and Diophantine Equations", D. R. Heath-Brown and B. Z. Moroz (eds.), Bonner Math. Schriften Nr. 360, Univ. Bonn, 2003, n. 25, 17pp.

[29] K. Matsumoto, *Functional equations for double zeta-functions*, Math. Proc. Cambridge Philos. Soc. **136** (2004), 1-7.

[30] K. Matsumoto and Y. Tanigawa, *The analytic continuation and the order estimate of multiple Dirichlet series*, J. Théorie des Nombres de Bordeaux **15** (2003), 267-274.

[31] K. Matsumoto and H. Tsumura, *Generalized multiple Dirichlet series and generalized multiple polylogarithms*, preprint.

[32] H. Mellin, *Eine Formel für den Logarithmus transcendenter Funktionen von endlichem Geschlecht*, Acta Soc. Sci. Fenn. **29**, no. 4 (1900), 49pp.

[33] H. Mellin, *Die Dirichlet'schen Reihen, die zahlentheoretischen Funktionen und die unendlichen Produkte von endlichem Geschlecht*, Acta Math. **28** (1904), 37-64.

[34] L. J. Mordell, *On the evaluation of some multiple series*, J. London Math. Soc. **33** (1958), 368-371.

[35] T. Shintani, *On evaluation of zeta functions of totally real algebraic number fields at non-negative integers*, J. Fac. Sci. Univ. Tokyo Sect. IA Math. **23** (1976), 393-417.

[36] T. Shintani, *On a Kronecker limit formula for real quadratic fields*, ibid. **24** (1977), 167-199.

[37] M. V. Subbarao and R. Sitaramachandrarao, *On some infinite series of L. J. Mordell and their analogues*, Pacific J. Math. **119** (1985), 245-255.

[38] L. Tornheim, *Harmonic double series*, Amer. J. Math. **72** (1950), 303-314.

[39] H. Tsumura, *On some combinatorial relations for Tornheim's double series*, Acta Arith. **105** (2002), 239-252.

[40] H. Tsumura, *On alternating analogues of Tornheim's double series*, Proc. Amer. Math. Soc. **131** (2003), 3633-3641.

[41] H. Tsumura, *Evaluation formulas for Tornheim's type of alternative double series*, Math. Comput. **73** (2003), 251-258.

[42] H. Tsumura, *Multiple harmonic series related to multiple Euler numbers*, J. Number Theory **106** (2004), 155-168.

[43] H. Tsumura, *On Mordell-Tornheim zeta values*, Proc. Amer. Math. Soc. **133** (2005), 2387-2393.

[44] H. Tsumura, *On a class of combinatorial relations for the Mordell-Tornheim triple series*, preprint.

[45] H. Tsumura, *On functional relations between the Mordell-Tornheim double zeta functions and the Riemann zeta function*, preprint.

[46] H. Tsumura, *Certain functional relations for the double harmonic series related to the double Euler numbers*, J. Austral. Math. Soc. to appear.

[47] Maoxiang Wu, *On analytic continuation of Mordell-Tornheim and Apostol-Vu L-functions* (in Japanese), Master Thesis, Nagoya University, 2003.

[48] D. Zagier, *Values of zeta functions and their applications*, in "First European Congress of Mathematics, Vol.II Invited Lectures (Part 2)", A. Joseph et al. (eds.), Progress in Math. Vol. 120, Birkhäuser, 1994, pp. 497-512.

[49] Jianqiang Zhao, *Analytic continuation of multiple zeta-functions*, Proc. Amer. Math. Soc. **128** (2000), 1275-1283.

CUBIC FIELDS AND MORDELL CURVES

Katsuya Miyake
Department of Mathematical Science, Faculty of Science and Engineering, Waseda University, 3-4-1 Okubo, Shinjuku-ku, Tokyo, Japan, 169-8555
miyakek@aoni.waseda.jp

Abstract The purpose of this article is to describe a relationship between several cubic polynomials and elliptic curves, and show a clearer view on it than that in the former half of our previous work [Mi-2003]. For a monic irreducible cubic polynomial $P(u)$ in u over $\mathbb{Q}$, the curve $E = E(P(u))$ defined by the equation $w^3 = P(u)$ is an elliptic curve whose j-invariant is equal to 0. We describe the set $E[\mathbb{Q}]$ of all rational points of E over $\mathbb{Q}$ by use of a root ξ of $P(u)$ as

$$\mathcal{W}(\xi) = \{\alpha = q\xi + r \mid \mathrm{N}_{K/\mathbb{Q}}(\alpha) = 1, q, r \in \mathbb{Q}\}.$$

Then we show that the short form of E is a Mordell curve, $y^2 = x^3 + k$, with a certain rational number k determined by the coefficients of $P(u)$. It is also pointed out that $E(P(u))$ is essentially dependent on the polynomial $P(u)$ rather than the cubic field $\mathbb{Q}(\xi)$ even though $E[\mathbb{Q}]$ is completely described by the subset $\mathcal{W}(\xi)$ of the cubic field.

Keywords: Elliptic Curves, Cubic Fields, Mordell Curves.

2000 Mathematics Subject Classification: Primary 11G05, 11R16

1. Introduction

In this paper we study those elliptic curves which are defined over the rational number field $\mathbb{Q}$ and whose rational points over $\mathbb{Q}$ are described by certain subsets of the associated cubic fields. The curve $E = E(P(u))$ we are interested in is defined by the equation

$$E : w^3 = P(u) := u^3 + au^2 + bu + c$$

where $P(u)$ is an irreducible cubic polynomial in u over $\mathbb{Q}$. One of the points at infinity is rational over $\mathbb{Q}$ (although other two are not). Hence

The author was partly supported by the Grant-in-Aid for Scientific Research (C) (2) No. 14540037, Japan Society for the Promotion of Science, while he prepared this work.

Number Theory: Tradition and Modernization, pp. 175–183
W. Zhang and Y. Tanigawa, eds.

E is an elliptic curve defined over $\mathbb{Q}$. Let $E[\mathbb{Q}]$ denote the set of all rational points of E over $\mathbb{Q}$. One of our main interests should be the Mordell-Weil rank of $E(P(u))$ which is an isogeny-invariant. We will not, however, go into its detailed study in this article.

Take a root ξ of $P(u)$ in the complex number field $\mathbb{C}$, and let $K = \mathbb{Q}(\xi)$ be the cubic field generated by ξ. Put

$$\mathcal{W}(\xi) = \{\alpha = q\xi + r \mid \mathrm{N}_{K/\mathbb{Q}}(\alpha) = 1, q, r \in \mathbb{Q}\}.$$

Then there is a bijective map from $\mathcal{W}(\xi)$ onto $E[\mathbb{Q}]$ which maps $1 \in \mathcal{W}(\xi)$ to the point at infinity in $E[\mathbb{Q}]$. Note that the subset $\mathcal{W}(\xi)$ of $K^\times$ is not well determined by the cubic field K itself but depends on ξ or, we may say, on the polynomial $P(u)$; indeed, if we take another root η of $P(u)$, the isomorphism between the cubic fields K and $\mathbb{Q}(\eta)$ bijectively maps $\mathcal{W}(\xi)$ to $\mathcal{W}(\eta)$. In this way we can completely describe the set of all rational points of E over $\mathbb{Q}$ by the subset $\mathcal{W}(\xi)$ of the cubic field K.

It is clear that $\mathcal{W}(\xi) = \mathcal{W}(s\xi + t)$ for $s, t \in \mathbb{Q}, s \neq 0$. Thus in Section 3 we reduce the curve $E(P(u))$ to two typical elliptic curves $F_0(a)$ and $E_0(b)$ which are defined as

$$\begin{aligned} F_0(a) &: w^3 = u^3 + a, \\ E_0(b) &: w^3 = R(b; u) := u^3 + bu + b, \end{aligned}$$

and parametrized respectively by $a, b \in \mathbb{Q}^\times$. The first $F_0(a)$ is isomorphic to the well known pure cubic twist $X^3 + Y^3 + aZ^3 = 0$ of the cubic Fermat curve. Its short form is a Mordell curve $M_5(a)$ defined as

$$M_5(a) : y^2 = x^3 - 2^4 3^3 a^2$$

(cf. e.g. Cassels [Ca-50]). In Section 6 we will show our main result as Theorem 1; it states that a short form of $E_0(b)$ is

$$M_7(B) : y^2 = x^3 - B^2(B + 3), \quad B = \frac{4b}{9}$$

and that this has a rational point of infinite order except two special values of the parameter. (We use the indices of curves here in such a way as they are to be consistent with the ones in the previous paper [Mi-2003].)

As is well known, every cubic field is generated by a root of $R(b; u) := u^3 + bu + b$ for a suitable choice of $b \in \mathbb{Q}$ (cf. Section 2). If we restrict ourselves to cyclic cubic fields, then

$$Q(s; u) := u^3 - (s - 3)u^2 - su - 1$$

is generic for them over $\mathbb{Q}$ (cf. e.g. Jensen, Ledet and Yui [JLY-2002]). Here we only note the fact that every cyclic cubic field is generated by a root of $Q(s;u)$ for a suitable $s \in \mathbb{Q}$, and the converse also holds as far as $Q(s;u)$ is irreducible over $\mathbb{Q}$. A short form of the elliptic curve

$$E_4(s) : w^3 = Q(s;u) = u^3 - (s-3)u^2 - su - 1$$

is

$$M_8(s) : y^2 = x^3 + 2^4(s^2 - 3s + 9)^2$$

(see Section 5, Proposition 7). We also determine the subfamily of $E_0(b)$ corresponding to $E_4(s)$ in Section 5, Proposition 6.

2. Miscellaneous Preliminaries

Let $P(u) = eu^3 + au^2 + bu + c$ be an irreducible cubic polynomial in u over $\mathbb{Q}$, and $E = E(P(u))$ is a curve defined by the equation

$$E(P(u)) : w^3 = P(u) = eu^3 + au^2 + bu + c.$$

Proposition 1. *One of the points of E at infinity is rational over $\mathbb{Q}$ if and only if the leading coefficient e of $P(u)$ is a cube in $\mathbb{Q}$ (other than 0). In this case, other two points of E at infinity are not rational over $\mathbb{Q}$ but over $\mathbb{Q}(\sqrt{-3})$.*

This is almost obvious, and the proof is omitted.

As we stated in the introduction, we suppose in this paper that one of the points of $E(P(u))$ at infinity is rational over $\mathbb{Q}$. Then by Proposition 1 we have $e = r^3, r \in \mathbb{Q}^\times$. Replacing the variable w in the defining equation of E by rw, we may hereafter assume $e = 1$.

Let ξ be a root of $P(u)$ in $\mathbb{C}$, and $K = \mathbb{Q}(\xi)$ be the cubic field generated by ξ. Furthermore, let $\mathcal{W}(\xi)$ be the subset of $K^\times$ defined in Section 1.

Proposition 2. *There exists a bijective map from $\mathcal{W}(\xi)$ onto $E[\mathbb{Q}]$ which maps $1 \in \mathcal{W}(\xi)$ to the point at infinity in $E[\mathbb{Q}]$.*

Proof. Let ξ, ξ' and ξ'' be the three roots of $P(u)$ in $\mathbb{C}$. Then we have

$$P(u) = (u - \xi)(u - \xi')(u - \xi'').$$

Hence for $k \in \mathbb{Q}$, we have $P(k) = \mathrm{N}_{K/\mathbb{Q}}(k - \xi)$. For $\alpha = q\xi + r \in \mathcal{W}(\xi)$ with $q \neq 0$, put $u = -r/q$ and $w = -1/q$. Then we have $w^3 = P(u)$ because $\mathrm{N}_{K/\mathbb{Q}}(\alpha) = 1$; that is, we have a point $(u, w) = (-r/q, -1/q)$ of $E[\mathbb{Q}]$. For $\alpha = r \in \mathbb{Q}$, $\mathrm{N}_{K/\mathbb{Q}}(\alpha) = r^3 = 1$ if and only if $\alpha = 1$. Define a map $\rho : \mathcal{W}(\xi) \longrightarrow E[\mathbb{Q}]$ by $\rho(\alpha) := (-r/q, -1/q)$ for $\alpha = q\xi + r \in \mathbb{Q}$ with $q \neq 0$ and $\rho(1) :=$ the point at infinity in $E[\mathbb{Q}]$. It is clear that ρ is well defined and injective. There is only one point at infinity in $E[\mathbb{Q}]$. Let $(u, w) = (k, l), k, l \in \mathbb{Q}$ be a point of $E[\mathbb{Q}]$. Then we have $l^3 = P(k)$ and hence $l \neq 0$ because $P(u)$ is irreducible over $\mathbb{Q}$. Therefore $(k, l) = \rho(\alpha)$ with $\alpha = (-1/l)\xi + k/l \in \mathcal{W}(\xi)$. The proof is completed. □

Lemma 1. $\mathcal{W}(\xi) = \mathcal{W}(s\xi + t)$ *for* $s, t \in \mathbb{Q}, s \neq 0$.

The proof, being almost obvious, is omitted.

3. The curves $F_0(a)$ and $E_0(b)$

Let us see that our curve $E(P(u))$ for $P(u) = u^3 + au^2 + bu + c$ is isomorphically reduced to either $F_0(a)$ or $E_0(b)$ defined in Section 1 without changing the set $\mathcal{W}(\xi)$.

First put $x = u + a/3$. Then we have $P(u) = x^3 + b'x + c'$ with

$$b' = -a^2/3 + b,$$
$$c' = 2a^3/27 - ab/3 + c.$$

Since $P(u)$ is irreducible over $\mathbb{Q}$, c' is not equal to 0. If $b' = 0$, then we have the curve

$$F_0(a) : w^3 = u^3 + a;$$

here we replaced x and c' by the letters u and a, respectively, for simplicity.

Suppose now $b' \neq 0$. Put $y = \frac{b'}{c'} \cdot x$. Then we see

$$\left(\frac{b'}{c'}\right)^3 \cdot P(u) = y^3 + \frac{b'^3}{c'^2} \cdot y + \frac{b'^3}{c'^2}.$$

We have, therefore, the curve

$$E_0(b) : w^3 = R(b; u) = u^3 + bu + b;$$

here we replaced $\frac{b'}{c'} \cdot w$, y and $\frac{b'^3}{c'^2}$ by the letters w, u and b, respectively, for simplicity. We easily see by Lemma 1 in the preceding section that the original $\mathcal{W}(\xi)$ coincides with either $\mathcal{W}(\sqrt[3]{a})$ for $F_0(a)$ or $\mathcal{W}(\eta)$ for $E_0(b)$ where η is a root of $R(b; u)$.

Thus we obtained the following proposition.

Proposition 3. *The elliptic curve* $E(P(u))$ *is isomorphic over* $\mathbb{Q}$ *to either* $F_0(a)$ *or* $E_0(b)$ *for some* a *or* b *in* $\mathbb{Q}$ *with the same subset* $\mathcal{W}(\xi)$ *of the cubic field* $\mathbb{Q}(\xi)$.

4. The curve $E_1(c)$

In our previous work [Mi-2003], we studied a family of elliptic curves

$$E_1(c) : w^3 = P(c; u) := u^3 + u^2 + c$$

where $P(c;u)$ is an irreducible cubic polynomial in $\mathbb{Q}[u]$. This polynomial is related to $R(b;u) := u^3 + bu + b$ as follows: as before let ξ be the fixed root of $R(b;u)$. Then ξ^{-1} is a root of $P(c;u)$ with $c = b^{-1}$. This relation may give us a morphism between the two algebraic curves $E_0(b)$ and $E_1(c)$. If we try to pick up such one, however, it may not be defined over $\mathbb{Q}$ as we do not have a natural correspondence between $\mathcal{W}(\xi)$ and $\mathcal{W}(\xi^{-1})$. Hence we should adopt the reduction process in the preceding section to obtain a homomorphism of elliptic curves defined over $\mathbb{Q}$.

Proposition 4. *Suppose $27c + 2 \neq 0$. Then $E_1(c)$ is isomorphic to $E_0(b)$ for $b = -27/(27c+2)^2$ over $\mathbb{Q}$.*

Proof. In this case, indeed, the defining equation

$$w^3 = u^3 + u^2 + c$$

of $E_1(c)$ is equivalent to

$$\left(\frac{-9w}{27c+2}\right)^3 = \left(\frac{-3(3u+1)}{27c+2}\right)^3 + b\left(\frac{-3(3u+1)}{27c+2}\right) + b$$

as is easily checked. □

Remark 1. In case of $27c + 2 = 0$, the cubic polynomial $P(c;u) = u^3 + u^2 + c$ is reducible over $\mathbb{Q}$.

It should be noted that c may not always be obtained from b in $\mathbb{Q}$ through the relation $b = -27/(27c+2)^2$. Hence the family of elliptic curves $E_1(c), c \in \mathbb{Q}$, covers a strict subfamily of $E_0(b), b \in \mathbb{Q}$.

We restate a few results on $E_1(c)$ given in [Mi-2003, Prop. 1] for the convenience of readers.

Proposition 5. (i) *The short form of $E_1(c)$ for $c \in \mathbb{Q} - \{0, -4/27\}$ is $M_1(c)$ which is defined by the equation*

$$M_1(c) : y^2 = x^3 - 2^4c(27c+4).$$

(ii) *$M_1(c)$ has a rational point*

$$(x_0, y_0) = (2^2 3^4 c^2 + 2^4 3c + 4/9,\ 2^3 3^6 c^3 + 2^4 3^4 c^2 + 60c - 8/27).$$

This is not a torsion point for $c \neq -2/27$, and hence the Mordell-Weil rank is greater than or equal to 1 if $c \neq -2/27$.
(iii) *The exceptional curve $M_1(-2/27) : y^2 = x^3 + (4/3)^3$ has Mordell-Weil rank $= 0$. Its torsion group is of order 2 and generated by the point $(-4/3, 0)$.*

5. The curve $E_4(s)$ associated to cyclic cubic fields

As we stated in Section 1, the polynomial

$$Q(s; u) := u^3 - (s-3)u^2 - su - 1$$

is generic for cyclic cubic fields over $\mathbb{Q}$ (cf. e.g. Jensen, Ledet and Yui [JLY-2002]). This implies that every cyclic cubic field is generated by a root of $Q(s; u)$ for a suitable $s \in \mathbb{Q}$, and the converse also holds as far as $Q(s; u)$ is irreducible over $\mathbb{Q}$. In this section we study the elliptic curve

$$E_4(s) : w^3 = Q(s; u) = u^3 - (s-3)u^2 - su - 1.$$

First we give a subfamily of $E_0(b)$ which is covered by $E_4(s)$.

Proposition 6. *Suppose $s \neq 3/2$. Then $E_4(s)$ is isomorphic to $E_0(b)$ for $b = -27(s^2 - 3s + 9)/(2s-3)^2$ over $\mathbb{Q}$.*

Proof. In this case, the defining equation

$$w^3 = u^3 - (s-3)u^2 - su - 1$$

of $E_4(s)$ is equivalent to

$$\left(\frac{9w}{2s-3}\right)^3 = \left(\frac{3(3u-s+3)}{2s-3}\right)^3 + b\left(\frac{3(3u-s+3)}{2s-3}\right) + b$$

as is easily checked. □

Remark 2. In case of $s = 3/2$, the cubic polynomial $Q(s; u) = u^3 - (s-3)u^2 - su - 1$ is reducible over $\mathbb{Q}$.

Proposition 7. *The short form of $E_4(s)$ is*

$$M_8(s) : y^2 = x^3 + 2^4(s^2 - 3s + 9)^2.$$

Let θ be a root of $Q(s; u)$ and $K = \mathbb{Q}(\theta)$. Then $-\theta - 1$ belongs to

$$\mathcal{W}(\theta) = \{\alpha = a\theta + b \mid \mathrm{N}_{K/\mathbb{Q}}(\alpha) = 1, a, b \in \mathbb{Q}\}.$$

The rational point $(w,u) = (1,-1)$ *of* $E_4(s)$ *corresponding to* $-\theta - 1$ *is mapped to the rational point*

$$(x, y) = (0,\ 4(s^2 - 3s + 9))$$

of $M_8(s)$. *This is a torsion point of order* 3.

The proof may be obtained in a standard manner and is omitted (cf. Section 6 below).

6. The short form of $E_0(b)$

In this final section, we show our main result on $E_0(b)$ which corresponds to Proposition 5 on $E_1(c)$. The author owes to M. Imaoka for a portion of the calculation.

Theorem 1. (i) *The short form of* $E_0(b)$ *for* $b \in \mathbb{Q} - \{0, -27/4\}$ *is* $M_7(B), B = 4b/9 \in \mathbb{Q} - \{0, -3\}$, *which is defined by the equation*

$$M_7(B) : y^2 = x^3 - B^2(B+3), \quad B = \frac{4b}{9}$$

(ii) $M_7(B)$ *has a rational point*

$$(x_0, y_0) = (B + 4,\ 3B + 8).$$

This is not a torsion point unless either $B = -4$ *or* $B = -8/3$, *and hence its Mordell-Weil rank is greater than or equal to* 1 *if* $B \neq -4, -8/3$.

(iii) *The exceptional curves* $M_7(-4) : y^2 = x^3 + 16$ *and* $M_7(-8/3) : y^2 = x^3 - (4/3)^3$ *have Mordell-Weil rank* $= 0$. *Their torsion groups are of order* 3 *and of* 2, *respectively, and generated by the point* (x_0, y_0) $(= (0,-4)$ *and* $= (4/3, 0)$, *respectively).*

Proof. (i) Put $W = w - u$. Then we obtain

$$3Wu^2 + (3W^2 - b)u = -W^3 + b$$

from the equation $w^3 = u^3 + bu + b$. To make a perfect square out of the left-hand side, multiply both sides by $12W$, and get

$$(6Wu + 3W^2 - b)^2 = -3W^4 - 6bW^2 + 12bW + b^2.$$

The constant term of the right-hand side is a square. Hence by multiplying both sides by b^2/W^4, we obtain

$$\begin{aligned} Y^2 &= (X-3)^4 + 12(X-3)^3 - 6b(X-3)^2 - 3b^2 \\ &= X^4 - 6(b+9)X^2 + 36(b+6)X - 3(b+9)^2, \end{aligned}$$

where we put

$$Y = \frac{b(6Wu + 3W^2 - b)}{W^2} = -\frac{b^2}{W^2} + \frac{6bu}{W} + 3b$$
$$X = \frac{b}{W} + 3.$$

Then we utilize the method of Mordell [Mo-69, Ch.10, Th.2]. If we put $\alpha = b + 9, \beta = 9(b + 6)$, and $\gamma = -3(b + 9)^2$ then we have $3\alpha^2 + \gamma = 0$ and $-\alpha\gamma - \beta^2 + \alpha^3 = b^2(4b + 27)$. Therefore if we put

$$\begin{cases} x = 2X^2 + 2Y - 2\alpha, \\ y = 4X(X^2 + Y - 3\alpha) + 4\gamma, \end{cases}$$

we easily obtain

$$y^2 = x^3 - (4b)^2(4b + 27).$$

If we divide both sides of the equation by 27^2, then we have

$$\left(\frac{y}{27}\right)^2 = \left(\frac{x}{9}\right)^3 - \left(\frac{4b}{9}\right)^2 \cdot \left(\frac{4b}{9} + 3\right).$$

Hence we obtain (i) of the theorem by replacing $x/9$ and $y/27$ by x and y, respectively.

(ii) The former half is confirmed by direct calculation. We prove the latter half. Put $P = (x_0, y_0)$ for simplicity and suppose that it is of finite order. We know by Fueter [Fu-30] (also see [GPZ-98, Prop.1]) that the torsion part of the rational points of a Mordell curve is cyclic and of order 1, 2, 3 or 6; furthermore, a rational point on it is of order 2 or 3 if and only if its y-coordinate is equal to 0 or its x-coordinate is equal to 0, respectively. We see immediately that $y_0 = 0$ if and only if $B = -8/3$; in this case, we have $M(-8/3) : y^2 = x^3 - (4/3)^3$ and the point P is of order 2. It is also clear that $x_0 = 0$ if and only if $B = -4$; in this case, we have $M(-4) : y^2 = x^3 + 16$ and the order of P is 3.

Let us now check if P might be of order 6. Suppose that $x_0 \neq 0$ and $y_0 \neq 0$. The x-coordinate $x(2P)$ of $2P = P + P$ on $M_7(B)$ is given by

$$x(2P) = \frac{x_0({x_0}^3 - 8 \cdot B^2(B + 3))}{4 \cdot {y_0}^2};$$

cf. e.g. Silverman and Tate [ST-92]. We know $x_0 \neq 0$. Hence $x(2P) = 0$ if and only if

$${x_0}^3 - 8 \cdot B^2(B + 3) = -7B^3 - 12B^2 + 48B + 64 = 0,$$

and hence if and only if

$$(7B)^3 + 12(7B)^2 - 48 \cdot 7 \cdot (7B) - 64 \cdot 7^2 = 0.$$

Since the algebraic integer $7B$ is contained in $\mathbb{Q}$, it is a rational integer which divides the constant term $-64 \cdot 7^2$. Such a divisor is, however, unable to make the cubic polynomial equal to 0 as is easily checked. Therefore we conclude that the point P is not of order 6. This completes the proof of (ii).

(iii) It is sufficient to show that the Mordell-Weil ranks of the curves $M(-8/3) : y^2 = x^3 - (4/3)^3$ and $M(-4) : y^2 = x^3 + 16$ are equal to 0. The latter curve is on Table 1 of Cassels [Ca-50]; as for the former, it is isomorphic to $y^2 = x^3 - 3^3 \cdot 2^6$ which is also on the table.

The theorem is now completely proved. □

References

[Ca-50] J. W. S. Cassels, *The Rational Solutions of the Diophantine Equation, $y^2 = x^3 - D$*, Acta Math. **82** (1950), 243–273.

[Fu-30] R. Fueter, *Ueber kubische diophantische Gleichungen*, Comm. Math. Helv. **2** (1930), 69–89.

[GPZ-98] J. Gebel, A. Pethő and H. G. Zimmer, *On Mordell's Equation*, Comp. Math. **110** (1998), 335–367.

[JLY-2002] C. U. Jensen, A. Ledet and N. Yui, *Generic Polynomials, Constructive Aspects of the Inverse Galois Problem*, Cambridge U. P., Cambridge, 2002.

[Mi-2003] K. Miyake, *Some Families of Mordell Curves associated to Cubic Fields*, Jour. Comp. and Applied Math. Sciences **160** (2003), 217–231.

[Mo-14] L. J. Mordell, *The Diophantine Equation $y^2 - k = x^3$*, Proc. London Math. Soc. **13** (1914), 60–80.

[Mo-69] L. J. Mordell, *Diophantine Equations*, Acad. Press, London and New York, 1969.

[Si-86] J. H. Silverman, *The Arithmetic of Elliptic Curves*, Springer-Verlag, New York, 1986.

[ST-92] J. H. Silverman and John Tate, *Rational Points on Elliptic Curves*, Springer-Verlag, New York, 1992.

TOWARDS THE RECIPROCITY OF QUARTIC THETA-WEYL SUMS, AND BEYOND

Yoshinobu Nakai
Department of Mathematics, Faculty of Education and Human Sciences, University of Yamanashi, Kofu, Yamanashi, 400-8510, Japan
nakai@edu.yamanashi.ac.jp

Abstract In the van der Corput method, one of the three principal methods of exponential sums, one treats the vdC reciprocal function $f^*(y) = f(x_y) - yx_y$ $(f'(x_y) = y)$. In the case where $f(x)$ is a polynomial of degree K, i.e. the Weyl sum, one encounters a situation similar to the elliptic transformation and the quadratic and cubic polynomial cases were successfully treated by the author. The main idea is to introduce the decomposition $x = n^{K-1} + m$ and think of n as the global variable of the function $f(n) = \frac{K-1}{K}\alpha x^{\frac{K}{K-1}} F((\Xi))$. Then the k-th vdC reciprocal function $f^*(y)$ given by (2.9) is essentially of the similar form to f (in terms of n_1, $y = n_1^{K-1} + m_1$) and the length of interval of summation remains the same order, establishing the k-th reciprocation (under elliptic transformations). The behavior of f^* under parabolic transformation is postponed for later researches. Instead, the inductive representation of f^* is also given, with the concrete examples of the quartic case.

Keywords: the van der Corput method, the reciprocal function of van der Corput type with respect to the global variable n, theta-Weyl sums, Weyl sums, finite theta series, elliptic transformations

2000 Mathematics Subject Classification: 11L15

Throughout in what follows we use the following
Notation

$\mathbb{N}$ – the set of all positive integers
$\mathbb{Z}$ – the set of all integers
$\mathbb{R}$ – the set of all real numbers
$e(\xi) = \exp(2\pi\sqrt{-1}\xi)$ – the complex exponential function $(\xi \in \mathbb{R})$
N – a large variable, appearing as the limit of summation
N' – a variable associated to $N, 0 \le N' \ll N$, used as the length of the interval of summation

Number Theory: Tradition and Modernization, pp. 185–204
W. Zhang and Y. Tanigawa, eds.

k – a non-negative integer used in the context of the k-th iterate of "theta reciprocation"

$(\xi) = (\xi_0, \xi_1, \cdots, \xi_k) \in \mathbb{R}^{k+1}$ - a $(k+1)$ dimensional vector

K – a fixed integer indicating the degree of the polynomial-like function $f(n)$

α – the leading coefficient of $f(n)$, used as the variable of integration (in "circle method")

For a real-valued function f with the positive second derivative, n_y is defined by

$$f'(n_y) = y, \quad y \in (f'(N), f'(N+N')], \quad (n_y \in (N, N+N']),$$

$f^* = f^*(y) = f(n_y) - yn_y$ - the van der Corput reciprocal function in y (the vdC reciprocal function for short)

For a differentiable function $H = H((\Xi)) = H(\Xi_0, \Xi_1, \cdots, \Xi_k)$ and $\kappa > 0$, let

$$\mathfrak{G} = \mathfrak{G}_\kappa(H) = H - \kappa^{-1} \sum_{h=0}^{k} \left(\frac{\partial}{\partial \Xi_h} H \right) \Xi_h,$$

$$G = G((\Xi)) = \mathfrak{G}_{\frac{K}{K-1}}(F),$$

$$\mathfrak{L}(H) = \mathfrak{G}_1 \left(H((\Xi)) - H(0, 0, \cdots, 0) \right),$$

$$1 + X = \frac{(K-1)\alpha}{y} x_y, \quad \text{with} \quad x_y = n_y^{K-1} + m$$

$$Q = \frac{m}{x}, \quad \widetilde{Q} = \frac{(K-1)\alpha}{y} m,$$

$$\Xi_h = \frac{\xi_h}{x}, \quad \widetilde{\Xi}_h = \frac{(K-1)\alpha}{y} \xi_h, \quad 0 \le h \le k,$$

where m comes from the decomposition $x = n^{K-1} + m$ with $0 \le m < (n+1)^{K-1} - n^{K-1}$.

$\Psi(t) = \Psi_K(t)$ is the unique function determined by (2.6):

$$(\Psi_K(t) - t)^{K-2} (\Psi_K(t) + (K-2)t) = 1 \quad \text{and} \quad \Psi_K(0) = 1.$$

$$T = T(X) = \frac{1}{K-1} \Big\{ \widetilde{Q} + (1+X) \Big(G^{K-1} \Big(\frac{(\widetilde{\Xi})}{1+X} \Big) - 1 \Big) \Big\},$$

$$\begin{aligned} \widetilde{\Lambda}_{j-1} &= T(\widetilde{X}_{j-1}) \\ &= \frac{1}{K-1} \Big\{ \widetilde{Q} + (1+\widetilde{X}_{j-1}) \Big(G^{K-1} \Big(\frac{(\widetilde{\Xi})}{1+\widetilde{X}_{j-1}} \Big) - 1 \Big) \Big\}, \end{aligned}$$

$$1+\widetilde{X}_j-\widetilde{Q}=\widetilde{\Psi}_K(\widetilde{\Lambda}_{j-1})-\widetilde{\Lambda}_{j-1}, \quad (j\geq 0)$$
$$\widetilde{\Lambda}_{-1}=\widetilde{\Lambda}=T(0), \quad \widetilde{X}_{-1}=0,$$
$$1+\widetilde{X}_0-\widetilde{Q}=\Psi_K(\widetilde{\Lambda}_{-1})-\widetilde{\Lambda}_{-1}.$$

1. Introduction

One of the fundamental tools in analytic number theory invented in the 20th century is the method of exponential sums (or sometimes referred to as trigonometrical sums). There are three mutually related treatments of these sums known due to H. Weyl, I. M. Vinogradov, and J. G. van der Corput, respectively (cf. e.g. [2] or [10]). We are concerned with the third type of treatment, the van der Corput method, whose main ingredient is the following lemma.

Lemma 1. *Let $f(n)$ be a real-valued function on $[N, N+N']$ with the positive second derivative. For $y \in (f'(N)-0, f'(N+N')+0)$, we define n_y and f^* as in Notation: $f'(n_y)=y$ and $f^*=f^*(y)=f(n_y)-yn_y$. Then we have*

$$\sum_{N<n\leq N+N'} e(f(n)) = e\Big(\frac{1}{8}\Big)\sum_{y} f''(n_y)^{-\frac{1}{2}} e(f^*(y)) + O(\cdot),$$

where y runs through integer values in the mapped interval $(f'(N)-\eta; f'(N+N')+\eta]$ with $0<\eta<\frac{1}{10}$.

Lemma 1 is often applied, with f a polynomial (in which case the exponential sum is often called the "Weyl sum"), to an integral with respect to the leading coefficient of f up to a constant factor or sometimes applied repeatedly k times, i.e. the k-th iterate.

If $f(n)=\frac{K-1}{K}\alpha n^K$ for $K\in\mathbb{N}$ $(K>1)$, $\alpha>0$ and $N\leq n\leq N+N' \,(\ll N)$, then

$$f^*(y) = -\frac{K-1}{K}\cdot\big((K-1)\alpha\big)^{-\frac{1}{K-1}}\cdot y^{\frac{K}{K-1}} \tag{1.1}$$

and

$$(K-1)\alpha N^{K-1}\leq y\leq (K-1)\alpha(N+N')^{K-1}. \tag{1.2}$$

Hence for $K=2$, we have

$$f(n)=\frac{1}{2}\alpha n^2 \quad\text{and}\quad f^*(y)=-\frac{1}{2}\alpha^{-1}y^2, \tag{1.3}$$

so that both sums appearing in Lemma 1 are finite theta series and Lemma 1 may be thought of as claiming the theta reciprocation (i.e.

the elliptic transformation under the action of $\begin{pmatrix} 0 & -1 \\ 1 & 0 \end{pmatrix}$) sending α into $-\alpha^{-1}$. This situation may be termed as the "theta-Weyl" sum, meaning that the Weyl sum admits a treatment as a (finite) theta series.

Thus for $K = 2$ we have a (quadratic) theta-Weyl sum which we define by

$$\theta(\beta^{-1}, t; N, N + N') = \sum_{N<n\leq N+N'} e\Big(\frac{1}{2}\beta^{-1}(n+t)^2\Big)$$

with a slight change of notation, where $\beta > 0$, $t \in \mathbb{R}$ (cf. [6]; see also [7], Chap. 2 and [1]). If $\beta > 1$, then decomposing β into $\beta = a + \beta'^{-1}$ with $a = [\beta]$ (the integral part of β) and $\beta' > 1$, we have a relation between $\theta(\beta, t; N, N + N')$ and $\theta(\beta'^{-1}, t'; N, N + N')$ where t' admits a number-theoretic expression. This corresponds to the parabolic transformation of β under the action of $\begin{pmatrix} 1 & -a \\ 0 & 1 \end{pmatrix}$, which we shall keep off in this paper.

In view of (1.2) we distinguish two cases $\beta'^{-1}N' \ll 1$ or $\beta'^{-1}N' \gg 1$, and in the former case we appeal to another lemma of van der Corput while in the latter case, Lemma 1 in conjunction with (1.3) gives a θ-reciprocity-like formula between $\theta(\beta'^{-1}, t'; N, N')$ and $\theta(\beta', t''; \beta'^{-1}N, \beta'^{-1}(N + N'))$.

We may then take it for granted that the quadratic theta-Weyl sum is satisfactorily treated, since all variables and parameters contained are interpreted in a good number-theoretic manner, in terms of the regular continued fraction expansion of β.

The estimation of the error term being important in applying Lemma 1, we shall confine ourselves to the elliptic transformation of α, i,e, the vdC function f^* with the admissible error term $O(x^{\frac{K}{K-1}}N^{-K-1})$ and the y-interval $(f'(N), f'(N + N')]$.

However, (1.2) shows that the length of the y-interval is $\asymp N^{K-2}N'$, which is longer than N' if $K \geq 3$, invalidating the naïve theta-series like treatment.

It took the author quite a long time to overcome this difficulty and hit on the following natural setup: Viewing n as the global variable ranging over $(N, N+N']$ we write $x = n^{K-1}+m$, where $n^{K-1} \leq x = n^{K-1}+m < (n+1)^{K-1}$, or $0 \leq m < (n+1)^{K-1} - n^{K-1} \ll n^{K-2}$ and we view m as a local variable. Then as n ranges over $(N, N + N']$ we have $x \asymp N^{K-1}$ and as (1.2) implies (cf. also Lemma 2 below), $y \asymp N^{K-1}$, so that they are well-balanced.

With this in mind and with the purpose of incorporating the k-th reciprocation, we consider

$$f(n) = \frac{K-1}{K}\alpha x^{\frac{K}{K-1}} F((\Xi)), \tag{1.4}$$

where $x = n^{K-1} + m$ as above and $F((\Xi)) = F(\Xi_0, \ldots, \Xi_k)$ is a function in $(k+1)$ variables, having $(K+1)$-th derivatives for $|\Xi_0|, \ldots, |\Xi_k| \ll N^{-1}$, and $F((0)) = 1$.

In what follows we let $\xi_0, \ldots, \xi_k$ be real parameters satisfying $|\xi_h| \ll N^{K-2}$ $(h = 0, \ldots, k)$, so that according to Notation, $\Xi_h = \frac{\xi_h}{x}$ lies in the range $|\Xi_h| \ll N^{-1}$.

2. The van der Corput reciprocal function f^*

We are to find the vdC reciprocal function f^* to f given by (1.4). First we shall find n_y in Lemma 1.

Lemma 2. *Viewing n as a real variable ($x = n^{K-1} + m$), we have*

$$\frac{\partial}{\partial n} f(n) = (K-1)\alpha x (1-Q)^{\frac{K-2}{K-1}} G((\Xi)), \tag{2.1}$$

where, as in Notation, $Q = x^{-1}m, \Xi_h = x^{-1}\xi_h, (0 \le h \le k)$, and

$$G = G((\Xi)) = \mathfrak{G}_{\frac{K}{K-1}}(F(\Xi)) = F((\Xi)) - \frac{K-1}{K}\sum_{h=0}^{k} \Xi_h\Big(\frac{\partial}{\partial \Xi_h}F\Big).$$

We may choose an $n_y \in [N, N+N']$ such that

$$\frac{\partial}{\partial n} f(n)\big|_{n=n_y} = y, \tag{2.2}$$

or

$$\frac{(K-1)\alpha}{y} x_y \Big(1 - \frac{m}{x_y}\Big)^{\frac{K-2}{K-1}} G\Big(\frac{(\xi_h)}{x_y}\Big) = 1, \tag{2.3}$$

where we put

$$x_y = n_y^{K-1} + m. \tag{2.4}$$

Proof. (2.1) follows by formal differentiation with respect to n and factoring out $(K-1)\alpha x n^{K-2} x^{-\frac{K-2}{K-1}}$, thereby noting that

$$n^{K-2} x^{-\frac{K-2}{K-1}} = (1-Q)^{\frac{K-2}{K-1}}.$$

Secondly, as n and m range over the interval $[N, N+N']$ and $[0, O(n^{K-2})]$, respectively, x range over $(N^{K-1}, (N+N')^{K-1})$ whence so does $\frac{\partial}{\partial n} f(n)$, because $G = 1 + O(N^{-1})$. Hence, for any given $y \asymp N^{K-1}$, we may find an $n_y \in [N, N+N']$ satisfying (2.2). (2.1) and (2.2) together imply (2.3), completing the proof. □

Now appealing to Notation, we have the data $\frac{(K-1)\alpha}{y} x_y = 1 + X$, $\frac{m}{x_y} = \frac{\widetilde{Q}}{1+X}$, $\frac{(\xi_h)}{x_y} = \frac{(\widetilde{\Xi})}{1+X}$. Hence (2.3) raised to the the $(K-1)$-th power with these data reads

$$(1+X)(1+X-\widetilde{Q})^{K-2} G^{K-1}(\frac{\widetilde{\Xi}}{1+X}) = 1. \tag{2.5}$$

Note that by (2.4), $X \ll N^{-1}$ and $\widetilde{Q} \ll N^{-1}$.

Now as in Notation we introduce the unique function $\Psi(t) = \Psi_K(t)$ determined by the functional equation

$$(\Psi_K(t) - t)^{K-2}(\Psi_K(t) + (K-2)t) = 1, \quad \Psi_K(0) = 1 \tag{2.6}$$

and put

$$T = T(X) = \frac{1}{K-1}\Big(\widetilde{Q} + (1+X)\Big(G^{K-1}\Big(\frac{(\widetilde{\Xi})}{1+X}\Big) - 1\Big)\Big), \tag{2.7}$$

where $T \ll N^{-1}$.

We may now state

Lemma 3. *Notation being as above, we have*

$$1 + X - \widetilde{Q} = \Psi_K(T) - T. \tag{2.8}$$

Proof. Since $1 + X - \widetilde{Q} + T$ satisfies (2.6) with $t = T$, (2.8) follows by uniqueness of the solution. □

We may now state an expression for the vdC reciprocal function.

Lemma 4. *The vdC reciprocal function f^* of (1.4) is given in the first instance by*

$$\begin{aligned} f^*(y) = &-\frac{K-1}{K}((K-1)\alpha)^{-\frac{1}{K-1}} y^{\frac{K}{K-1}} \\ &\times \Big\{\frac{K}{K-1}(1+X-\widetilde{Q})^{\frac{1}{K-1}} - \frac{1}{K-1}(1+X-\widetilde{Q})^{-\frac{K(K-2)}{K-1}} \\ &- \frac{1}{K-1}(1+X)^{\frac{K}{K-1}}(F - G^K)\Big(\frac{(\widetilde{\Xi})}{1+X}\Big)\Big\}. \end{aligned} \tag{2.9}$$

Proof. In the definition of f^*,

$$f^*(y) = \frac{K-1}{K}\alpha x_y^{\frac{K}{K-1}} F\left(\frac{(\xi)}{x_y}\right) - y n_y,$$

we rewrite the second term as

$$yn_y = y(x_y - m)^{\frac{1}{K-1}} = y^{\frac{K}{K-1}} \left(\frac{x_y - m}{y}\right)^{\frac{1}{K-1}}$$
$$= y^{\frac{K}{K-1}}((K-1)\alpha)^{-\frac{1}{K-1}}(1 + X - \widetilde{Q})^{\frac{1}{K-1}}$$

and factor out $-\frac{K-1}{K}((K-1)\alpha)^{-\frac{1}{K-1}} y^{\frac{K}{K-1}}$ to obtain

$$f^*(y) = -\frac{K-1}{K}((K-1)\alpha)^{-\frac{1}{K-1}} y^{\frac{K}{K-1}} \left\{ \frac{K}{K-1}(1 + X - \widetilde{Q})^{\frac{1}{K-1}} \right.$$
$$- \frac{1}{K-1}\Big((1+X)G^{K-1}\Big(\frac{(\widetilde{\Xi})}{1+X}\Big)\Big)^{\frac{K}{K-1}}$$
$$\left. - \frac{1}{K-1}(1+X)^{\frac{K}{K-1}}(F - G^K)\Big(\frac{(\widetilde{\Xi})}{1+X}\Big) \right\}$$

after slight transformation. Using (2.5) we see that the second term in the third factor become $-\frac{1}{K-1}(1 + X - \widetilde{Q})^{-\frac{K(K-2)}{K-1}}$, completing the proof. □

Although they are not in perfect conformity because of different processes, it is instructive and worth while reviewing our previous result for the cubic case [8].

First recall the notations in this special case.

$$N < n \le N + N', \quad x = n^2 + m, \quad 0 \le m \le 2n,$$
$$|\xi_h| \ll N, \quad |\Xi_h| = \left|\frac{\xi_h}{x_y}\right| \ll N^{-1}, \quad 0 \le h \le k.$$

The vdC reciprocal function f^* of

$$f(n) = f(n; m; \xi_0, \xi_1, \cdots, \xi_k)$$
$$= \frac{2}{3}\alpha(n^2+m)^{\frac{3}{2}} F\left(\frac{\xi_0}{n^2+m}, \frac{\xi_1}{n^2+m}, \cdots, \frac{\xi_k}{n^2+m}\right)$$
$$= \frac{2}{3}\alpha x^{\frac{3}{2}} F\left(\frac{(\xi)}{x}\right)$$

is

$$f^*(y)$$
$$= -\frac{2}{3}(2\alpha)^{-\frac{1}{2}} y^{\frac{3}{2}} \left\{ \mathrm{tr}\frac{1+\sqrt{-1}}{2}\left(1 + \frac{\sqrt{-1}}{2}\left(\frac{2\alpha m}{y} + \left(G^2\Big(\frac{2\alpha}{y}(\xi)\Big) - 1\right)\right)\right)^{\frac{3}{2}} \right.$$

$$-\frac{1}{2}(F-G^3)\Big(\frac{2\alpha}{y}(\xi)\Big)+O(N^{-4})\Big\},$$

where $y=n_1^2+m_1$, $0\le m_1\le 2n_1$, $\sqrt{2\alpha}N\le n_1\le\sqrt{2\alpha}(N+N')$, and

$$\begin{aligned}&\mathrm{tr}\frac{1+\sqrt{-1}}{2}(1+U+\sqrt{-1}V)^{\frac{3}{2}}\\&=\frac{1+\sqrt{-1}}{2}(1+U+\sqrt{-1}V)^{\frac{3}{2}}+\frac{1-\sqrt{-1}}{2}(1+U-\sqrt{-1}V)^{\frac{3}{2}}.\end{aligned}$$

for $U,V\in\mathbb{R}, U,V\ll N^{-1}$.

This $\frac{2\alpha m}{y}$ ($\widetilde{Q}$ in Lemma 4) works as the Ξ_{k+1} in the $(k+1)$-st reciprocation.

We close this section by stating some properties of the function $\Psi_K(t)$.

$$\Psi_K'(t)=\frac{(K-2)t}{\Psi_K(t)+(K-3)t}, \tag{2.10}$$

$$\begin{aligned}\Psi_K(t)=&1+\frac{K-2}{2}t^2-(K-2)\cdot\frac{K-3}{3}t^3\\&+(K-2)\cdot\frac{2K-5}{2}\cdot\frac{K-4}{4}t^4\\&-(K-2)(K-3)\cdot\frac{3K-7}{3}\cdot\frac{K-5}{5}t^5\\&+(K-2)\cdot\frac{2K-7}{2}\cdot\frac{3K-8}{3}\cdot\frac{4K-9}{4}\frac{K-6}{6}t^6+\cdots.\end{aligned} \tag{2.11}$$

It seems that the coefficient of t^K is 0.

First a few functions are

$$\Psi_2(t)=1,\quad \Psi_3=\sqrt{1+t^2} \tag{2.12}$$

and

$$\Psi_4(t)=\sqrt[3]{\frac{1}{2}(1-2t^3)+\frac{1}{2}\sqrt{1-4t^3}}+\sqrt[3]{\frac{1}{2}(1-2t^3)-\frac{1}{2}\sqrt{1-4t^3}}.$$

3. Inductive representation for the vdC reciprocal function

To prove Proposition 1 below which provides an inductive representation for f^*, we first establish Lemma 5 below giving the same for the variable $x_y=n_y^{k-1}+m$ under the notation:

$$\widetilde{\Lambda}=T(0)=\frac{1}{K-1}(\widetilde{Q}+(\widetilde{G}^{K-1}-1)),$$

where

$$\widetilde{G} = G((\widetilde{\Xi}));$$

$$\widetilde{\Lambda}_{-1} = \widetilde{\Lambda},\ \widetilde{X}_{-1} = 0,\ 1 + \widetilde{X}_0 - \widetilde{Q} = \Psi_K(\widetilde{\Lambda}) - \widetilde{\Lambda},$$

in general $\widetilde{\Lambda}_{j-1}$ and $\widetilde{X}_j$ are defined by

$$\begin{aligned}\widetilde{\Lambda}_{j-1} &= T(\widetilde{X}_{j-1}) \\ &= \frac{1}{K-1}\Big\{\widetilde{Q} + (1+\widetilde{X}_{j-1})\Big(G^{K-1}\Big(\frac{(\widetilde{\Xi})}{1+\widetilde{X}_{j-1}}\Big) - 1\Big)\Big\},\end{aligned} \tag{3.1}$$

$$1 + \widetilde{X}_j - \widetilde{Q} = \Psi_K(\widetilde{\Lambda}_{j-1}) - \widetilde{\Lambda}_{j-1}, \quad j = 0, 1, 2, \cdots$$

Lemma 5. *For $j \geq 0$ we have*

$$\begin{aligned}1 + X - \widetilde{Q} &= \Psi_K(\widetilde{\Lambda}_{j-1}) - \widetilde{\Lambda}_{j-1} + E_{2j+3} \\ &= 1 + \widetilde{X}_j - \widetilde{Q} + E_{2j+3},\end{aligned} \tag{3.2}$$

where $E_{2j+3} = O(N^{-(2j+3)})$.

Proof. On writing $\frac{1}{1+X} = 1 - \frac{X}{1+X}$, we have, by Taylor's formula,

$$\begin{aligned}&G^{K-1}\Big(\frac{(\widetilde{\Xi})}{1+X}\Big) = G^{K-1}\Big((\widetilde{\Xi}) - \frac{X}{1+X}(\widetilde{\Xi})\Big) \\ &= \widetilde{G}^{K-1} - \sum_h \frac{X}{1+X}\widetilde{\Xi}_h \cdot \Big(\frac{\partial}{\partial \Xi_h}G^{K-1}\Big)((\widetilde{\Xi})) \\ &+ \frac{1}{2}\sum_{h_1,h_2} \frac{-X}{1+X}\widetilde{\Xi}_{h_1} \cdot \frac{-X}{1+X}\widetilde{\Xi}_{h_2}\Big(\frac{\partial}{\partial \Xi_{h_1}}\frac{\partial}{\partial \Xi_{h_2}}G^{K-1}\Big)\Big((\widetilde{\Xi})\Big(1 - \delta\frac{X}{1+X}\Big)\Big),\end{aligned}$$

with suitable $0 < \delta < 1$. Multiplying this by $1 + X$, subtracting $1 + X$, and using the notation

$$\mathfrak{L}(G^{K-1}(\widetilde{\Xi})) = \widetilde{G}^{K-1} - 1 - \sum_{h=0}^{k} \widetilde{\Xi}_h\Big(\frac{\partial}{\partial \Xi_h}G^{K-1}\Big),$$

we deduce that

$$\begin{aligned}&(1+X)\Big(G^{K-1}\Big(\frac{(\widetilde{\Xi})}{1+X}\Big) - 1\Big) = \widetilde{G}^{K-1} - 1 + \mathfrak{L}(G^{K-1})((\widetilde{\Xi}))X \\ &+ \frac{1}{2}\sum_{h_1,h_2} \widetilde{\Xi}_{h_1}\widetilde{\Xi}_{h_2}\Big(\frac{\partial}{\partial \Xi_{h_1}}\frac{\partial}{\partial \Xi_{h_2}}G^{K-1}\Big)\Big((\widetilde{\Xi})\Big(1 - \delta\frac{X}{1+X}\Big)\Big) \cdot \frac{X^2}{1+X}.\end{aligned} \tag{3.3}$$

Substituting the data $X = O(N^{-1})$, $\widetilde{\Xi}_h = O(N^{-1})$, and $\mathfrak{L}(G^{K-1}(\widetilde{\Xi})) = O(N^{-2})$, we conclude that

$$(1+X)\Big(G^{K-1}\Big(\frac{(\widetilde{\Xi})}{1+X}\Big)-1\Big) = \widetilde{G}^{K-1} - 1 + O(N^{-3}),$$

which proves (3.2) in the case $j = 0$.

Now suppose (3.2) is true for $j \geq 0$. Then by writing $1 + X = 1+\widetilde{X}_j+E_{2j+3}$ and $\frac{1}{1+\widetilde{X}_j+E_{2j+3}} = \frac{1}{1+\widetilde{X}_j} - \frac{E_{2j+3}}{(1+\widetilde{X}_j)(1+\widetilde{X}_j+E_{2j+3})}$, and arguing as above (with $\frac{X}{1+X}$ replaced by $\frac{E_{2j+3}}{(1+\widetilde{X}_j)(1+\widetilde{X}_j+E_{2j+3})}$), we obtain

$$\begin{aligned}
&(1+X)\Big(G^{K-1}\Big(\frac{(\widetilde{\Xi})}{1+X}\Big)-1\Big)\\
&= (1+\widetilde{X}_j+E_{2j+3})\Big\{G^{K-1}\Big(\frac{(\widetilde{\Xi})}{1+\widetilde{X}_j}\Big) - 1 - \frac{E_{2j+3}}{1+\widetilde{X}_j+E_{2j+3}}\\
&\quad\times\sum_h \frac{\widetilde{\Xi}_h}{1+\widetilde{X}_j}\Big(\frac{\partial}{\partial\Xi_h}G^{K-1}\Big)\Big(\frac{(\widetilde{\Xi})}{1+\widetilde{X}_j}\Big) + O(N^{-2}|E_{2j+3}|^2)\Big\}\\
&= (1+\widetilde{X}_j)\Big(G^{K-1}\Big(\frac{(\widetilde{\Xi})}{1+\widetilde{X}_j}\Big)-1\Big) + E_{2j+3}\cdot\mathfrak{L}(G^{K-1})\Big(\frac{(\widetilde{\Xi})}{1+\widetilde{X}_j}\Big)\\
&\quad + O\Big(N^{-2}|E_{2j+3}|^2\Big).
\end{aligned}$$

This reduces further, on account of the induction hypothesis, to

$$\begin{aligned}
&(1+X)\Big(G^{K-1}\Big(\frac{(\widetilde{\Xi})}{1+X}\Big)-1\Big)\\
&= (1+\widetilde{X_j})\Big(G^{K-1}\Big(\frac{(\widetilde{\Xi})}{1+X_j}\Big)-1\Big) + O\Big(N^{-(2(j+1)+3)}\Big). \qquad (3.4)
\end{aligned}$$

Substituting (3.4) in (2.7), we conclude (3.2) with $j+1$ in place of j, completing the proof. □

Note that to apply Lemma with admissible error term, we must choose j so that $2j + 3 \geq K + 1$, or $j \geq \frac{1}{2}(K-1)$; hence for $K = 3, 4$, we must take $j \geq 1$.

We are now in a position to state the main result of the paper.

Theorem 1. *The vdC reciprocal function f^* of f given by* (1.4) *admits the inductive representation with admissible error term:*

$$f^*(y) = -\frac{K-1}{K}((K-1)\alpha)^{-\frac{1}{K-1}}\, y^{\frac{K}{K-1}}\Big\{\frac{K}{K-1}\Big(\Psi_K(\widetilde{\Lambda}_{j-1}) - \widetilde{\Lambda}_{j-1}\Big)^{\frac{1}{K-1}}$$

$$- \frac{1}{K-1}\Big(\Psi_K(\widetilde{\Lambda}_{j-1}) + (K-2)\widetilde{\Lambda}_{j-1}\Big)^{\frac{K}{K-1}}$$

$$- \frac{1}{K-1}\Big(1+\widetilde{X}_{j-1}\Big)^{\frac{K}{K-1}}(F-G^K)\Big(\frac{(\widetilde{\Xi})}{1+\widetilde{X}_{j-1}}\Big) + O(N^{-(2j+5)})\Big\}, \tag{3.5}$$

where $j \ge \max(1, \frac{K-4}{2})$ *and for the* $(k+1)$*-st reciprocation* y *is to be decomposed as* $y = {n_1}^{K-1} + m_1$ *(as in* (2.4)), $((K-1)\alpha)^{\frac{1}{K-1}}N < n_1 \le ((K-1)\alpha)^{\frac{1}{K-1}}(N+N')$ *and* ${n_1}^{K-1} \le {n_1}^{K-1} + m_1 < (n_1+1)^{K-1}$, *the* O*-constants not depending on* N, m, ξ_h $(h = 0, 1, \cdots, k)$.

The proof is elementary but long, and we shall not state it here. Instead we shall prove Proposition 1, which gives an exact formula for $f^*(y)$ as well as an induction representation with a weaker error term:

Proposition 1. *The vdC recportial function* f^* *admits the exact representation*

$$f^*(y) = -\frac{K-1}{K}((K-1)\alpha)^{-\frac{1}{K-1}}y^{\frac{K}{K-1}}$$
$$\times \Big\{\frac{K}{K-1}\big(\Psi_K(\widetilde{\Lambda}_{j-1}) - \widetilde{\Lambda}_{j-1} + E_{2j+3}\big)^{\frac{1}{K-1}}$$
$$- \frac{1}{K-1}\big(\Psi_K(\widetilde{\Lambda}_{j-1}) - \widetilde{\Lambda}_{j-1} + E_{2j+3}\big)^{-\frac{K(K-2)}{K-1}}$$
$$- \frac{1}{K-1}\big(1+\widetilde{X}_{j-1} + E_{2(j-1)+3}\big)^{\frac{K}{K-1}}$$
$$\times (F-G^K)\Big(\frac{(\widetilde{\Xi})}{1+\widetilde{X}_{j-1}+E_{2(j-1)+3}}\Big)\Big\}. \tag{3.6}$$

Moreover, (3.5) *holds with* $O\left(N^{-(2j+3)}\right)$, *if* $j \ge \max\left(1, \frac{K-2}{2}\right)$.

Proof. (3.6) follows from Lemmas 4 and 5, and the second assertion, (3.5) with weaker error term, follows from (3.6) by order estimates of E_{2j+3} and $E_{2(j-1)+3}$ in Lemma 5, (2.6) and the fact that $F - G^K$ begins with the second order term, which in turn is a consequence of

$$\mathfrak{G}_{\frac{K}{K-1}}(F-G^K) = -G\mathfrak{L}(G^{K-1}). \tag{3.7}$$

This completes the proof. □

In applying Theorem 1, the dependence of the implied constant on $\alpha, k, j, F, K,$ etc. must be made explicit, and we shall do it elsewhere.

We close this section by stating some formulas on whose grounds the intermediate order term cancel each other in the proof of Theorem 1.

Lemma 6. *We have for* $j \geq -1$

$$\widetilde{X}_{j+1} - \widetilde{X}_j = O(N^{-(2j+3)}), \tag{3.8}$$

$$\begin{aligned}&(1 - \Psi_K'(t))\big((\Psi_K(t) - t) + (K-2)(\Psi_K(t) + (K-2)t)\big)\\&= (K-1)(\Psi_K(t) - t),\end{aligned} \tag{3.9}$$

and

$$\begin{aligned}&(1 + \widetilde{X}_j)(\Psi_K(\widetilde{\Lambda}_{j-1}) - \widetilde{\Lambda}_{j-1})^{K-2} G^{K-1}\Big(\frac{(\widetilde{\Xi})}{1 + \widetilde{X}_{j-1}}\Big)\\&= 1 + (\widetilde{X}_j - \widetilde{X}_{j-1})\Big(G^{K-1}\Big(\frac{(\widetilde{\Xi})}{1 + \widetilde{X}_{j-1}}\Big) - 1\Big)\Big(\Psi_K(\widetilde{\Lambda}_{j-1}) - \widetilde{\Lambda}_{j-1}\Big)^{K-2}.\end{aligned} \tag{3.10}$$

Proof. (3.8) follows from Lemma 5 for $j \geq 0$. (3.9) follows from (2.6) and (2.10).

To deduce (3.10), we first derive

$$\begin{aligned}&(1 + \widetilde{X}_{j-1})\Big(G^{K-1}\Big(\frac{(\widetilde{\Xi})}{1 + \widetilde{X}_{j-1}}\Big) - 1\Big) + \widetilde{X}_j\\&= -1 + \big(\Psi_K(\widetilde{\Lambda}_{j-1}) - \widetilde{\Lambda}_{j-1}\big)^{-(K-2)}.\end{aligned} \tag{3.11}$$

The left-hand side of (3.10) is equal to

$$\begin{aligned}&(\widetilde{X}_j - \widetilde{X}_{j-1})\Big(G^{K-1}\Big(\frac{(\widetilde{\Xi})}{1 + \widetilde{X}_{j-1}}\Big) - 1\Big)\Big(\Psi_K(\widetilde{\Lambda}_{j-1}) - \widetilde{\Lambda}_{j-1}\Big)^{K-2}\\&+ \Big(\Psi_K(\widetilde{\Lambda}_{j-1}) - \widetilde{\Lambda}_{j-1}\Big)^{K-2}(1 + \widetilde{X}_{j-1})\Big(G^{K-1}\Big(\frac{(\widetilde{\Xi})}{1 + \widetilde{X}_{j-1}}\Big) - 1\Big)\\&+ \Big(\Psi_K(\widetilde{\Lambda}_{j-1}) - \widetilde{\Lambda}_{j-1}\Big)^{K-2}(1 + \widetilde{X}_j),\end{aligned}$$

which turns out to be equal to the right-hand side of (3.10) by (3.11). □

4. Refinement of the expression and the quartic Theta - Weyl sums

By the same reasoning as used in the proof of Lemma 5, we may prove

Lemma 7. *For $j \geq 1$,*

$$\widetilde{\Lambda}_{j-1} = \widetilde{\Lambda}_{j-2} + \frac{1}{K-1} \cdot \frac{1+\widetilde{X}_{j-2}}{1+\widetilde{X}_{j-1}} \cdot (\widetilde{X}_{j-1} - \widetilde{X}_{j-2}) \\ \times \mathfrak{L}(\widetilde{G}^{K-1})\Big(\frac{(\widetilde{\Xi})}{1+\widetilde{X}_{j-2}}\Big) + O(N^{-(2j+3)}).$$

Lemma 8. *For $K \geq 4$ and $j \geq 1$ we have*

$$\frac{1}{K-2}\left\{\Psi_K\left(\frac{1}{K-1}\Big(\widetilde{X}_{j-1} - \widetilde{X}_{j-2}\Big)\right)^{\frac{K}{K-1}} \right. \\ \left. \times (F - G^K)\left(\frac{(\widetilde{\Xi})}{\Psi_K(\frac{1}{K-1}(\widetilde{X}_{j-1} - \widetilde{X}_{j-2}))}\right) - \Big(F((\widetilde{\Xi})) - G^K((\widetilde{\Xi}))\Big)\right\} \\ = -\frac{1}{K-1}\Big(F((\widetilde{\Xi})) - G^K((\widetilde{\Xi}))\Big)\frac{K-2}{2}\Big(\frac{1}{K-1}(\widetilde{X}_{j-1} - \widetilde{X}_{j-2})\Big)^2 \\ + O\Big(N^{-2}(\widetilde{X}_{j-1} - \widetilde{X}_{j-2})^3\Big) + O\Big(N^{-3}(\widetilde{X}_{j-1} - \widetilde{X}_{j-2})^2\Big).$$

Using these and more elaborate argument, we may prove the following refinement of Theorem 1.

Proposition 2. *For $j \geq 1$ we have*

$$f^*(y) = -\frac{K-1}{K}((K-1)\alpha)^{-\frac{1}{K-1}} y^{\frac{K}{K-1}} \left\{ \frac{K}{K-1}(\Psi_K(\widetilde{\Lambda}_{j-2}) - \widetilde{\Lambda}_{j-2})^{\frac{1}{K-1}} \right. \\ - \frac{1}{K-1}(\Psi_k(\widetilde{\Lambda}_{j-2}) + (K-2)\widetilde{\Lambda}_{j-2})^{\frac{K}{K-1}} \\ - \frac{1}{K-1}(1+\widetilde{X}_{j-2})^{\frac{K}{K-1}}(F - G^K)\Big(\frac{(\widetilde{\Xi})}{1+\widetilde{X}_{j-2}}\Big) \\ + \frac{1}{K-2}\left(\Psi_K\Big(\frac{1}{K-1}(\widetilde{X}_{j-1} - \widetilde{X}_{j-2})\Big)^{\frac{K}{K-1}} \right. \\ \left. \times (F - G^K)\Big(\frac{(\widetilde{\Xi})}{\Psi_K(\frac{1}{K-1}(\widetilde{X}_{j-1} - \widetilde{X}_{j-2}))}\Big) - (\widetilde{F} - \widetilde{G}^K)\right) \\ - \frac{K}{K-1}\frac{1}{K-1}(\widetilde{X}_{j-1} - \widetilde{X}_{j-2})\Big(1 - (1+\widetilde{X}_{j-2})^{\frac{1}{K-1}}\Big) \\ \times \mathfrak{G}_{\frac{K}{K-1}}(F - G^K)\Big(\frac{(\widetilde{\Xi})}{1+\widetilde{X}_{j-2}}\Big) + O\left(N^{-(2j+5)}\right)$$

$$+O\left(N^{-2}(\widetilde{X}_{j-1}-\widetilde{X}_{j-2})^3\right)+O\left(N^{-3}(\widetilde{X}_{j-1}-\widetilde{X}_{j-2})^2\right)\Bigg\}, \tag{4.1}$$

where $\widetilde{F}=F((\widetilde{\Xi}))$.

Remark 1. (i) What we reported in the China-Japan seminar in Xi'an is

$$\begin{aligned} f^*(y) = -\frac{K-1}{K}((K-1)\alpha)^{-\frac{1}{K-1}}y^{\frac{K}{K-1}}\Bigg\{&\frac{K}{K-1}(\Psi_K(\widetilde{\Lambda})-\widetilde{\Lambda})^{\frac{1}{K-1}} \\ &-\frac{1}{K-1}\left(\Psi_K(\widetilde{\Lambda})+(K-2)\widetilde{\Lambda}\right)^{\frac{K}{K-1}}-\frac{1}{K-1}(F-G^K)((\widetilde{\Xi})) \\ &\times\left(1+\frac{K-2}{2(K-1)^2}(\widetilde{Q}-\widetilde{\Lambda})^2\right)+O(N^{-5})\Bigg\}. \end{aligned}$$

If we apply Lemma 8, this becomes (4.1) with $j=1$.

(ii) If we need the error term $O(N^{-(2j+3)})$ or $O(N^{-(2j+5)})$, then $j\geq 1$ or $j\geq 2$ is sufficient, respectively.

(iii) We expect to obtain the expression of f^* with the error term $O(N^{-(K+1)})$ for $K\geq 5$ in which case the "main term" is of the form suggested as in the following Theorem 2.

Hereafter, we shall dwell on the proof of the following theorem which is the case $j=1$, $K=4$ of Proposition 2 with the "main term" expressed in more symmetrical form.

Theorem 2. *The vdC reciprocal function of*

$$f(n)=\frac{3}{4}\alpha x^{\frac{4}{3}}F\left(\frac{(\xi)}{x}\right)$$

is

$$\begin{aligned} f^*(y) = -\frac{3}{4}(3\alpha)^{-\frac{1}{3}}y^{\frac{4}{3}}\Bigg\{&\rho(1+\rho\widetilde{\Lambda})^{\frac{4}{3}}+\bar{\rho}(1+\bar{\rho}\widetilde{\Lambda})^{\frac{4}{3}}-\frac{1}{9}\sum_{j \bmod 3}\omega^j(1+\omega^j\widetilde{\Lambda}^2)^{\frac{2}{3}} \\ &-\frac{1}{3}(\widetilde{F}-\widetilde{G}^4)+\frac{1}{2}\left(\Psi_4(\frac{1}{3}\widetilde{X}_0)^{\frac{4}{3}}(F-G^4)\Big(\frac{(\widetilde{\Xi})}{\Psi_4(\frac{1}{3}\widetilde{X}_0)}\Big)-(\widetilde{F}-\widetilde{G}^4)\right) \\ &+O(N^{-5})\Bigg\} \end{aligned}$$

where $\widetilde{F}=F((\widetilde{\Xi}))$ *and* $\rho=\frac{1}{2}(1+\sqrt{-3})$ *and* $\omega=\frac{1}{2}(-1+\sqrt{-3})$ *are primitive* 6-*th and* 3-*rd root of* 1, *respectively.*

Proof. Since for $j = 1$, the term containing $\mathfrak{G}_{\frac{K}{K-1}}$ disappears from (4.1) on account of $\widetilde{X}_{-1} = 0$, it is enough to express the main term

$$M_4(t) = \frac{4}{3}(\Psi_4(t) - t)^{\frac{1}{3}} - \frac{1}{3}(\Psi_4(t) + 2t)^{\frac{4}{3}} \tag{4.2}$$

in the form given above. We appeal to the Taylor expansion

$$M_4(t) = 1 - \frac{4}{3}t - \frac{4}{9}t^2 + \frac{4}{81}t^3 + \frac{14}{3^5}t^4 + O(t^5), \tag{4.3}$$

which we wish to express as $\sum_{h=0}^{5} a_h c_h t^h + O(t^6)$. Using the orthogonality of $\frac{1}{6}\sum_{j=0}^{5} \rho^{lj}$ for $l \in \mathbb{Z}$, we further decompose

$$\sum_{h=0}^{5} a_h c_h t^h = \frac{1}{6}\sum_{j=0}^{5} A_j B_j(t), \tag{4.4}$$

where

$$A_j = \sum_{h=0}^{5} a_h \rho^{-hj}, \quad B_j(t) = \sum_{k=0}^{5} c_k (\rho^j t)^k. \tag{4.5}$$

We determine c_k as the Taylor coefficient of $(1+t)^{\frac{4}{3}}$, i.e.

$$c_0 = 1, \quad c_1 = \frac{4}{3}, \quad c_2 = \frac{2}{9}, \quad c_3 = -\frac{4}{81}, \quad c_4 = \frac{5}{3^5}, \quad c_5 = -\frac{8}{3^6},$$

whence

$$B_j(t) = (1 + \rho^j t)^{\frac{4}{3}} + O(t^6) \tag{4.6}$$

We determine a_h's by equating (4.3) and (4.4), therewith substituting the values of c_k: $a_0 = 1, a_1 = -1, a_2 = -2, a_3 = -1, a_4 = 1 + \frac{9}{5}, a_5$ to be specified presently. Substituting these values, we get, after some transformation,

$$A_j = (1 - \rho^{-j} - 2\rho^{-2j})(1 - (-1)^j) + (-1)^j \frac{9}{5}\rho^{-j} + (-1)^j (a_5 - 2)\rho^{-2j}.$$

Choose $a_5 = 2$ and classify the values of j mod 6 to obtain

$$\begin{aligned} A_j = & \frac{9}{5} \quad (j \equiv 0 \mod 6), \quad \frac{9}{5}\rho^4 \quad (j \equiv 2 \mod 6), \quad \frac{9}{5} \quad (j \equiv 3 \mod 6), \\ & \frac{9}{5}\bar{\rho}^4 \quad (j \equiv 4 \mod 6), \quad 2(1 - \bar{\rho} - 2\bar{\rho}^2) + \frac{9}{5}\rho^2 \quad (j \equiv 1 \mod 6), \\ & 2(1 - \rho - 2\rho^2) + \frac{9}{5}\bar{\rho}^2 \quad (j \equiv 5 \mod 6). \end{aligned} \tag{4.7}$$

Noting $1-\bar{\rho}-2\bar{\rho}^2=3\rho$, we see that

$$\frac{1}{6}A_j=\rho+\frac{3}{10}\rho^2 \quad (j\equiv 1 \mod 6),$$

and similarly for $j\equiv 5 \mod 6$.

Substituting (4.6) and (4.7) in $M_4(t)=\frac{1}{6}\sum_{j=0}^{5}A_jB_j(t)+O(t^5)$, we deduce that

$$\begin{aligned}M_4(t)&=\rho(1+\rho t)^{\frac{4}{3}}+\bar{\rho}(1+\bar{\rho}t)^{\frac{4}{3}}+\frac{3}{10}\Big((1+t)^{\frac{4}{3}}+\rho^2(1+\rho t)^{\frac{4}{3}}\\&\quad+\rho^4(1+\rho^2t)^{\frac{4}{3}}+(1-t)^{\frac{4}{3}}+\bar{\rho}^2(1+\bar{\rho}t)^{\frac{4}{3}}+\bar{\rho}^4(1+\bar{\rho}^2t)^{\frac{4}{3}}\Big)+O(t^5)\\&=\rho(1+\rho t)^{\frac{4}{3}}+\bar{\rho}(1+\bar{\rho}t)^{\frac{4}{3}}+\frac{1}{3^3}t^4+O(t^5).\end{aligned} \tag{4.8}$$

Finally, replacing the term $\frac{1}{3^3}t^4$ by $-\frac{1}{3^2}\sum_{j=0}^{2}\omega^j(1+\omega^jt^2)^{\frac{2}{3}}$ with an error $O(t^5)$, we conclude that $M_4(t)$ is of the form given in Theorem 2. □

Remark 2. (i) In order to treat general quartic polynomials, it is enough to consider the function of the form

$$f(n)=\frac{3}{4}\alpha\left(x^{\frac{4}{3}}F(\frac{(\xi)}{x})+c_1x+L((\xi))+c_2x^{\frac{2}{3}}F_2\Big(\frac{(\xi)}{x}\Big)\right)$$

where c_1 and c_2 are constants, $L((\xi))$ is a linear function in (ξ) and F_2 is a similar function as F. This will be studied elsewhere.

(ii) As we remarked after the proof of Proposition 1, we need to make the dependence of error terms on parameters, especially on α, explicit for practical applications of Theorem 2 (like Theorem 1).

To be entitled to call our process theta-Weyl sum, we need to study the parabolic transformation ([8, §6]) of the "k"-ic continued fraction expansion ([8, §2]). These tasks are to be conducted elsewhere.

Appendix: Three methods for treating exponential sums $\sum_{X<n\leq X+N} e^{2\pi i F(n)}$ or Weyl sums $\sum_{X<n\leq X+N} e^{2\pi i \alpha f(n)}$

	The Weyl-Hardy-Littlewood method	**The van der Corput method**	**The Vinogradov method**
Inequalities	Schwarz inequality; triangular inequality	Estimates from above using the mean value thorem $\int_X^x f(x)g(x)dx = f(X)\int_X^\xi g(x)dx$ in integral calculus ($f(x)$ is positive and monotone decreasing)	Hölder inequality (both for sums and integrals); double sum methods
tools	differencing (polarization in case of polynomials)	Euler's summation formula (Poisson formula in finite form), esp., $< x >= -\sum_{\nu=1}^{\infty} \frac{\sin 2\pi\nu x}{\pi\nu}$, the saw-tooth Fourier series	Newton formula for the relation between power sums and fundamental symmetric functions; the number of integer-solutions of a system of linear equations as L_1-norm
the most effectively applied objects	Weyl sums	sums with F such that F'' is of constant sign, or F' is monotone	Weyl sums with the core polynomial having real coefficients; also effective for exponential sums

contd. on next page

results obtained first	local type	local type	mean value type
alteration of summation intervals at each step of application	inside $\|(\cdots)\|$, the interval remains invariant, with summation over consecutive integers;outside $\|(\cdots)\|$, the length of interval increasing geometrically, where $\|(\cdots)\|$ means the inner sum of $\sum_{y_1,\ldots,y_H}\left\|\sum_x e(K(K-1)\cdots(K-H+1)\alpha x^{K-H}y_1\cdots y_H)\right\|$ and outside means the sum over $y_1,\ldots,y_H$ in the case $f=\alpha x^K$	changes into F'(the previous interval), summation is extended over consecutive integers	inside $\|(\cdots)\|$, the length of intervals changes from X into $X^{1-1/K}$ outside, the length of intervals increases of order $X^{1/K}\cdot(\cdots)$, where $(\cdots)$ means a similar sum as in W-H-L; division into equal subintervals (Vinogradov) or selection as arithmetic progressions (Linnik-Karatsuba), both putting subintervals together using parallel translation of subintervals
estimate of Diophantine approximation used (all linear type)	upper bounds for $\sharp\left\{x\in\mathbb{Z}\,\middle\|\,\begin{array}{l}X<n\le X+N,\\ \|\alpha x\|\le\frac{1}{U}\end{array}\right\}$	the order of F'' in the form $\|F''(x)\|\asymp\lambda$ on $[X,X+N]$ (in paractice, use is made of $\|F^{(K+1)}(x)\|\asymp\lambda_{K+1}$ using differencing)	upper bounds for $\sharp\left\{x\in\mathbb{Z}\,\middle\|\,\begin{array}{l}X<n\le X+N,\\ \|\alpha x\|\le\frac{1}{U}\end{array}\right\}$, $\|F^{(K+1)}(x)\|\asymp\lambda_{K+1}$ on $[X,X+N]$

contd. on next page

result local/mean	$O_{(K)}(N^{1-\frac{1}{2^{K-1}}})\big/ l > 2^K$	the same as W-H-L for Weyl sums, more efficient if used as a theta series	$O_{(K)}(N^{1-\frac{c}{K^2 \log K}})\big/ l \gg K^2 \log K$
impression	the most general and necessarily the most inefficient	undeveloped save for quadratic polynomials	the most effective, but one defect is that polynomials with generic real coefficients must be considered because of the translation of intervals referred to above

References

[1] H. Fiedler, W. Jurkat and O. Körner, *Asymptotic expansions of finite theta series*, Acta Arith. **32** (1977), 129–146.

[2] S. W. Graham and G. Kolesnik, *Van der Corput's method of exponential sums*, London Mathematical Society Lecture Note Series 126, Cambridge Univ. Press, London, 1991.

[3] J.-I. Igusa, *Lectures on forms of higher degree*, Tata Institute of Fundamental Research Lectures on Mathematics and Physics 59, Tata Inst., Bombay, 1978.

[4] T. Kubota, *On an analogy to the Poisson summation formula for generalised Fourier transformation*, J. Reine Angew. Math. **268/269** (1974), 180–189.

[5] W. Maier, *Transformation der kubischen Thetafunktionen*, Math. Ann. **111** (1935), 183–196.

[6] Y.-N. Nakai, *On a θ-Weyl sum*, Nagoya Math. J. **52** (1973), 163–172. Errata, ibid. **60** (1976), 217.

[7] Y.-N. Nakai, *On Diophantine inequalities of real indefinite quadratic forms of additive type in four variables*, Advanced Studies in Pure Mathematics **13**, (1988), Investigations in Number Theory, 25-170, Kinokuniya Compa.LTD., Tokyo, Japan. (This series is now published by Math. Soc. Japan).

[8] Y.-N. Nakai, *A penultimate step toward cubic theta-Weyl sums*, Number Theoretic Methods, Future trends, ed. by S. Kanemitsu and C.-H. Jia, (2002), 311-338, Kluwer Acad. Publishers.

[9] W. Raab, *Kubischen und biquadratische Thetafunktionen* I *und* II, Sizungsber. Österreich. Akad. Wiss. Mat-Natur. Kl. **188** (1979), 47-77 and 231–246.

[10] E. C. Titchmarsh, *The theory of the Riemann zeta-function*, Oxford Univ. Press, 1951, 2nd ed. 1986 (edited by D. R. Heath-Brown).

EXPLICIT CONGRUENCES FOR EULER POLYNOMIALS

Zhi-Wei Sun
Department of Mathematics, Nanjing University
Nanjing 210093, People's Republic of China
zwsun@nju.edu.cn

Abstract In this paper we establish some explicit congruences for Euler polynomials modulo a general positive integer. As a consequence, if $a, m \in \mathbb{Z}$ and $2 \nmid m$ then

$$\frac{m^{k+1}}{2}E_k\left(\frac{x+a}{m}\right) - \frac{(-1)^a}{2}E_k(x) \in \mathbb{Z}[x] \quad \text{for every } k = 0, 1, 2, \ldots,$$

which may be regarded as a refinement of a multiplication formula.

Keywords: Euler polynomial, Congruence, q-adic number.

2000 Mathematics Subject Classification: Primary 11B68; Secondary 11A07, 11S05.

1. Introduction

Congruences for Bernoulli numbers have been a very intriguing objective of research since the time of L. Euler, and they recently got revived in connection with p-adic interpolation of L-functions. Congruences for Euler numbers, being cognates of Bernoulli numbers, have also received much attention from the same point of view of p-adic interpolation. In [S4] the author determined Euler numbers modulo powers of two, while Euler numbers modulo any odd integer are essentially trivial.

As a natural further step, we are led to consider congruences among Bernoulli and Euler polynomials, the latter of which will be our main concern in this paper. We prove the integrity of coefficients of $f_k(x; a, m)$ (defined by (1.7)), which are related to the summands in the multiplica-

Supported by the National Science Fund for Distinguished Young Scholars (No. 10425103) and the Key Program of NSF (No. 10331020) in China.

Number Theory: Tradition and Modernization, pp. 205–218
W. Zhang and Y. Tanigawa, eds.

tion formula (1.6) for Euler polynomials, and establish number-theoretic generalizations thereof (Theorems 1.2 and 2.1).

Hereafter, the labelled formulae with star indicate those known ones which have their counterparts for Bernoulli or Euler polynomials. Hopefully, these will serve also as a basic table for these polynomials (for more information, the reader is referred to [AS], [E], [S1]). In referring to them, we omit the star symbol.

Euler numbers $E_0, E_1, E_2, \ldots$ are defined by

$$E_0 = 1 \quad \text{and} \quad \sum_{\substack{k=0 \\ 2|k}}^{n} \binom{n}{k} E_{n-k} = 0 \quad \text{for } n \in \mathbb{N} = \{1, 2, 3, \cdots\}. \tag{1.1*}$$

It is well known that they are integers and odd-numbered ones E_1, $E_3, E_5, \cdots$ are all zero.

For each $n \in \mathbb{N}_0 = \{0, 1, 2, \ldots\}$, the Euler polynomial $E_n(x)$ of degree n is given by

$$E_n(x) = \sum_{k=0}^{n} \binom{n}{k} \frac{E_k}{2^k} \left(x - \frac{1}{2}\right)^{n-k} \tag{1.2*}$$

Note that

$$E_n = 2^n E_n(1/2). \tag{1.3*}$$

Here are basic properties of Euler polynomials:

$$E_n(x) + E_n(x+1) = 2x^n, \tag{1.4*}$$

$$E_n(x+y) = \sum_{k=0}^{n} \binom{n}{k} E_k(x) y^{n-k}, \tag{1.5*}$$

and

$$m^k \sum_{a=0}^{m-1} (-1)^a E_k\left(\frac{x+a}{m}\right) = E_k(x). \tag{1.6*}$$

From now on we always assume that q is `a fixed integer greater than one`, and let $\mathbb{Z}_q$ denote the ring of q-adic integers (see [M]). For $\alpha, \beta \in \mathbb{Z}_q$, by $\alpha \equiv \beta \pmod{q}$ we mean that $\alpha - \beta = q\gamma$ for some $\gamma \in \mathbb{Z}_q$. A rational number in $\mathbb{Z}_q$ is usually called a *q-integer.*

In this paper we aim at establishing some explicit congruences for Euler polynomials modulo a general positive integer.

We adopt some standard notations. For example, for a real number α, $\lfloor \alpha \rfloor$ stands for the greatest integer not exceeding α, and $\{\alpha\} = \alpha - \lfloor \alpha \rfloor$ the fractional part of α, (a, b) the greatest common divisor of $a, b \in \mathbb{Z}$, and $\Delta(P(x))$ the difference $P(x+1) - P(x)$ of a polynomial $P(x)$. For a prime p, $n \in \mathbb{N}_0$ and $a \in \mathbb{Z}$, we write $p^n \| a$ if $p^n \mid a$ and $p^{n+1} \nmid a$. For $a, b \in \mathbb{Z} \setminus \{0\}$, by $a \sim_2 b$ we mean that both $2^n \| a$ and $2^n \| b$ for a common $n \in \mathbb{N}_0$.

For convenience we also use the logical notations $\wedge$ (*and*), $\vee$ (*or*), $\Leftrightarrow$ (*if and only if*), and the special notation:

$$[A] = \begin{cases} 1 & \text{if } A \text{ holds,} \\ 0 & \text{otherwise.} \end{cases}$$

By (1.1) and (1.2) it is easy to get

$$E_0(x) = 1, \quad E_1(x) = x - \frac{1}{2} \quad \text{and} \quad E_2(x) = x^2 - x,$$

and verify that the polynomial

$$f_k(x) = f_k(x; a, m) = \frac{m^{k+1}}{2} E_k\left(\frac{x+a}{m}\right) - \frac{(-1)^a}{2} E_k(x) \tag{1.7}$$

has integral coefficients for $k = 0, 1, 2$.

This phenomenon is not contingent but universal as asserted by the following general theorem.

Theorem 1.1. *For each $k \in \mathbb{N}_0$ and $a, m \in \mathbb{Z}$ with $2 \nmid m$, we have*

$$f_k(x) = f_k(x; a, m) = \frac{m^{k+1}}{2} E_k\left(\frac{x+a}{m}\right) - \frac{(-1)^a}{2} E_k(x) \in \mathbb{Z}[x].$$

This result is remarkable in that (1.6) can be expressed as the vanishing arithmetic mean

$$\frac{1}{m} \sum_{a=0}^{m-1} (-1)^a f_k(x; a, m) = 0.$$

Theorem 1.1 follows from the following more general result whose proof will be given in Section 2.

Theorem 1.2. *Let $k \in \mathbb{N}_0$, $d, m \in \mathbb{N}$ and $d \mid m$. Let c be a real number, and let $P(x)$ denote the polynomial*

$$\frac{[2 \nmid q] + d\,[2 \mid q]}{2} m^k E_k\left(\frac{x}{m}\right)$$
$$- (-1)^{\lfloor \frac{c}{(d,q)} \rfloor} \frac{(d,q)^{k+1}}{2} \left(\frac{m}{d}\right)^k E_k\left(\frac{d}{(d,q)} \cdot \frac{x}{m} - \left\lfloor \frac{c}{(d,q)} \right\rfloor\right).$$

Then $P(x) \in \mathbb{Z}_q[x]$. Furthermore, if q is odd, then

$$P(x) \equiv \sum_{\substack{j=0 \\ 2\nmid (d-1)j+\lfloor \frac{c+jd}{q} \rfloor}}^{q-1} (-1)^j (x+jm)^k \pmod{q}; \tag{1.8}$$

if q is even, then

$$\begin{aligned} P(x) \equiv & \sum_{j=0}^{q-1} (-1)^{j-1}(x+jm)^k \left(\left\lfloor \frac{c+jd}{q} \right\rfloor + [2 \mid d \ \wedge \ 2 \mid j](-1)^{\lfloor \frac{c+jd}{q} \rfloor} \right) \\ & + \begin{cases} \frac{q}{2}[2 \nmid m \ \wedge \ 4 \mid d+1]\Delta(x^k) \pmod{q} & \text{if } 2 \mid k, \\ \frac{q}{2}[4 \mid d+1](\Delta(x^k) + [4 \mid q]\Delta(x^{k-1})) \pmod{q} & \text{if } 2 \nmid km, \\ \frac{q}{2}[d \sim_2 m]([2\|m] + [2\|q])x^{k-1} \pmod{q} & \text{if } 2 \nmid k(m-1). \end{cases} \end{aligned} \tag{1.9}$$

Now we derive various consequences of Theorem 1.2.

Corollary 1.1. *Let $k \in \mathbb{N}_0$ and $m \in \mathbb{N}$, and let x be a q-integer. If q is odd, then*

$$\begin{aligned} & \frac{m^k}{2} E_k\left(\frac{x}{m}\right) - (-1)^{\lfloor \frac{x}{(m,q)} \rfloor} \frac{(m,q)^{k+1}}{2} E_k\left(\left\{\frac{x}{(m,q)}\right\}\right) \\ \equiv & \sum_{\substack{j=0 \\ 2\nmid (m-1)j+\lfloor \frac{x+jm}{q} \rfloor}}^{q-1} (-1)^j (x+jm)^k \pmod{q}. \end{aligned}$$

If q is even, then

$$\begin{aligned} & \frac{m^{k+1}}{2} E_k\left(\frac{x}{m}\right) - (-1)^{\lfloor \frac{x}{(m,q)} \rfloor} \frac{(m,q)^{k+1}}{2} E_k\left(\left\{\frac{x}{(m,q)}\right\}\right) \\ \equiv & \sum_{j=0}^{q-1} (-1)^{j-1}(x+jm)^k \left(\left\lfloor \frac{x+jm}{q} \right\rfloor + [2 \mid m \ \wedge \ 2 \mid j](-1)^{\lfloor \frac{x+jm}{q} \rfloor} \right) \\ & + \begin{cases} \frac{q}{2}([2\|m] + [2\|q])x^{k-1} \pmod{q} & \text{if } 2 \nmid k \text{ and } 2 \mid m, \\ 0 \pmod{q} & \text{if } k = 0, \\ \frac{q}{2}[4 \mid m+1 \ \wedge \ (k=1 \ \vee \ 2 \mid k \ \vee \ 2\|q)] \pmod{q} & \text{otherwise.} \end{cases} \end{aligned}$$

Proof. Just apply Theorem 1.2 with $c = x$ and $d = m$, and note that $\frac{q}{2}x^k \equiv \frac{q}{2}x \pmod{q}$ if $k > 0$ (cf. [S3, Lemma 2.1]). □

Corollary 1.2. *Let $a \in \mathbb{Z}$, $k \in \mathbb{N}_0$, $m \in \mathbb{Z}^+$, $2 \mid q$ and $(m, q) = 1$. Then*

$$\begin{aligned} &f_k(x; a, m) \\ &\equiv \sum_{j=0}^{q-1} (-1)^{j-1} \left\lfloor \frac{a + jm}{q} \right\rfloor (x + a + jm)^k \\ &+ \frac{q}{2}[4 \mid m + 1] \left(\Delta(x^k) + [2 \nmid k \ \wedge \ 4 \mid q] \Delta(x^{k-1}) \right) \pmod{q}. \end{aligned}$$

Proof. Clearly $2 \nmid m$ and $\Delta((x+a)^k) - \Delta(x^k) \in 2\mathbb{Z}[x]$. Applying Theorem 1.2 with $c = a$, $d = m$, and x replaced by $x + a$, we obtain the desired congruence. □

Proof of Theorem 1.1. Suppose first that $m > 0$. Then by Corollary 1.2,

$$f_k(x; a, m) \in \mathbb{Z}_2[x].$$

If p is an odd prime, then $E_k(x)/2 \in \mathbb{Z}_p[x]$, and also by (1.5)

$$\frac{m^{k+1}}{2} E_k \left(\frac{x + a}{m} \right) = \frac{m}{2} \sum_{l=0}^{k} \binom{k}{l} E_l(0) m^k \left(\frac{x + a}{m} \right)^{k-l} \in \mathbb{Z}_p[x].$$

Thus every coefficient of the polynomial $f_k(x) = f_k(x; a, m)$ is a p-integer for any prime p, which amounts to $f_k(x) \in \mathbb{Z}[x]$.

For the negative modulus case $(-m > 0)$, using

$$E_k(1 - x) = (-1)^k E_k(x) \tag{1.10}^*$$

and $2 \nmid m$, we may express $f_k(x; a, -m)$ as

$$-\frac{m^{k+1}}{2} E_k \left(\frac{x + (a + m)}{m} \right) + \frac{(-1)^{a+m}}{2} E_k(x).$$

Thus the positive modulus case applies and the proof is complete. □

In the spirit of Sun [S3], Theorem 1.2 can also be used to deduce some general congruences of Kummer's type for Euler polynomials. However, in order not to make this paper too long, we will not go into details.

2. Proof of Theorem 1.2

We introduce the Bernoulli polynomials $B_n(x)$ ($n \in \mathbb{N}_0$) by the generating power series

$$\frac{ze^{xz}}{e^z-1} = \sum_{n=0}^{\infty} B_n(x)\frac{z^n}{n!}. \qquad (|z|<2\pi) \tag{2.1}^*$$

Their values $B_n(0)$ at $x=0$ are rational numbers, called Bernoulli numbers and denoted by B_n; it is well known that $B_{2k+1}=0$ for $k=1,2,3,\ldots$. Raabe's multiplication formula (counterpart of (1.6)) reads

$$m^{n-1}\sum_{r=0}^{m-1} B_n\left(\frac{x+r}{m}\right) = B_n(x) \text{ for any } m\in\mathbb{N}. \tag{2.2}^*$$

Other properties include

$$\Delta(B_n(x)) = nx^{n-1} \tag{2.3}^*$$

and

$$E_n(x) = \frac{2}{n+1}\left(B_{n+1}(x) - 2^{n+1}B_{n+1}\left(\frac{x}{2}\right)\right), \tag{2.4}^*$$

the last one links the Euler and the Bernoulli polynomials.

Lemma 2.1. *Let k be a positive integer and y be a real number. Then*

$$kE_{k-1}(x+\{y\}) = 2(-1)^{\lfloor y\rfloor}\left(B_k(x+\{y\}) - 2^kB_k\left(\frac{x}{2}+\left\{\frac{y}{2}\right\}\right)\right). \tag{2.5}$$

Proof. Observe that

$$\left\{\frac{y}{2}\right\}<\frac{1}{2} \iff \{y\}=2\left\{\frac{y}{2}\right\} \iff \lfloor y\rfloor = 2\left\lfloor\frac{y}{2}\right\rfloor \iff 2\mid\lfloor y\rfloor. \tag{2.6}$$

Hence, if $2\mid\lfloor y\rfloor$, the right hand side of (2.5) is

$$2\left(B_k(x+\{y\}) - 2^kB_k\left(\frac{x+\{y\}}{2}\right)\right),$$

which coincides with the left hand side of (2.5) in view of (2.4). Now, if $2\nmid\lfloor y\rfloor$, then the right hand side of (2.5) is

$$-2\left(B_k(x+\{y\}) - 2^kB_k\left(\frac{x+\{y\}+1}{2}\right)\right).$$

We may express $B_k\left(\frac{x+\{y\}+1}{2}\right)$ as $2^{1-k}B_k(x+\{y\}) - B_k\left(\frac{x+\{y\}}{2}\right)$ by (2.2). Then what remains is $-2\left(2^kB_k\left(\frac{x+\{y\}}{2}\right) - B_k(x+\{y\})\right)$ which equals the left hand side of (2.5). This completes the proof. □

Lemma 2.2. *Let $k \in \mathbb{N}_0$ and $m \in \mathbb{Z} \setminus \{0\}$. Then $(k+1)m^k E_k(x/m) \in \mathbb{Z}[x]$. Furthermore,*

$$(k+1)m^k E_k\left(\frac{x}{m}\right) - x^k - (x-km)\frac{(x+m)^k - x^k}{m} \in 2\mathbb{Z}[x].$$

Proof. First note that by (2.4),

$$(l+1)E_l(0) = 2(1-2^{l+1})B_{l+1}$$

for any $l \in \mathbb{N}_0$. Hence, if l is even then

$$(l+1)E_l(0) = [l=0];$$

while for odd l, we see that in the expression

$$(l+1)E_l(0) = 2^{l+1} - 1 - 2(2^{l+1}-1)\Bigg(B_{l+1} + \sum_{p-1 \mid l+1} \frac{1}{p}\Bigg) + 2\sum_{\substack{p\neq 2 \\ p-1\mid l+1}} \frac{2^{l+1}-1}{p},$$

the third and fourth terms are integers by the von Staudt–Clausen theorem (cf. [IR, pp. 233-236]) and Fermat's little theorem, respectively, so that $(l+1)E_l(0)$ is an odd integer.

Using (1.5) and writing $(k+1)\binom{k}{l}$ as $(l+1)\binom{k+1}{l+1}$, we find that

$$(k+1)m^k E_k\left(\frac{x}{m}\right) = \sum_{l=0}^{k} \binom{k+1}{l+1}(l+1)E_l(0)m^l x^{k-l}$$

lies in $\mathbb{Z}[x]$ and that

$$(k+1)m^k E_k\left(\frac{x}{m}\right) - \binom{k+1}{1}E_0(0)x^k - \sum_{\substack{l=1 \\ 2\nmid l}}^{k} \binom{k+1}{l+1} m^l x^{k-l} \in 2\mathbb{Z}[x].$$

Now, since

$$\begin{aligned}\sum_{\substack{l=1 \\ 2\nmid l}}^{k} \binom{k+1}{l+1} m^l x^{k-l} &= \frac{(x+m)^{k+1} + (x-m)^{k+1}}{2m} - \frac{x^{k+1}}{m} \\ &= \frac{(x+m)^{k+1} - x^{k+1}}{m} - \binom{k+1}{1}(x+m)^k\end{aligned}$$

$$-\sum_{j=2}^{k+1}\binom{k+1}{j}(-2m)^j(x+m)^{k+1-j},$$

the third term lying in $2\mathbb{Z}[x]$, we conclude that what we should subtract from $(k+1)m^kE_k\left(\frac{x}{m}\right)$ is

$$\begin{aligned}&\binom{k+1}{1}E_0(0)x^k+\frac{(x+m)^{k+1}-x^{k+1}}{m}-\binom{k+1}{1}(x+m)^k\\&=x^k+(x-km)\frac{(x+m)^k-x^k}{m},\end{aligned}$$

as asserted, and the proof is complete. □

Lemma 2.3 ([S3, Theorem 4.1]). *Let $k\in\mathbb{N}_0$, $d,m,n\in\mathbb{N}$, $d\mid n$, $m\mid qn$, and $2\nmid d$ or $2\nmid q$ or $2\mid\frac{qn}{m}$. Put $\bar d=(d,qn/m)$ and $\bar m=(m,qn/d)$. Then, for any real number y, the polynomial*

$$L(x,y)=\frac{1}{k+1}\left(dm^kB_{k+1}\left(\frac{x}{m}\right)-\bar d\bar m^kB_{k+1}\left(\frac{x}{\bar m}-\left\lfloor\frac{y}{\bar m}\right\rfloor\right)\right)\tag{2.7}$$

is in $\mathbb{Z}_q[x]$ and is congruent to

$$\begin{aligned}R(x,y)=&\sum_{j=0}^{qn/m-1}(x+jm)^k\left(\left\lfloor\frac{y+jm}{qn/d}\right\rfloor+\frac{1-d}{2}\right)-\frac{q}{3}[3\mid d]\frac{n}{d}\cdot\frac{qn}{m}kx^{k-1}\\&+\frac{q}{2}k[d\sim_2 n]\left(\left[2\mid n\ \wedge\ 2\|\frac{qn}{m}\right]x^{k-1}\right.\\&\qquad\left.+[2\nmid m\ \wedge\ 2\|n\ \wedge\ 2\|q]\,\Delta(x^{k-1})\right)\end{aligned}\tag{2.8}$$

modulo q.

Now we establish a result more general than Theorem 1.2 ((2.9) and (2.10) below are generalizations of (1.8) and (1.9), respectively).

Theorem 2.1. *Let $k\in\mathbb{N}_0$, $d,m,n\in\mathbb{N}$, $d\mid n$, $m\mid qn$, and $2\nmid d$ or $2\nmid q$ or $2\mid\frac{qn}{m}$. Put $\bar d=(d,qn/m)$ and $\bar m=(m,qn/d)$. Then for any real number y we have*

$$\begin{aligned}P_k(x,y):=&\frac{[2m\nmid qn]+d[2m\mid qn]}{2}m^kE_k\left(\frac{x}{m}\right)\\&-\frac{\bar d\bar m^k}{2}(-1)^{\lfloor y/\bar m\rfloor}E_k\left(\frac{x}{\bar m}-\left\lfloor\frac{y}{\bar m}\right\rfloor\right)\in\mathbb{Z}_q[x].\end{aligned}$$

Moreover, if qn/m is odd then

$$P_k(x,y) \equiv \sum_{\substack{j=0 \\ 2\nmid(d-1)j+\lfloor\frac{y+jm}{qn/d}\rfloor}}^{qn/m-1} (-1)^j(x+jm)^k + \frac{q}{2}[d \sim_2 n \ \wedge \ d \not\equiv 0,1 \pmod 4]kx^{k-1} \pmod q; \quad (2.9)$$

if qn/m is even then

$$P_k(x,y) \equiv \sum_{j=0}^{qn/m-1} (-1)^{j-1}(x+jm)^k\left(\left\lfloor\frac{y+jm}{qn/d}\right\rfloor + [2\mid d \ \wedge \ 2\mid j](-1)^{\lfloor\frac{y+jm}{qn/d}\rfloor}\right) + R_k(x) \pmod q, \quad (2.10)$$

where

$$R_k(x) = \begin{cases} \frac{q}{2}[d \sim_2 n \ \wedge \ d \not\equiv 0,1 \pmod 4 \ \wedge \ 2\nmid m]\Delta(x^k) & \text{if } 2\mid k, \\ \frac{q}{2}[d \sim_2 n \ \wedge \ d \not\equiv 0,1 \pmod 4](\Delta(x^k)+[4\mid q]\Delta(x^{k-1})) & \text{if } 2\nmid km, \\ \frac{q}{2}[d \sim_2 n]([d \not\equiv 0,1 \pmod 4]+[2\mid n \ \wedge \ 2\|(qn/m)])x^{k-1} & \text{if } 2\nmid k(m-1). \end{cases}$$

Proof. We observe that the $(k+1)$-degree terms in (2.7) cancel each other in view of $d/m = \bar{d}/\bar{m}$. Hence $L(x,y)$ is of degree at most k. Writing

$$L(x,y) = \sum_{i=0}^{k} a_i(y)x^i,$$

we see that $2^kL(\frac{x}{2},\frac{y}{2}) \in \mathbb{Z}_q[x]$, and similarly $2^kR(\frac{x}{2},\frac{y}{2}) \in \mathbb{Z}_q[x]$, and *a fortiori* that

$$L(x,y) - 2^{k+1}L\left(\frac{x}{2},\frac{y}{2}\right) \in \mathbb{Z}_q[x]$$

and

$$L(x,y) - 2^{k+1}L\left(\frac{x}{2},\frac{y}{2}\right) \equiv R(x,y) - 2^{k+1}R\left(\frac{x}{2},\frac{y}{2}\right) \pmod q. \quad (2.11)$$

By (2.7) we can express the left hand side of (2.11) in such a way that we may apply Lemma 2.1 to deduce that

$$\begin{aligned} &L(x,y)-2^{k+1}L\left(\frac{x}{2},\frac{y}{2}\right)\\ &=\frac{dm^k}{2}E_k\left(\frac{x}{m}\right)-(-1)^{\lfloor y/\bar m\rfloor}\frac{\bar d\bar m^k}{2}E_k\left(\frac{x}{\bar m}-\left\lfloor\frac{y}{\bar m}\right\rfloor\right). \end{aligned}\tag{2.12}$$

On the other hand, the right hand side of (2.11) is congruent to

$$\Sigma-\frac{q}{3}[3|d]\frac{n}{d}\cdot\frac{qn}{m}k\left(x^{k-1}-2^{k+1}\left(\frac{x}{2}\right)^{k-1}\right)+r(x)$$

modulo q, where

$$\begin{aligned} r(x)=\frac{q}{2}k[d\sim_2 n]\Big(&\left[2\mid n\ \wedge\ 2\|\frac{qn}{m}\right]x^{k-1}\\ &+[2\nmid m\ \wedge\ 2\|n\ \wedge\ 2\|q]\Delta(x^{k-1})\Big) \end{aligned}\tag{2.13}$$

and

$$\begin{aligned} \Sigma=\sum_{j=0}^{qn/m-1}\Bigg(&(x+jm)^k\left(\left\lfloor\frac{y+jm}{qn/d}\right\rfloor+\frac{1-d}{2}\right)\\ &-(x+2jm)^k\left(2\left\lfloor\frac{y+2jm}{2qn/d}\right\rfloor+1-d\right)\Bigg). \end{aligned}\tag{2.14}$$

Hence

$$R(x,y)-2^{k+1}R\left(\frac{x}{2},\frac{y}{2}\right)\equiv\Sigma+r(x)\pmod q.\tag{2.15}$$

By the counterpart of (2.6), the second term on the right hand side of (2.14) becomes

$$-\sum_{j=0}^{qn/m-1}(x+2jm)^k\left(\left\lfloor\frac{y+2jm}{qn/d}\right\rfloor-d+\left[2\mid\left\lfloor\frac{y+2jm}{qn/d}\right\rfloor\right]\right)$$

in which we shall divide the sum into two parts via midpoint. Then

$$\begin{aligned} \Sigma=&\sum_{j=0}^{qn/m-1}(x+jm)^k\left(\left\lfloor\frac{y+jm}{qn/d}\right\rfloor+\frac{1-d}{2}\right)\\ &-\sum_{\substack{i=0\\ 2|i}}^{qn/m-1}(x+im)^k\left(\left\lfloor\frac{y+im}{qn/d}\right\rfloor-d+\left[2\mid\left\lfloor\frac{y+im}{qn/d}\right\rfloor\right]\right) \end{aligned}$$

$$- \sum_{\substack{i=0 \\ 2|i+qn/m}}^{qn/m-1} \left(x + \left(i + \frac{qn}{m}\right) m\right)^k \left(\left\lfloor \frac{y + (i + qn/m)m}{qn/d} \right\rfloor - d + \left[2 \mid \left\lfloor \frac{y + im}{qn/d} + d \right\rfloor \right] \right).$$

Whence, writing

$$c_j(y) = \left\lfloor \frac{y + jm}{qn/d} \right\rfloor \tag{2.16}$$

for $j = 0, 1, \ldots$, we obtain

$$\begin{aligned} \Sigma \equiv & \sum_{j=0}^{qn/m-1} (x + jm)^k c_j(y) [2 \nmid j] + \frac{d-1}{2} \sum_{j=0}^{qn/m-1} (x + jm)^k (2 [2 \mid j] - 1) \\ & + \sum_{j=0}^{qn/m-1} (x + jm)^k ([2 \mid j \ \wedge \ 2 \nmid c_j(y)]) \\ & - \sum_{j=0}^{qn/m-1} (x + jm)^k c_j(y) \left[2 \mid \frac{qn}{m} + j \right] \\ & - \sum_{j=0}^{qn/m-1} (x + jm)^k \left[2 \mid \frac{qn}{m} + j \ \wedge \ 2 \mid c_j(y) + d \right] \pmod{q}. \end{aligned}$$

Recombination of terms yields

$$\Sigma \equiv \sum_{j=0}^{qn/m-1} (x + jm)^k c_j(y) \left([2 \nmid j] - \left[2 \mid \frac{qn}{m} - j \right] \right) + \Sigma_1 + \Sigma_2 \pmod{q},$$

where

$$\begin{aligned} \Sigma_1 = & \frac{d-1}{2} \sum_{j=0}^{qn/m-1} (-1)^j (x + jm)^k, \\ \Sigma_2 = & \sum_{j=0}^{qn/m-1} (x + jm)^k \Big([2 \mid j \ \wedge \ 2 \nmid c_j(y)] \\ & \qquad - \left[2 \mid \frac{qn}{m} - j \ \wedge \ 2 \mid c_j(y) + d \right] \Big). \end{aligned} \tag{2.17}$$

Thus, in view of the equality $[2\nmid j]-[2\mid \frac{qn}{m}-j]=[2\mid \frac{qn}{m}](-1)^{j-1}$, we deduce that

$$\Sigma\equiv\left[2\mid\frac{qn}{m}\right]\sum_{j=0}^{qn/m-1}(-1)^{j-1}(x+jm)^kc_j(y)+\Sigma_1+\Sigma_2\pmod q. \tag{2.18}$$

Writing

$$\Sigma_1=\frac{d-1}{2}m^k\sum_{j=0}^{qn/m-1}(-1)^j\left(\frac{x}{m}+j\right)^k$$

and applying (1.4) and (1.5) successively, we obtain

$$\begin{aligned}\Sigma_1&=\frac{d-1}{2}\cdot\frac{m^k}{2}\sum_{j=0}^{qn/m-1}\left((-1)^jE_k\left(\frac{x}{m}+j\right)-(-1)^{j+1}E_k\left(\frac{x}{m}+j+1\right)\right)\\&=\frac{d-1}{2}\cdot\frac{m^k}{2}\left(E_k\left(\frac{x}{m}\right)-(-1)^{qn/m}E_k\left(\frac{x}{m}+\frac{qn}{m}\right)\right)\\&=\frac{d-1}{2}\cdot\frac{m^k}{2}\left(E_k\left(\frac{x}{m}\right)-(-1)^{qn/m}\sum_{l=0}^{k}\binom{k}{l}E_{k-l}\left(\frac{x}{m}\right)\left(\frac{qn}{m}\right)^l\right);\end{aligned}$$

whence separating the term with $l=0$,

$$\begin{aligned}\Sigma_1=&\frac{d-1}{2}\cdot\left[2\nmid\frac{qn}{m}\right]m^kE_k\left(\frac{x}{m}\right)\\&-(-1)^{qn/m}\sum_{0<l\le k}\binom{k}{l-1}(k-l+1)m^{k-l}E_{k-l}\left(\frac{x}{m}\right)\frac{d-1}{2}\cdot\frac{n^lq^l}{2l}.\end{aligned} \tag{2.19}$$

Now, by [S3, Lemma 2.1], $q^{l-1}/l\equiv 0 \pmod q$ for $l>2$, so that if $0<l\le k$, then

$$\begin{aligned}\frac{d-1}{2}\cdot\frac{n^lq^l}{2l}&=\binom{d}{2}\frac{n^l}{d}\cdot\frac{q}{2}\cdot\frac{q^{l-1}}{l}\\&\equiv\frac{q}{2}[d\sim_2 n\ \wedge\ d\not\equiv 0,1 \pmod 4]\\&\qquad\times\big([l=1]+[l=2\ \wedge\ 2\nmid n\ \wedge\ 2\|q]\big)\pmod q,\end{aligned} \tag{2.20}$$

and in particular, only two terms with $l=1,2$ appear on the right hand side of (2.19) modulo q. Hence the sum on the right hand side of (2.19) is congruent to

$$\frac{d-1}{2}\cdot\frac{qn}{2}[k>0]km^{k-1}E_{k-1}\left(\frac{x}{m}\right)$$
$$+\frac{d-1}{2}\cdot\frac{n^2q^2}{4}[k>1]k(k-1)m^{k-2}E_{k-2}\left(\frac{x}{m}\right)$$

modulo q. Thus, by Lemma 2.2, Σ_1 is congruent to

$$\frac{d-1}{2}\left[2\nmid\frac{qn}{m}\right]m^kE_k\left(\frac{x}{m}\right)-(-1)^{qn/m}\frac{d-1}{2}$$
$$\times\left(\frac{nq}{2}[k>0]\left(x^{k-1}+(x-(k-1)m)\frac{(x+m)^{k-1}-x^{k-1}}{m}\right)\right.$$
$$\left.+\frac{n^2q^2}{4}[k>1]k\left(x^{k-2}+(x-(k-2)m)\frac{(x+m)^{k-2}-x^{k-2}}{m}\right)\right)$$

modulo q. By (2.20) with $l=1,2$, the second term of the above expression is congruent to $-(-1)^{qn/m}\bar{r}(x)$ modulo q, where

$$\begin{aligned}\bar{r}(x)=&\frac{q}{2}\left[d\sim_2 n\ \wedge\ d\not\equiv 0,1\ (\mathrm{mod}\ 4)\ \wedge\ 2\mid k\right]\frac{(x+m)^k-x^k}{m}\\&+\frac{q}{2}[d\sim_2 n\ \wedge\ d\not\equiv 0,1\ (\mathrm{mod}\ 4)\ \wedge\ 2\nmid k]\\&\qquad\times\frac{x(x+m)^{k-1}-(x+m)x^{k-1}}{m}\\&+\frac{q}{2}[2\|q\ \wedge\ 2\nmid kn\ \wedge\ 4\mid d+1]\frac{(x+m)^{k-1}-x^{k-1}}{m}.\end{aligned}\tag{2.21}$$

Thus

$$\Sigma_1\equiv\frac{d-1}{2}[2\nmid\frac{qn}{m}]m^kE_k\left(\frac{x}{m}\right)+\bar{r}(x)\pmod q.\tag{2.22}$$

Now, from (2.11)–(2.13), (2.18) and (2.22), it follows that

$$\frac{dm^k}{2}E_k\left(\frac{x}{m}\right)-(-1)^{\lfloor y/\bar{m}\rfloor}\frac{\bar{d}\bar{m}^k}{2}E_k\left(\frac{x}{\bar{m}}-\left\lfloor\frac{y}{\bar{m}}\right\rfloor\right)\equiv\Sigma+r(x)$$
$$\equiv\left[2\mid\frac{qn}{m}\right]\sum_{j=0}^{qn/m-1}(-1)^{j-1}(x+jm)^kc_j(y)$$
$$+\frac{d-1}{2}\left[2\nmid\frac{qn}{m}\right]m^kE_k\left(\frac{x}{m}\right)+r(x)+\bar{r}(x)$$
$$+\sum_{j=0}^{qn/m-1}(x+jm)^k\Big([2\mid j\ \wedge\ 2\nmid c_j(y)]$$

$$- \left[2 \mid \frac{qn}{m} - j \ \wedge \ 2 \mid c_j(y) + d\right]\Big) \pmod q.$$

With the help of the binomial theorem, we can easily verify that $r(x) + \bar{r}(x) \equiv R_k(x) \pmod q$. If qn/m is odd, then either $2 \nmid q$ or $2 \mid m$, whence

$$R_k(x) \equiv \frac{q}{2}[d \sim_2 n \ \wedge \ d \not\equiv 0, 1 \pmod 4]kx^{k-1} \pmod q.$$

and the desired results follow. □

Proof of Theorem 1.2. Just apply Theorem 2.1 with $n = m$ and $y = cm/d$.

References

[AS] M. Abramowitz and I. A. Stegun, *Handbook of Mathematical Functions*, Dover, New York, 1972.

[E] A. Erdélyi, W. Magnus, F. Oberhettinger and F. G. Tricomi, *Higher Transcendental Functions I*, McGraw-Hill, New York, Toronto and London, 1953.

[IR] K. Ireland and M. Rosen, *A Classical Introduction to Modern Number Theory* (Graduate texts in mathematics; 84), 2nd ed., Springer, New York, 1990.

[M] K. Mahler, *Introduction to p-adic Numbers and their Functions*, Cambridge Univ. Press, Cambridge, 1973.

[S1] Z. W. Sun, *Introduction to Bernoulli and Euler polynomials*, a talk given at Taiwan, 2002, `http://pweb.nju.edu.cn/zwsun/BerE.pdf`.

[S2] Z. W. Sun, *Combinatorial identities in dual sequences*, European J. Combin. **24**(2003), 709–718.

[S3] Z. W. Sun, *General congruences for Bernoulli polynomials*, Discrete Math. **262**(2003), 253–276.

[S4] Z. W. Sun, *On Euler numbers modulo powers of two*, J. Number Theory, 2005, in press.

SQUARE-FREE INTEGERS AS SUMS OF TWO SQUARES

Wenguang Zhai
School of Mathematical Sciences, Shandong Normal University, Jinan, Shandong, 250014, P.R.China
zhaiwg@hotmail.com

Abstract Let $r(n)$ denote the number of representations of the integer n as a sum of two squares, $\mu(n)$ the Möbius function and $P(x)$ the error term of the Gauss circle problem. Let $Q(x) := \sum_{n\le x}|\mu(n)|r(n)$. In this short note we shall prove that if the estimate $P(x) = O(x^{\theta})$ holds, then $Q(x+y) - Q(x) = Ay + O(yx^{-\varepsilon/2} + x^{\theta+\varepsilon})$, where A is a constant. In particular this asymptotic formula is true for $\theta = 131/416$. Our result improves Krätzel's previous result.

Keywords: Gauss circle problem, square-free number

2000 Mathematics Subject Classification: 11N37

1. Introduction

Let $r(n)$ denote the number of representations of the integer $n \ge 1$ as a sum of two squares and $\mu(n)$ the Möbius function. The celebrated Gauss circle problem is to determine the smallest exponent α for which the estimate for the error term $P(x) := \sum_{n\le x} r(n) - \pi x = O(x^{\alpha+\varepsilon})$ holds for every $\varepsilon > 0$. It was Gauss who proved that $P(x) = O(x^{1/2})$. The exponent $1/2$ was improved by many authors. The latest result is due to Huxley [2], who proved that

$$P(x) \ll x^{131/416}(\log x)^{26957/8320} \tag{1.1}$$

It is conjectured that $\alpha = 1/4$. For a historical survey on the circle problem, see E. Krätzel [6].

This work is supported by National Natural Science Foundation of China (Grant No. 10301018).

Number Theory: Tradition and Modernization, pp. 219–227
W. Zhang and Y. Tanigawa, eds.

Since $|\mu(n)|$ is the characteristic function of the set of square-free integers, the function $|\mu(n)|r(n)$ is the number of representation of a square-free integer as a sum of two squares.

Let

$$f(n) = \frac{1}{4}|\mu(n)|r(n).$$

We have (for $\sigma > 1$)

$$F(n) := \sum_{n=1}^{\infty} \frac{f(n)}{n^s} = (1 + 2^{-s}) \prod_{p \in \mathcal{P}_1} (1 + 2p^{-s}), \tag{1.2}$$

where $\mathcal{P}_1$ denotes the set of all primes which are congruent to 1 modulo 4. Then for the summatory function $Q(x)$ of $|\mu(n)|r(n)$, K.-H. Fischer [1] proved that

$$Q(x) := \sum_{n \leq x} |\mu(n)|r(n) = Ax + O(x^{1/2} \log x), \tag{1.3}$$

where

$$A = \operatorname*{Res}_{s=1} F(s).$$

The exponent $1/2$ in (1.3) cannot be reduced with the present knowledge of the zero-free region for $\zeta(s)$.

E. Krätzel [5] studied the short interval case, and proved that if

$$P(x) = O(x^\theta) \quad (\theta < 1/3) \tag{1.4}$$

and

$$D_3(x) := \sum_{n \leq x} d_3(n) = x(c_1 \log^2 x + c_2 \log x + c_3) + O(x^\delta) \quad (\delta \leq 1/2) \tag{1.5}$$

with some constants c_1, c_2, c_3, then the asymptotic formula

$$Q(x + y) - Q(x) = Ay + o(y) \tag{1.6}$$

holds for

$$x^{\frac{4-\delta(1+\theta)}{2(5-\theta-2\delta)}} \log^3 x \leq y = o(x).$$

With the present best known estimates $\delta = 43/96+\varepsilon$, $\theta = 131/416+\varepsilon$ (see Kolesnik [4] and Huxley [2], respectively), (1.6) is true for $y \geq x^{0.4501\cdots+\varepsilon}$. Since $\delta \geq 1/3$, $\theta \geq 1/4$, the limit of E. Krätzel's method is $y \geq x^{43/98+\varepsilon}$. Note that $43/98 = 0.4387\cdots$ and $131/416 = 0.3149\cdots$.

In this short note we shall use the convolution method to prove the following theorem.

Theorm. *Suppose (1.4) is true for some* $1/4 < \theta < 1/3$, *then we have*

$$Q(x+y) - Q(x) = Ay + O(yx^{-\varepsilon/2} + x^{\theta+\varepsilon}). \tag{1.7}$$

Corollary. *The asymptotic formula (1.6) is true for* $x^{131/416+2\varepsilon} \le y \le x$.

Notations. Throughout this note, ε always denotes a fixed sufficiently small positive constant. For any fixed integers $1 \le a_1 \le a_2 \le \cdots \le a_k$,

$$d(a_1, a_2, \cdots, a_k; n) := \sum_{n=n_1^{a_1}\cdots n_k^{a_k}} 1, \quad d_k(n) := d(1, 1, \cdots, 1; n).$$

$\zeta(s)$ is the Riemann zeta-function, and $L(s,\chi)$ is the Dirichlet L-function associated to the non-principal characters χ mod 4.

2. Proof of Theorem

We recall from (1.2) that

$$F(s) := \sum_{n=1}^{\infty} \frac{f(n)}{n^s} = (1+2^{-s}) \prod_{p\in\mathcal{P}_1} (1+2p^{-s}), \quad \Re s > 1, \tag{2.1}$$

where $f(n) = \frac{1}{4}|\mu(n)|r(n)$.

The following lemma plays a crucial role in the proof.

Lemma 1. *Suppose* $\Re s > 1$. *Then we have*

$$F(s) = \frac{\zeta(s)L(s,\chi)\zeta(3s)L(3s,\chi)}{\zeta^2(2s)L(2s,\chi)\zeta(4s)L^2(4s,\chi)} M(s), \tag{2.2}$$

where $\zeta(s)$ *is the Riemann zeta-function,* $L(s,\chi)$ *is the Dirichlet* L-*function, respectively as above and* $M(s)$ *is a certain Dirichlet series which is absolutely convergent for* $\Re s > 1/5$.

Proof. By the well-known Euler product representations, we have

$$\zeta(s) = \prod_p (1-p^{-s})^{-1} = (1-2^{-s})^{-1} \prod_{p\in\mathcal{P}_1} (1-p^{-s})^{-1} \prod_{p\in\mathcal{P}_3} (1-p^{-s})^{-1} \tag{2.3}$$

and

$$L(s,\chi) = \prod_p (1-\chi(p)p^{-s})^{-1} = \prod_{p\in\mathcal{P}_1} (1-p^{-s})^{-1} \prod_{p\in\mathcal{P}_3} (1+p^{-s})^{-1}, \tag{2.4}$$

where $\mathcal{P}_3$ signifies the set of all primes which are congruent to 3 mod 4. Hence

$$\prod_{p\in\mathcal{P}_1}(1-p^{-s})^{-2} = (1-2^{-s})\zeta(s)L(s,\chi)\prod_{p\in\mathcal{P}_3}(1-p^{-2s}). \tag{2.5}$$

Now for $|u| < 1/2$, we note that

$$1+2u = (1-u)^{-2}(1-u^2)^3(1-u^3)^{-2}(1-u^4)^3E(u) \tag{2.6}$$

with $E(u) = 1 + O(u^5)$. Hence we may rewrite (2.1) as

$$\begin{aligned} F(s) = (1+2^{-s})&\prod_{p\in\mathcal{P}_1}(1-p^{-s})^{-2}\prod_{p\in\mathcal{P}_1}(1-p^{-2s})^3\prod_{p\in\mathcal{P}_1}(1-p^{-4s})^3 \\ &\times\prod_{p\in\mathcal{P}_1}(1-p^{-3s})^{-2}M_1(s), \end{aligned} \tag{2.7}$$

where $M_1(s) := \prod_{p\in\mathcal{P}_1} E(p^{-s})$, which has a Dirichlet series expansion, absolutely convergent for $\Re s > 1/5$.

Proof of (2.2) amounts to substituting (2.5) (its powers) and replace the infinite products by (2.3) and (2.4). In fact we have

$$\begin{aligned} &(1+2^{-s})\prod_{p\in\mathcal{P}_1}(1-p^{-s})^{-2}\prod_{p\in\mathcal{P}_1}(1-p^{-2s})^3\prod_{p\in\mathcal{P}_1}(1-p^{-4s})^3 \\ &= (1-2^{-2s})\zeta(s)L(s,\chi)\prod_{p\in\mathcal{P}_3}(1-p^{-2s})\prod_{p\in\mathcal{P}_1}(1-p^{-2s})^3\prod_{p\in\mathcal{P}_1}(1-p^{-4s})^3 \\ &= \frac{\zeta(s)L(s,\chi)}{\zeta(2s)}\prod_{p\in\mathcal{P}_1}(1-p^{-2s})^2\prod_{p\in\mathcal{P}_1}(1-p^{-4s})^3 \\ &= (1-2^{-2s})^{-1}\frac{\zeta(s)L(s,\chi)}{\zeta^2(2s)L(2s,\chi)}\prod_{p\in\mathcal{P}_3}(1-p^{-4s})^{-1}\prod_{p\in\mathcal{P}_1}(1-p^{-4s})^3 \\ &= (1-2^{-2s})^{-1}(1-2^{-4s})\frac{\zeta(s)L(s,\chi)\zeta(4s)}{\zeta^2(2s)L(2s,\chi)}\prod_{p\in\mathcal{P}_1}(1-p^{-4s})^4 \\ &= (1-2^{-2s})^{-1}(1-2^{-4s})^{-1} \\ &\quad\times\frac{\zeta(s)L(s,\chi)}{\zeta^2(2s)L(2s,\chi)\zeta(4s)L^2(4s,\chi)}\prod_{p\in\mathcal{P}_3}(1-p^{-8s})^{-2} \end{aligned} \tag{2.8}$$

and

$$\prod_{p\in\mathcal{P}_1}(1-p^{-3s})^{-2} = (1-2^{-3s})\zeta(3s)L(3s,\chi)\prod_{p\in\mathcal{P}_3}(1-p^{-6s}). \tag{2.9}$$

Hence we get (2.2) with

$$M(s) = \frac{(1-2^{-3s})M_1(s)}{(1-2^{-2s})(1-2^{-4s})} \prod_{p\in\mathcal{P}_3} (1-p^{-8s})^{-2} \prod_{p\in\mathcal{P}_3} (1-p^{-6s}),$$

which has a Dirichlet series expansion, absolutely convergent for $\Re s > 1/5$. □

By Lemma 1, we have for $\Re s > 1$,

$$F(s) = F_1(s)F_2(s)F_3(s), \tag{2.10}$$

where

$$F_1(s) = \sum_{n=1}^{\infty} \frac{f_1(n)}{n^s} = \zeta(s)L(s,\chi)\zeta^{-1}(4s)L^{-2}(4s,\chi)M(s) \tag{2.11}$$

$$F_2(s) = \sum_{n=1}^{\infty} \frac{f_2(n)}{n^s} = \zeta^{-2}(s)L^{-1}(s,\chi), \tag{2.12}$$

and

$$F_3(s) = \sum_{n=1}^{\infty} \frac{f_3(n)}{n^s} = \zeta(s)L(s,\chi) = \frac{1}{4}\sum_{n=1}^{\infty} \frac{r(n)}{n^s}. \tag{2.13}$$

Then we have

Lemma 2. *If (1.4) holds, then we have*

$$A(x) := \sum_{n\le x} f_1(n) = A_1 x + O(x^{\theta+\varepsilon}), \tag{2.14}$$

where

$$A_1 = \operatorname*{Res}_{s=1} F_1(s).$$

Proof. We introduce the notation which we will use in this proof only.

$$\sum_{n=1}^{\infty} \frac{\alpha(n)}{n^s} = M(s), \quad \sum_{n=1}^{\infty} \frac{\beta(n)}{n^s} = \zeta^{-1}(4s)L^{-2}(4s,\chi)M(s)$$

and

$$\sum_{n=1}^{\infty} \frac{\gamma(n)}{n^s} = \zeta^{-1}(s)L^{-2}(s,\chi).$$

Then

$$f_1(n) = \sum_{ml=n} f_3(m)\beta(l), \quad \beta(n) = \sum_{m^4 l=n} \gamma(m)\alpha(l)$$

and

$$\gamma(n) = \sum_{n=n_1n_2n_3} \mu(n_1)\mu(n_2)\mu(n_3)\chi(n_2)\chi(n_3) \ll d_3(n).$$

By Perron's formula we see that

$$\sum_{n\le x} |\alpha(n)| \ll x^{1/5+\varepsilon}. \tag{2.15}$$

Hence

$$\begin{aligned}\sum_{n\le x} |\beta(n)| &\ll \sum_{m^4 l\le x} |\alpha(l)| d_3(m)\\ &= \sum_{l\le x} |\alpha(l)| D_3(x^{1/4} l^{-1/4})\\ &\ll x^{1/4}\log^2 x \sum_{l\le x} |\alpha(l)| l^{-1/4}\\ &\ll x^{1/4}\log^2 x\end{aligned}$$

by partial summation, (1.5), and (2.15).

Now we may appeal to Ivić [3] Theorem 14.1 to conclude (2.14). Or we may directly apply the hyperbola method as follows.

$$\begin{aligned}\sum_{n\le x} f_1(n) &= \sum_{ml\le x} \frac{1}{4} r(m)\beta(l)\\ &= \sum_{m\le x^\varepsilon} \frac{1}{4} r(m) \sum_{l\le \frac{x}{m}} \beta(l) + \sum_{l\le x^{1-\varepsilon}} \beta(l) \sum_{m\le \frac{x}{l}} \frac{1}{4} r(m)\\ &\quad - \Big(\sum_{m\le x^\varepsilon} \frac{1}{4} r(m)\Big)\Big(\sum_{l\le x^{1-\varepsilon}} \beta(l)\Big)\\ &= \frac{1}{4}\pi x \sum_{l\le x^{1-\varepsilon}} \frac{\beta(l)}{l} + O\Big(x^\theta \sum_{l\le x^{1-\varepsilon}} l^{-\theta} |\beta(l)|\Big)\\ &\quad + O\Big(x^{1/4} \sum_{m\le x^\varepsilon} \frac{r(m)}{m^{1/4}}\Big) + O\left(x^\varepsilon \cdot x^{(1-\varepsilon)/4}\right)\\ &= \frac{1}{4}\pi c x + O\left(x^{\theta+\varepsilon}\right),\end{aligned}$$

where $c = \int_1^\infty \left(\sum_{n\le y} \beta(n)\right) y^{-2}dy$.

Since the generating function for $f_1(s)$ is $F_1(s)$ having a simple pole at $s = 1$, the main term $\frac{1}{4}\pi cx$ must coincide in view of Tauberian argument or Perron's formula: $\frac{1}{4}\pi c = A_1 = \mathrm{Res}_{s=1} F_1(s)$. □

Finally we prepare a lemma for estimating the error term.

Lemma 3. *Let $k \ge 2$ be a fixed integer , $1 < y \le x$ be large real numbers and*

$$\mathcal{A}(x,y;k,\varepsilon) := \sum_{\substack{x<nm^k\le x+y \\ m>x^\varepsilon}} 1.$$

Then we have

$$\mathcal{A}(x,y;k,\varepsilon) \ll yx^{-\varepsilon} + x^{1/4}. \tag{2.16}$$

Proof. Without loss of generality, we may suppose $k = 2$. It is clear that

$$\mathcal{A}(x,y;2,\varepsilon) = D(1,2;x+y) - D(1,2;x) - \sum\nolimits_0,$$

where

$$D(1,2;u) := \sum_{n\le u} d(1,2;n)$$

and

$$\sum\nolimits_0 = \sum_{\substack{x<nm^2\le x+y \\ m>x^\varepsilon}} 1.$$

From Richert [7] we have

$$D(1,2;u) = \zeta(2)u + \zeta(1/2)u^{1/2} + O(u^{2/9}\log u),$$

which we apply with the error estimate $O(x^{1/4})$. For $\sum_0$, we have

$$\begin{aligned}\sum\nolimits_0 &= \sum_{m\le x^\varepsilon} \left(\left[\frac{x+y}{m^2}\right] - \left[\frac{x}{m^2}\right]\right) \\ &= y\sum_{m\le x^\varepsilon} m^{-2} + O(x^\varepsilon) \\ &= \zeta(2)y + O(yx^{-\varepsilon} + x^\varepsilon).\end{aligned}$$

Lemma 3 follows from the above formulas. □

Now we prove our Theorem. By (2.10) –(2.13),

$$f(n) = \sum_{n=n_1n_2^2n_3^3} f_1(n_1)f_2(n_2)f_3(n_3)$$

and

$$f_i(n) \ll n^{\varepsilon^2}, \quad i = 1, 2, 3. \tag{2.17}$$

Hence we have

$$\begin{aligned} Q(x+y) - Q(x) &= 4 \sum_{x<n\le x+y} f(n) \\ &= 4 \sum_{x<n_1 n_2^2 n_3^3 \le x+y} f_1(n_1) f_2(n_2) f_3(n_3) \\ &= 4 \sum\nolimits_1 + O\left(\sum\nolimits_2 + \sum\nolimits_3\right), \end{aligned} \tag{2.18}$$

where

$$\sum\nolimits_1 = \sum_{\substack{n_2 \le x^\varepsilon \\ n_3 \le x^\varepsilon}} f_2(n_2) f_3(n_3) \sum_{\frac{x}{n_2^2 n_3^3} < n_1 \le \frac{x+y}{n_2^2 n_3^3}} f_1(n_1),$$

$$\sum\nolimits_2 = \sum_{\substack{x<n_1 n_2^2 n_3^3 \le x+y \\ n_2 > x^\varepsilon}} |f_1(n_1) f_2(n_2) f_3(n_3)|,$$

$$\sum\nolimits_3 = \sum_{\substack{x<n_1 n_2^2 n_3^3 \le x+y \\ n_3 > x^\varepsilon}} |f_1(n_1) f_2(n_2) f_3(n_3)|.$$

By Lemma 2 we have

$$\begin{aligned} \sum\nolimits_1 &= \sum_{\substack{n_2 \le x^\varepsilon \\ n_3 \le x^\varepsilon}} f_2(n_2) f_3(n_3) \left(\frac{A_1 y}{n_2^2 n_3^3} + O\left(\left(\frac{x}{n_2^2 n_3^3}\right)^\theta\right)\right) \\ &= Ay + O(yx^{-\varepsilon/2} + x^{\theta+\varepsilon}), \end{aligned} \tag{2.19}$$

where as in the proof of Lemma 2, the coefficient of the main term

$$A_1 \sum_{n_2=1}^{\infty} \sum_{n_3=1}^{\infty} \frac{f_2(n_2)}{n_2^2} \frac{f_3(n_3)}{n_3^3}$$

must coincide with $A = \operatorname{Res}_{s=1} F(s)$.

Also

$$\begin{aligned} \sum\nolimits_2 &\ll \sum_{\substack{x<n_1 n_2^2 n_3^3 \le x+y \\ n_2 > x^\varepsilon}} (n_1 n_2 n_3)^{\varepsilon^2} \\ &\ll x^{\varepsilon^2} \sum_{\substack{x<mn_2^2 \le x+y \\ n_2 > x^\varepsilon}} d(1,3;m) \end{aligned}$$

on writing $n_1 n_3^3 = n$. Since $d(1,3;m) \ll m^{\varepsilon^2}$ we have by Lemma 3

$$\begin{aligned}\sum\nolimits_2 &\ll x^{2\varepsilon^2}\mathcal{A}(x,y;2,\varepsilon)\\ &\ll yx^{2\varepsilon^2-\varepsilon} + x^{1/4+2\varepsilon^2}\\ &\ll yx^{-\varepsilon/2} + x^{1/4+\varepsilon}.\end{aligned} \tag{2.20}$$

Similarly, we have

$$\sum\nolimits_3 \ll yx^{-\varepsilon/2} + x^{1/4+\varepsilon}. \tag{2.21}$$

Now our Theorem follows from (2.18)-(2.21).

Acknowledgments

The author deeply thanks Professor Kanemitsu, who kindly read the manuscript of the paper and proposed many helpful suggestions.

References

[1] K.-H. Fischer, *Über die Anzahl der Gitterpunkte auf Kreisen mit quadratfreien Radien-quadraten*, Arch. Math. **33** (1979), 150–154.

[2] M. N. Huxley, *Exponential sums and Lattice points III*, Proc. London Math. Soc. **87**(3) (2003), 591–609.

[3] A. Ivić, *The Riemann zeta-function*, John Wiley & Sons, New York, 1985.

[4] G. Kolesnik, *On the estimation of multiple exponential sums*, in *Recent Progress in Analytic Number Theory*, Symposium Durham 1979 (Vol.1), Academic, London, 231–246.

[5] E. Krätzel, *Squarefree numbers as sums of two squares*, Arch. Math. **39** (1982), 28–31.

[6] E. Krätzel, *Lattice points*, Deutsch. Verlag Wiss. Berlin, 1988.

[7] H.-E. Richert, *Über die Anzahl Abelscher Gruppen gegebener Ordnung I*, Math. Z. **56** (1952),21–32; II. ibid.**58** (1953), 71–84.

SOME APPLICATIONS OF L- FUNCTIONS TO THE MEAN VALUE OF THE DEDEKIND SUMS AND COCHRANE SUMS *

ZHANG Wenpeng
Department of Mathematics, Northwest University
Xi'an, Shaanxi, P.R.China
wpzhang@nwu.edu.cn

Abstract In this paper, the applications of L- functions to the mean value of Dedekind sums and Cochrane sums are described, and a few asymptotic formulae are presented.

Keywords: L- functions, Dedekind sums, Cochrane sums

2000 Mathematics Subject Classification: 11M20; 11F20

1. Mean value properties of L-functions

In the following we let q denote a fixed modulus, $q \geq 2$ and χ a Dirichlet character mod q. The Dirichlet L-function $L(s,\chi)$ associated to χ is defined by

$$L(s,\chi) = \sum_{n=1}^{\infty} \chi(n) n^{-s},$$

absolutely convergent for $\Re s > 1$ if $\chi = \chi^0$, the principal character mod q, and convergent for $\Re s \geq 0$ if $\chi \neq \chi^0$.

They play a very important role in number theory, e.g. in the proof of Dirichlet's prime number theorem, in Dirichlet's class number formula, in the solution of D. H. Lehmer's problem on the parity of residues ([12]) and so on.

We summarize the mean value results on L-functions (for odd characters, $\chi(-1) = -1$) in §1 and the corresponding mean value results on Dedekind and Cochrane sums in §2 and §3, respectively, as the consequences of the results in §1, in conjunction with close relationships

*This work is supported by the N.S.F. (10271093, 60472068) of P.R.China

Number Theory: Tradition and Modernization, pp 229–237
W. Zhang and Y. Tanigawa, eds.

between L-functions and the afore-mentioned sums, to be stated in the respective section. We shall focus on achievements of our own school.

Let $e(y) = e^{2\pi iy}$ and let $G(\chi, n)$ denote the Gauss sum

$$G(\chi, n) = \sum_{a=1}^{q} \chi(a) e\left(\frac{na}{q}\right)$$

for $n \geq 1$ and χ mod q. For $n = 1$ we write

$$\tau(\chi) = G(\chi, 1) = \sum_{a=1}^{q} \chi(a) e\left(\frac{a}{q}\right).$$

Define the saw-tooth function $((x))$ by

$$((x)) = \begin{cases} x - [x] - \frac{1}{2}, & \text{if } x \text{ is not an integer,} \\ 0, & \text{if } x \text{ is an integer,} \end{cases}$$

with $[x]$ denoting the integer part of x.

For any primitive character χ modulo q, H. Walum [10] established the identity

$$\sum_{a=1}^{q} \chi(a) \left(\left(\frac{a}{q}\right)\right) = \begin{cases} -\frac{\tau(\chi)}{\pi i} L(1, \overline{\chi}), & \text{if } \chi(-1) = -1, \\ 0, & \text{if } \chi(-1) = 1, \end{cases} \tag{1.1}$$

where $\overline{\chi}(n) = \overline{\chi(n)}$ means the complex conjugate of $\chi(n)$. For $q = p$ (a prime), he deduced the beautiful exact formula

$$\sum_{\substack{\chi \bmod p \\ \chi(-1)=-1}} |L(1, \chi)|^2 = \frac{\pi^2}{12} \frac{(p-1)^2(p-2)}{p^2}. \tag{1.2}$$

For general q, S. Louboutin [8] and the author [13] obtained the following generalization of (1.2):

Proposition 1.1.

$$\sum_{\substack{\chi \bmod q \\ \chi(-1)=-1}} |L(1, \chi)|^2 = \frac{\pi^2}{12} \cdot \frac{\phi^2(q)}{q^2} \left(q \prod_{p|q} \left(1 + \frac{1}{p}\right) - 3 \right),$$

where $\phi(q)$ is the Euler function.

There are many mean value theorems on the Dirichlet L-functions. We state some of them.

First,

Proposition 1.2 ([4]). *Let u and v be integers with $(u,v)=d\geq 2$, χ_u^0 and χ_v^0 be the principal characters modulo u and v respectively. Then we have for $\Re s\geq 1$*

$$\sum_{\substack{\chi \bmod d\\ \chi(-1)=-1}} \left|L(s,\chi\chi_u^0)\right|^2 \left|L(s,\chi\chi_v^0)\right|^2$$
$$=\frac{\zeta^4(2s)}{2\zeta(4s)}\phi(d)\frac{\prod_{p|uv}\frac{(p^{2s}-1)^2}{p^{2s}(p^{2s}+1)}}{\prod_{p|d}\frac{p^{2s}}{p^{2s}-1}}+O\left(\frac{\phi(d)}{d^s}\exp\left(\frac{3\ln m}{\ln\ln m}\right)\right)$$

and

$$\sum_{\substack{\chi \bmod d\\ \chi(-1)=1}} \left|L(s,\chi\chi_u^0)\right|^2 \left|L(s,\chi\chi_v^0)\right|^2$$
$$=\frac{\zeta^4(2s)}{2\zeta(4s)}\phi(d)\frac{\prod_{p|uv}\frac{(p^{2s}-1)^2}{p^{2s}(p^{2s}+1)}}{\prod_{p|d}\frac{p^{2s}}{p^{2s}-1}}+O\left(\frac{\phi(d)}{d^s}\exp\left(\frac{3\ln m}{\ln\ln m}\right)\right).$$

The special case $s=1$ is given in [14].

Secondly,

Proposition 1.3 ([5]). *Let $q=uv$, where $(u,v)=1$, u be a square-full number or $u=1$, v be a square-free number. Then for any positive integers n and m we have*

$$\sum_{d|v}(ud)^m \sideset{}{^*}\sum_{\substack{\chi \bmod ud\\ \chi(-1)=1}}\left[\sum_{t|\frac{v}{d}}\frac{\overline{\chi}(t)\mu(t)\phi(t)}{t^n}\right]^m L^m(n,\overline{\chi})$$
$$=\frac{q^{m-1}\phi^2(q)}{2}\prod_{p\|q}\left(1-\frac{p^{m-1}-1}{p^{m-1}(p-1)^2}\right)+O\left(q^{m+\epsilon}\right)$$

and

$$\sum_{d|v}(ud)^m \sideset{}{^*}\sum_{\substack{\chi \bmod ud\\ \chi(-1)=-1}}\left[\sum_{t|\frac{v}{d}}\frac{\overline{\chi}(t)\mu(t)\phi(t)}{t^n}\right]^m L^m(n,\overline{\chi})$$
$$=\frac{q^{m-1}\phi^2(q)}{2}\prod_{p\|q}\left(1-\frac{p^{m-1}-1}{p^{m-1}(p-1)^2}\right)+O\left(q^{m+\epsilon}\right),$$

where $\sum\limits_{\substack{\chi \bmod m \\ \chi(-1)=\pm 1}}^{*}$ *means the summation over even or odd primitive characters mod m respectively.*

Proposition 1.4 ([18]). *Let $q \geq 3$ be an odd modulus. Then we have the asymptotic formulae*

$$\sum_{\substack{\chi \bmod q \\ \chi(-1)=-1}} |L(1,\chi)|^4 = \frac{5\pi^4}{144}\phi(q)\prod_{p|q}\frac{(p^2-1)^3}{p^4(p^2+1)} + O\left(\exp\left(\frac{3\ln q}{\ln\ln q}\right)\right),$$

$$\sum_{\substack{\chi \bmod q \\ \chi(-1)=-1}} \chi(2)\,|L(1,\chi)|^4 = \frac{\pi^4}{36}\phi(q)\prod_{p|q}\frac{(p^2-1)^3}{p^4(p^2+1)} + O\left(\exp\left(\frac{3\ln q}{\ln\ln q}\right)\right),$$

$$\sum_{\substack{\chi \bmod q \\ \chi(-1)=-1}} \chi(4)\,|L(1,\chi)|^4 = \frac{11\pi^4}{576}\phi(q)\prod_{p|q}\frac{(p^2-1)^3}{p^4(p^2+1)} + O\left(\exp\left(\frac{3\ln q}{\ln\ln q}\right)\right).$$

2. Mean value of Dedekind sums

For any integer h, the Dedekind sum is defined by

$$s(h,q) = \sum_{a=1}^{q}\left(\left(\frac{a}{q}\right)\right)\left(\left(\frac{ha}{q}\right)\right), \tag{2.1}$$

whose reciprocity law amounts to the transformation formula for the Dedekind eta function [9]. In the same paper [10] referred to above, H. Walum showed that for a prime p,

$$\sum_{\substack{\chi \bmod p \\ \chi(-1)=-1}} |L(1,\chi)|^4 = \frac{\pi^4(p-1)}{p^2}\sum_{h=1}^{p}|s(h,p)|^2,$$

which suggests the existence of some connections between Dedekind sums and L-functions.

J. B. Conrey, E. Fransen, R. Klein and C. Scott [3] studied the higher power mean value of the Dedekind sums and proved the following by elementary methods:

$$\sum_{h=1}^{q}{}' s^{2m}(h,q) = f_m(q)\left(\frac{q}{12}\right)^{2m} + O\left(\left(q^{9/5} + q^{2m-1+1/(m+1)}\right)\log^3 q\right), \tag{2.2}$$

where $\sum_h{}'$ denotes the summation over h relatively prime to q, $(h,q)=1$, and $f_m(k)$ is defined by the Dirichlet series

$$\sum_{k=1}^{\infty} \frac{f_m(k)}{k^s} = 2 \cdot \frac{\zeta^2(2m)}{\zeta(4m)} \cdot \frac{\zeta(s+4m-1)}{\zeta^2(s+2m)} \cdot \zeta(s),$$

$\zeta(s)$ denoting the Riemann zeta-function.

There are improvement over (2.2), due to Chaohua Jia [2] in the case $m>1$ and due to the author [14] in the case $m=1$, the latter reading

$$\sum_{h=1}^{q}{}' s^2(h,q) = \frac{5}{144} q\phi(q) \cdot \frac{\prod\limits_{p^\alpha \| q} \left(\left(1+\frac{1}{p}\right)^2 - \frac{1}{p^{3\alpha+1}} \right)}{\prod\limits_{p|q} \left(1+\frac{1}{p}+\frac{1}{p^2}\right)} + O\left(q \exp\left(\frac{4\ln q}{\ln\ln q} \right) \right) \tag{2.3}$$

with $p^\alpha \parallel q$ indicating that $p^\alpha | q$ but $p^{\alpha+1} \nmid q$. This depends on the following identity [13] expressing the relation between the L-functions and the Dedekind sums, alluded to in §1:

$$s(h,q) = \frac{1}{\pi^2 q} \sum_{d|q} \frac{d^2}{\phi(d)} \sum_{\substack{\chi \bmod d \\ \chi(-1)=-1}} \chi(h) \left| L(1,\chi) \right|^2. \tag{2.4}$$

Moreover, let

$$s_2(h,q) = \sum_{a=1}^{q} (-1)^a \left(\left(\frac{a}{q}\right)\right) \left(\left(\frac{ha}{q}\right)\right), \tag{2.5}$$

which is related to the Dedekind sum through

$$s_2(h,q) = 2s(h,q/2) - s(h,q)$$

for even q. By this identity, we were able to obtain an analogue of (2.3):

Proposition 2.1 ([16]). *Let $q=2^\beta M$ with $\beta \geq 1$ and $2 \nmid M$. Then we have*

$$\sum_{h=1}^{q}{}' |s_2(h,q)|^2 = \frac{5}{112} q\phi(q) \left(\frac{3}{5} - \frac{2}{2^{3\beta}} \right) \prod_{p^\alpha \| M} \frac{\left[\left(1+\frac{1}{p}\right)^2 - \frac{1}{p^{3\alpha+1}} \right]}{\left(1+\frac{1}{p}+\frac{1}{p^2}\right)} + O\left(q \exp\left(\frac{4\ln q}{\ln\ln q} \right) \right).$$

Huaning Liu [6, 7] obtained asymptotic formulae for the $2m$-th power mean value of sums analogous to Dedekind sums.

Furthermore, we have obtained the following results suggesting the relationships between Dedekind sums and the Hurwitz zeta function $\zeta(\frac{1}{2}, \frac{a}{q})$ and those between the sum $s_1(h,q)$ and the Ramanaujan sum $R_q(h)$, respectively:

$$s_1(h,q) = \sum_{a=1}^{q-1} (-1)^{a+1+[ha/q]} \tag{2.6}$$

$$R_q(h) = \sum_{a=1}^{q}{}' \, \mathrm{e}\left(\frac{ha}{q}\right). \tag{2.7}$$

Proposition 2.2 ([15]). *For $q \geq 3$ and any fixed positive integer m, we have the asymptotic formula*

$$\sum_{a=1}^{q}{}' \zeta^2\left(\frac{1}{2}, \frac{a}{q}\right) s^{2m}(a,q) = \frac{q^{2m+1}}{(12)^{2m}} \zeta(2m+1) \prod_{p|q} \left(1 - \frac{1}{p^{2m+1}}\right) + O\left(q^{2m+1/2} \exp\left(\frac{3 \ln q}{\ln \ln q}\right)\right).$$

Proposition 2.3 ([19]). *For $q \geq 3$ an odd modulus, we have the asymptotic formula*

$$\sum_{h=1}^{q}{}' R_q(2h+1) s_1(2h,q) = \phi^2(q) \prod_{p|q} \left(1 + \frac{1}{p}\right) \prod_{p||q} \left(1 + \frac{1}{(p+1)(p-1)^2}\right) + O\left(q^{1+\epsilon}\right).$$

3. Mean value of Cochrane sums

For an arbitrary integer h, we define the Cochrane sum in analogy to (2.1):

$$C(h,q) = \sum_{a=1}^{q}{}' \left(\left(\frac{a}{q}\right)\right) \left(\left(\frac{\bar{a}h}{q}\right)\right) \tag{3.1}$$

the sum being taken over all a, relatively prime to q, and $\bar{a}$ denoting the inverse of a mod q.

We have a counterpart of (2.4) [21] expressing the relation between the Cochrane sum and L-functions:

$$C(h,q) = \frac{-1}{\pi^2\phi(q)} \sum_{m|q} \sum_{\substack{\chi \bmod m \\ \chi(-1)=-1}}^{*} \overline{\chi}(h) \left(\prod_{\substack{p|q \\ p\nmid m}} (1-\chi(p)) \right)^2 \tau^2(\chi) L^2(1,\overline{\chi}). \tag{3.2}$$

Formula (3.2) enables us to obtain an upper bound for, and a mean value of, the Cochrane sum:

Proposition 3.1 ([21]). *For any integer h with $(h,q)=1$, we have the estimate*

$$|C(h,q)| \ll \sqrt{q}d(q)\ln^2 q,$$

where $d(q)=\sum_{d|q} 1$ is the divisor function.

Proposition 3.2 ([21]). *For $q=p$ an odd prime,*

$$\sum_{h=1}^{p-1} C^2(h,p) = \frac{5}{144}p^2 + O\left(p \exp\left(\frac{4\ln p}{\ln\ln p}\right)\right).$$

Now, corresponding to the k-dimensional analogue of (2.1) due to L. Carlitz [1], we may introduce the k-dimensional Cochrane sums

$$C(h,k,q) = \sum_{a_1=1}^{q}{}' \sum_{a_2=1}^{q}{}' \cdots \sum_{a_k=1}^{q}{}' \left(\left(\frac{a_1}{q}\right)\right)\left(\left(\frac{a_2}{q}\right)\right)\cdots\left(\left(\frac{a_k}{q}\right)\right)\left(\left(\frac{h\overline{a_1 a_2 \cdots a_k}}{q}\right)\right), \tag{3.3}$$

the case $k=1$ being (3.1).

Xu Zhefeng and the author [11] obtained the k-dimensional analogue of (3.2), which in turn is one of (2.4), and succeeded in proving counterpart of Proposition 3.1 and 3.2:

$$C(h,k,q) = \frac{i^{k+1}}{\pi^{k+1}\phi(q)} \sum_{m|q} \sum_{\substack{\chi \bmod m \\ \chi(-1)=-1}}^{*} \overline{\chi}(h) \left(\prod\nolimits_{\substack{p|q \\ p\nmid m}} (1-\chi(p)) \right)^{k+1} \tau^{k+1}(\chi) L^{k+1}(1,\overline{\chi}), \tag{3.4}$$

Proposition 3.3 ([11]). *For any integer h, $(h,q)=1$ and any fixed positive integer k with $(q,k(k+1))=1$, we have*

$$|C(h,k,q)| \ll \frac{2^{(k+1)^2}}{\pi^{k+1}} q^{\frac{k}{2}} d(q) \left(2^{k+2}k\right)^{\omega(q)} \ln^{k+1} q,$$

where $\omega(q)$ denotes the number of all distinct prime divisors of q.

Proposition 3.4 ([11]). *For $q=p^\alpha$ a prime power, we have*

$$\sum_{h=1}^{p^\alpha}{}' |C(h,k,p^\alpha)|^2 = \frac{\zeta^{2k+1}(2)\left[\left(1-\frac{1}{p}\right)p^{(\alpha+1)(k+2)} + p^{2k+3} - p^{k+2}\right]}{2\pi^{2k+2}p^\alpha(p^{k+2}-1)}$$
$$\times \left(\frac{p^2-1}{p^2}\right)^{2k+1} \prod_{p_1\neq p}\left(1-\frac{1-C_{2k}^k}{p_1^2}\right) + O_k\left(p^{\alpha k+\epsilon}\right),$$

where ϵ is any fixed positive number, $\prod_{p_1\neq p}$ denotes the product over all primes p_1 with $p_1\neq p$, $C_m^n = m!/n!(m-n)!$.

The definition (3.1) gives an impression that there may be some relation between the Cochrane sum and the classical Kloosterman sum $K(h,n;q)$ defined by

$$K(h,n;q) = \sum_{b=1}^{k}{}' e\left(\frac{hb+n\bar{b}}{q}\right). \tag{3.5}$$

This impression is strengthened by the following mean value result of Cochrane sums with Kloosterman sum weights:

Proposition 3.5 ([17]). *We have*

$$\sum_{h=1}^{q}{}' K(h,1;q)C(h,q) = \frac{-1}{2\pi^2} q\phi(q) \prod_{p\|q}\left(1-\frac{1}{p(p-1)}\right) + O\left(q^{\frac{3}{2}+\epsilon}\right)$$

for any $\epsilon>0$.

The proof rests on the identity (3.4) and some involved discussion of calculation. The author and Huaning Liu [20] improved the error estimate in Proposition 3.4 to $O(q^{1+\epsilon})$ by established the identity

$$C(h,q) = \frac{-1}{\pi^2\phi(q)} \sum_{\substack{\chi \bmod q\\ \chi(-1)=-1}} \overline{\chi}(h) \left(\sum_{n=1}^{\infty}\frac{G(\chi,n)}{n}\right)^2.$$

References

[1] L. Carlitz, *A note on generalized Dedekind sums*, Duke Math. J. **21** (1954), 399–404.

[2] Chaohua Jia, *On the Mean Values of Dedekind sums*, J. Number Theory **87** (2001), 173–188.

[3] J. B. Conrey, E. Fransen, R. Klein and C. Scott, *Mean values of Dedekind sums*, J. Number Theory **56** (1996), 214–226.

[4] Hongyan Liu and Wenpeng Zhang, *On the mean square value of a generalized Cochrane sum*, Soochow J. Math. **30** (2004), 165–175.

[5] Huaning Liu and Wenpeng Zhang, *On the hybrid mean value of Gauss sums and generalized Bernoulli numbers*, Proc. Japan Acad. Ser. A- Math. Sci. **80** (2004), 113–115.

[6] Huaning Liu and Wenpeng Zhang, *On certain Hardy sums and their 2m-th power mean*, Osaka J. Math. **41** (2004), 745–758.

[7] Huaning Liu and Wenpeng Zhang, *On the 2m-th power mean of a sum analogous to Dedekind sums*, Acta Math. Hungarica **106** (2005), 67–81.

[8] S. Louboutin, *Quelques formules exactes pour des moyennes de fonctions L de Dirichlet*, Canad. Math. Bull. **36** (1993), 190–196.

[9] H. Rademacher and E. Grosswald, *Dedekind sum*, MAA 1972.

[10] H. Walum, *An exact formula for an average of L-series*, Illinois J. Math. **26** (1982), 1–3.

[11] Zhefeng Xu and Wenpeng Zhang, *On the order of the high-dimensional Cochrane sum and its mean value*, J. Number Theory (in press).

[12] Wenpeng Zhang, *On a problem of D.H.Lehmer and its generalization*, Compositio Math. **86** (1991), 307–316.

[13] Wenpeng Zhang, *On the mean values of Dedekind sums*, J. de Theorie des Nombres de Bordeaux **8** (1996), 429–442.

[14] Wenpeng Zhang, *A note on the mean square value of the Dedekind Sums*, Acta Math. Hungarica **86** (2000), 275–289.

[15] Wenpeng Zhang, *On the hybrid mean value of Dedekind sums and Hurwitz zeta-function*, Acta Arith. **92** (2000), 141–152.

[16] Wenpeng Zhang, *A sums analogous to Dedekind sums and its mean value formula*, J. Number Theory **89** (2001), 1–13.

[17] Wenpeng Zhang, *On a Cochrane sum and its hybrid mean value formula (II)*, J. Math. Anal. and Appl. **276** (2003), 446–457.

[18] Wenpeng Zhang, *A problem of D. H. Lehmer and its mean square value formula*, Japan. J. Math. **29** (2003), 109–116.

[19] Wenpeng Zhang, *On a hybrid mean value of certain Hardy sums and Ramanujan sum*, Osaka J. Math. **40** (2003), 365–373.

[20] Wenpeng Zhang and Huaning Liu, *A Note on the Cochrane Sum and its Hybrid Mean Value Formula*, J. Math. Anal. and Appl. **288** (2003), 646–659.

[21] Wenpeng Zhang and Yuan Yi, *On the upper bound estimate of Cochrane sums*, Soochow J. Math. **28** (2002), 297–304.

Printed in the United States
108232LV00003B/83/A

9 780387 304144